# SKYRMIONS
## A Theory of Nuclei

# SKYRMIONS
## A Theory of Nuclei

### Nicholas S Manton
University of Cambridge, UK

 **World Scientific**

NEW JERSEY · LONDON · SINGAPORE · BEIJING · SHANGHAI · HONG KONG · TAIPEI · CHENNAI · TOKYO

*Published by*

World Scientific Publishing Europe Ltd.
57 Shelton Street, Covent Garden, London WC2H 9HE
*Head office:* 5 Toh Tuck Link, Singapore 596224
*USA office:* 27 Warren Street, Suite 401-402, Hackensack, NJ 07601

**Library of Congress Cataloging-in-Publication Data**
Names: Manton, Nicholas, 1952– author.
Title: Skyrmions : a theory of nuclei / Nicholas S. Manton.
Description: Hackensack, New Jersey : World Scientific, [2022] |
    Includes bibliographical references and index.
Identifiers: LCCN 2021062774 | ISBN 9781800612471 (hardcover) |
    ISBN 9781800612488 (ebook for institutions) | ISBN 9781800612495 (ebook for individuals)
Subjects: LCSH: Skyrme model. | Atomic structure. | Topology. | Nuclear physics--History. |
    Quantum theory--History.
Classification: LCC QC793.3.S8 M36 2022 | DDC 539.7/4--dc23/eng20220218
LC record available at https://lccn.loc.gov/2021062774

**British Library Cataloguing-in-Publication Data**
A catalogue record for this book is available from the British Library.

Copyright © 2022 by World Scientific Publishing Europe Ltd.

*All rights reserved. This book, or parts thereof, may not be reproduced in any form or by any means, electronic or mechanical, including photocopying, recording or any information storage and retrieval system now known or to be invented, without written permission from the Publisher.*

For photocopying of material in this volume, please pay a copying fee through the Copyright Clearance Center, Inc., 222 Rosewood Drive, Danvers, MA 01923, USA. In this case permission to photocopy is not required from the publisher.

For any available supplementary material, please visit
https://www.worldscientific.com/worldscibooks/10.1142/Q0368#t=suppl

# Foreword

### by Prof. K. K. Phua
### Chairman and Editor-in-Chief, World Scientific

World Scientific is honoured to publish *Skyrmions – A Theory of Nuclei* by Professor Nick Manton of Cambridge University to celebrate the centenary in December 2022 of the late Professor Tony Skyrme's birth.

I got my Ph.D. from the Department of Mathematical Physics at the University of Birmingham in early 1970, and had very close contact and some very good discussions with Professor Skyrme about different areas of research in physics.

I remember very clearly that Professor Skyrme came and discussed with me whether $SU(3)$ and his work on Skyrmions are in any way related or connected. We had some good discussion about this problem (which is considered in Chapter 14 of this book) and frankly speaking, I learnt and benefitted a lot from him.

Professor Skyrme was quite familiar with science and research in Malaysia and Singapore. He was Head of the Mathematics Department at the University of Malaya for about three years in the early 1960s and visited Singapore a few times. Hence, he got to know the lecturers and students in Malaysia and Singapore well, which is one of the reasons why the Departments of Physics and Mathematics at the University of Birmingham attracted quite a large number of Asian students from Singapore, Malaysia and Hong Kong.

In summary, Professor Skyrme was one of the greatest physicists of the 20th century. He was a deep thinker and had many original ideas.

# Preface

I would like to thank J. Martin Speight for encouraging me to write this book. For a while, I had too many duties and distractions, but an opportunity to start it was the closure of my university department in March 2020, part of wider lockdowns and the need to work from home because of the Coronavirus pandemic.

Much of this book is based on my published papers from 1986 onwards, often written jointly with my Ph.D. students and other collaborators. I would particularly like to thank Trevor Samols, Robert Leese, Bernd Schroers, Paul Sutcliffe, Richard Battye, Kim Baskerville, Conor Houghton, Patrick Irwin, Steffen Krusch, Bernard Piette, Olga Manko, Stephen Wood, Theodora Ioannidou, Dankrad Feist, P. H. Chris Lau, Chris Halcrow, Chris King, Dave Foster, Mareike Haberichter, Jonathan Rawlinson and Derek Harland for their contributions, and for discussions about Skyrmions over very many years. They originally carried out numerous analytical and numerical calculations whose results are presented here, and they produced most of the figures. I also wish to record my gratitude for the inspiring collaboration and discussions with the late Sir Michael Atiyah, whose interest in Skyrmions and related ideas in theoretical physics ran from about 1987 until the end of his life in 2019.

Part of this book is a reworking of some chapters in *Topological Solitons*, the CUP monograph I wrote jointly with Paul Sutcliffe. I thank Paul and Cambridge University Press for permission to reuse this material. I am also most grateful to Paul, and also to Chris Lau and Chris Halcrow, for reading substantial parts of a draft of this book, and for suggesting several improvements and corrections, many of which have been implemented. Last but not least, I warmly thank Shi Ying Koe and Rozita Osman at World

Scientific for the efficient and helpful way they have turned my manuscript into a book ready for publication.

This book is dedicated to the memory of Tony Skyrme, the centenary of whose birth falls in 2022. I was fortunate to have met him once.

# Contents

*Foreword*     v

*Preface*     vii

1. Introduction     1

2. Fields and Particles     11
   - 2.1 The Classical Notions of Particles and Fields . . . . . . .     11
     - 2.1.1 Quantum theory . . . . . . . . . . . . . . . . .     13
   - 2.2 Quantum Field Theory and the Standard Model . . . . .     14
   - 2.3 Pions and Nucleons . . . . . . . . . . . . . . . . . . . . .     16
   - 2.4 Nuclei and Isospin . . . . . . . . . . . . . . . . . . . . .     21
   - 2.5 Effective Field Theory . . . . . . . . . . . . . . . . . . .     24
   - 2.6 Skyrme's Effective Field Theory . . . . . . . . . . . . . .     27

3. Lagrangians and Symmetries     29
   - 3.1 Finite-Dimensional Systems . . . . . . . . . . . . . . . .     29
   - 3.2 Symmetries and Conservation Laws . . . . . . . . . . . .     32
   - 3.3 Lagrangian Field Theory . . . . . . . . . . . . . . . . . .     33
   - 3.4 Noether's Theorem in Field Theory . . . . . . . . . . . .     37

4. Skyrme Theory     41
   - 4.1 $SU(2)$ . . . . . . . . . . . . . . . . . . . . . . . . . . . .     41
   - 4.2 The Skyrme Lagrangian and Field Equation . . . . . . .     44
   - 4.3 Skyrme Field Topology . . . . . . . . . . . . . . . . . . .     47
     - 4.3.1 Topological degree of a map . . . . . . . . . . .     48

|      | 4.4   | Skyrme Field Energy                                      | 51  |
|------|-------|----------------------------------------------------------|-----|
|      |       | 4.4.1   Elastic strain formulation                       | 52  |
|      | 4.5   | Hedgehog Skyrmions                                       | 54  |
|      | 4.6   | Visualising Skyrmions                                    | 56  |
|      | 4.7   | Asymptotic Interactions of Hedgehogs                     | 58  |
|      | 4.8   | Adding a Pion Mass Term                                  | 60  |
| 5.   | Quantization of Skyrmions                                        | 63  |
|      | 5.1   | Quantization of Skyrme Fields                            | 63  |
|      | 5.2   | Topology and Quantization                                | 66  |
|      | 5.3   | Rigid-Body Quantization                                  | 67  |
|      | 5.4   | Quantized Hedgehog Skyrmion – Proton and Neutron         | 69  |
|      | 5.5   | Classical Interpretation of Quantized Skyrmion States    | 74  |
|      | 5.6   | The Need for Fermionic Quantization                      | 77  |
| 6.   | Skyrmions with Higher $B$ – Massless Pions                      | 81  |
|      | 6.1   | Skyrmions with Baryon Numbers $B \leq 8$                 | 82  |
|      | 6.2   | The Rational Map Ansatz                                  | 86  |
|      | 6.3   | Symmetric Rational Maps                                  | 90  |
|      |       | 6.3.1   Platonic symmetries                              | 92  |
|      | 6.4   | Skyrmions from Rational Maps                             | 96  |
|      | 6.5   | Skyrmions up to $B = 22$                                 | 99  |
|      | 6.6   | Rigorous Investigation of Skyrmions                      | 103 |
|      | 6.7   | The Skyrmion Crystal                                     | 106 |
| 7.   | Rigid-Body Skyrmion Quantization                                 | 111 |
|      | 7.1   | Collective Coordinate Quantization                       | 111 |
|      | 7.2   | Rational Maps and Finkelstein–Rubinstein Signs           | 115 |
|      | 7.3   | Parity of States                                         | 117 |
|      | 7.4   | The Quantized Toroidal $B = 2$ Skyrmion                  | 119 |
|      | 7.5   | The Quantized $B = 3$ Skyrmion                           | 122 |
|      | 7.6   | The $B = 4$ Skyrmion and the $\alpha$-Particle           | 124 |
|      | 7.7   | $B = 5$ and $B = 7$                                      | 126 |
|      | 7.8   | Quantization of the $B = 6$ Skyrmion                     | 128 |
| 8.   | Skyrmions with Higher $B$ – Massive Pions                       | 131 |
|      | 8.1   | Massive Pions                                            | 131 |
|      | 8.2   | The Double Rational Map Ansatz                           | 132 |

| | | | |
|---|---|---|---|
| 8.3 | Skyrmions from $B = 8$ to $B = 32$ | | 135 |
| | 8.3.1 $B = 8$ | | 135 |
| | 8.3.2 $B = 12$ | | 136 |
| | 8.3.3 $B = 16$ | | 139 |
| | 8.3.4 $B = 24$ and $B = 32$ | | 141 |
| 8.4 | Geometrical Construction of Rational Maps | | 144 |
| | 8.4.1 $B = 24$ to $B = 31$ solutions by corner cutting | | 147 |
| 8.5 | Rational Maps with $O_h$ and $T_d$ Symmetry | | 149 |
| 8.6 | Skyrmions up to Baryon Number 256 | | 151 |
| 8.7 | Summary | | 154 |

## 9. Quantized Skyrmions with Even $B \leq 12$    157

| | | |
|---|---|---|
| 9.1 | Masses, Charge Radii and Calibration | 157 |
| 9.2 | $B = 4$ | 160 |
| 9.3 | $B = 6$ | 161 |
| 9.4 | $B = 8$ | 164 |
| 9.5 | $B = 10$ | 167 |
| 9.6 | $B = 12$ | 171 |
| | 9.6.1 Quantizing the $D_{3h}$-symmetric Skyrmion | 172 |
| | 9.6.2 Comparison with experimental data | 174 |
| | 9.6.3 The chain Skyrmion and the Hoyle band | 176 |
| | 9.6.4 Matter radii | 179 |

## 10. Skyrmion Deformations and Vibrations    181

| | | |
|---|---|---|
| 10.1 | The Need to Consider Vibrations | 181 |
| 10.2 | The $\alpha$-Particle and its Vibrational Excitations | 185 |
| 10.3 | Deformations of the $B = 7$ Skyrmion | 189 |
| 10.4 | $B = 12$ Skyrmion Deformations and Carbon-12 | 190 |
| | 10.4.1 A multiphonon model for Carbon-12 | 193 |

## 11. Modelling Oxygen-16    199

| | | |
|---|---|---|
| 11.1 | Introduction | 199 |
| 11.2 | The Vibrational E-Manifold | 201 |
| | 11.2.1 The Hamiltonian and quantum states | 203 |
| 11.3 | E-Manifold States | 204 |
| | 11.3.1 E-phonons | 207 |
| | 11.3.2 Rovibrational states | 209 |

11.4 Beyond E-Vibrations . . . . . . . . . . . . . . . . . . . . . . . 211
11.5 The Complete Oxygen-16 Energy Spectrum . . . . . . . . 214

## 12. Modelling Calcium-40    221

12.1 Tetrahedral Structure of Calcium-40 . . . . . . . . . . . . 221
12.2 Rovibrational Bands . . . . . . . . . . . . . . . . . . . . . . 225
12.3 Interpreting the Calcium-40 Spectrum . . . . . . . . . . . 226
12.4 Summary . . . . . . . . . . . . . . . . . . . . . . . . . . . . 230

## 13. Electromagnetic Transition Strengths    233

13.1 $B(E2)$ Transition Strengths in Skyrme Theory . . . . . . 234
    13.1.1 Beryllium-8 . . . . . . . . . . . . . . . . . . . . . . . 239
    13.1.2 Carbon-12 and the Hoyle state . . . . . . . . . . . 239
    13.1.3 Oxygen-16 . . . . . . . . . . . . . . . . . . . . . . . 240
    13.1.4 Neon-20, Magnesium-24 and Sulphur-32 . . . . . . 240
13.2 Beryllium-12 . . . . . . . . . . . . . . . . . . . . . . . . . . 242
13.3 Further Transitions . . . . . . . . . . . . . . . . . . . . . . 244
13.4 Summary . . . . . . . . . . . . . . . . . . . . . . . . . . . . 246

## 14. Variants of Skyrme Theory    249

14.1 Adding a Sextic Term . . . . . . . . . . . . . . . . . . . . . 252
    14.1.1 The lightly-bound Skyrme model . . . . . . . . . . 253
14.2 The BPS Skyrme Model . . . . . . . . . . . . . . . . . . . 254
14.3 Including Heavier Mesons . . . . . . . . . . . . . . . . . . 258
    14.3.1 Skyrme model with $\rho$ mesons . . . . . . . . . . . . 258
    14.3.2 Skyrme model with $\omega$ mesons . . . . . . . . . . . . 261
14.4 The $SU(3)$ Extension of Skyrme Theory . . . . . . . . . . 263

## 15. The Sakai–Sugimoto Model    269

15.1 Skyrmions from Instantons . . . . . . . . . . . . . . . . . . 275
15.2 An Expansion for 4-Dimensional Gauge Potentials . . . . 280
15.3 Baryon Resonances in the Sakai–Sugimoto Model . . . . . 285

*Bibliography*    291

*Index*    305

*About the Author*    309

# Chapter 1

# Introduction

What is a particle? Is it pointlike, or does it have a finite size? How does a particle exert forces on other particles, and are these forces affected by the particle's finite size? What is responsible for a particle's spin? Is the spin a fundamental attribute, as seems inevitable for a pointlike particle, or does the spin arise from quantization of rotational motion about an axis, something that is possible for a particle with spatial structure?

These are hard questions and they don't have a single answer. Indeed, the way these questions have been answered has changed substantially over time. This book is about one type of answer that applies specifically to protons and neutrons, the particles that exist close together as constituents of the nuclei in the cores of atoms, and to the composite nuclei made from them. In an electrically neutral atom, the nucleus is orbited by electrons, but when studying the physics of nuclei we can almost entirely ignore the electrons. So this book presents a theory of nuclei – the theory of Skyrmions. It is an unconventional theory, and somewhat controversial. We will explain its motivation in terms of more conventional ideas about particles and fields, but will need to go into considerable mathematical detail in order to formulate it carefully and understand it, both classically and quantum mechanically.

Skyrme theory is a field theory, whose classical field equation is a nonlinear partial differential equation, and whose most important solutions for us are the Skyrmions. Much work over several decades has gone into finding the Skyrmion solutions, many of which have great geometrical beauty. To relate these solutions to physics we need to introduce some quantum mechanical ideas, and this is the way to get models of the proton and neutron, and of larger nuclei. The calculated results match known properties of several nuclei with reasonable accuracy, and the theory makes further predictions concerning nuclear excitation spectra that could be tested experimentally. The theory runs into problems with some nuclei. This may be because the analysis is hard and not yet complete, rather than for a more fundamental reason. Support for this view is that a decade or more ago, far fewer nuclei were understood using Skyrme theory, but subsequent work has resolved earlier difficulties. We will argue that the theory can compete strongly with other theories of nuclei that have been developed over the past 100 years or so.

The theory of Skyrmions was formulated in a group of papers written around 1960 by Tony H. R. Skyrme [215, 216]. It is usually referred to as the Skyrme model, to indicate a more provisional character. But the theory is now 60 years old and has been studied and developed by more than one generation of physicists (with some overlap). So calling it Skyrme theory, or the theory of Skyrmions, now seems justified.

A Skyrmion is a static solution of a classical field equation describing the 3-component pion field and its nonlinear self-interactions. The pion particles are the lightest of the so-called mesons, the particles that interact strongly with protons and neutrons. No explicit proton or neutron fields are introduced; instead, the proton and neutron are quantum states of the Skyrmion constructed out of the pion field. It is not controversial that there is a long-range pion field surrounding a proton or neutron. This tail field is responsible for forces between protons and neutrons. What is remarkable in Skyrme theory is that this long-range tail has no source at its centre. It is entirely self-sustaining as a result of the nonlinearity of the pion field equation. The pion tail field is analogous to the grin of the Cheshire Cat – as one probes the pion field closer up, one finds there is no cat.

The most novel aspect of a Skyrmion is that it is stable because of its topological character. Topology is the branch of mathematics that deals with properties of a system that do not change under any continuous deformation. A Skyrmion has the property that its pion field cannot be continuously deformed into the vacuum field – it has a topological twist that cannot be destroyed. This means that the Skyrmion is not simply a pion, which is a quantized wavelike disturbance of the pion field, close to the vacuum and topologically trivial. It is a different type of particle.

The topological twist is conserved during any time-evolution of the field, no matter how energetic. This captures a key property of protons and neutrons and all their interactions, the conservation of total baryon number. A proton and a neutron each have one unit of baryon number, and the baryon number of a nucleus is the sum of its proton and neutron numbers, so it is an integer. In Skyrme theory, a Skyrmion's degree of topological twisting is characterised by an integer too, and this is the Skyrmion's conserved baryon number. The proton and neutron are distinct quantum states of the basic Skyrmion with unit baryon number.

A Skyrmion is an example in 3-dimensional space of the more general notion of a topological soliton [184]. A topological soliton is a localised but smooth classical solution of a nonlinear field equation that doesn't need a source, and has finite energy. Far from its centre the field approaches the vacuum, but around its centre the soliton has a topological twist, which stabilises it. The soliton therefore behaves as a classical model of a particle of finite size. Usually, if the soliton is at rest then the field configuration is time-independent. This is the case for a Skyrmion. A soliton can also be in motion, and in a relativistic field theory like Skyrme theory the soliton is free to move through space at any speed less than the speed of light. The soliton is Lorentz-contracted in its direction of motion if its speed is large.

In fact, the most remarkable aspect of Skyrme theory is the way it unifies particles and fields. A Skyrmion represents a particle of matter – a material object – yet it is a solution of a classical field equation. Fields are usually interpreted dynamically, as giving rise to forces, waves and other transitory processes, and matter particles are usually the sources for fields, or provide the background media where fields operate. Several other types of soliton, e.g. vortices in superconductors, localised waves in fluids, and optical solitons, require a background material medium, but a Skyrmion exists as a particle in empty space.

The basic Skyrmion is found to have a spherically symmetric energy density, like a billiard ball, but with a smoother and softer radial profile. Like a billiard ball, the Skyrmion can classically rotate. While at rest, it has a definite orientation in space determined by the internal structure of the pion field components, so rotational motion is physically possible and increases the kinetic energy. This rotational motion needs to be quantized, and the second remarkable aspect of the basic Skyrmion is that in its lowest-energy quantum state it has spin $\frac{1}{2}\hbar$, where $\hbar$ is Planck's constant (below, we will just say spin $\frac{1}{2}$). This half-integer spin has a topological explanation.

It is usually thought that a quantized classical rotor can only have an integer spin. It was one of Skyrme's insights that half-integer quantization of a Skyrmion's spin is possible. This makes the model of a proton or neutron as a quantized Skyrmion unconventional. Normally, the spin of these particles is either regarded as fundamental and treated algebraically using a 2- or 4-component spinor, or regarded as arising from the spin of the confined quarks inside. In Skyrme theory, it is the quantization of the Skyrmion's orientational degrees of freedom that is responsible for the spin.

Work on Skyrmions, until the present, can be divided roughly into four phases. First, there was the pioneering work of Skyrme. It made little impression, mainly because theoretical nuclear and particle physicists were not in the habit of solving nonlinear classical field equations, and the available computational resources were too limited to do this easily. A little later, Finkelstein and Rubinstein studied topological aspects of the quantization of Skyrmions, and established a spin-statistics relationship for Skyrmions [87]. They showed that a Skyrmion with spin $\frac{1}{2}$ is inevitably a fermion, whereas if the Skyrmion had been quantized to have integer spin, then it would be a boson. The existence of a relatively simple topological explanation for the connection between quantum mechanical spin and statistics in Skyrme theory is one of the theory's surprising strengths. The usual arguments for the spin-statistics relation depend on subtleties of relativistic field correlation functions, as discussed long ago by Streater and Wightman [220], building on an argument of Pauli, but the topic is complicated and relies on certain assumptions [76]. Other attempts to prove the spin-statistics relation in non-relativistic quantum mechanics also rely on certain assumptions [38]. The explanation from Skyrme theory is much more attractive, although it could be argued that it relies on the assumptions of Skyrme theory itself – in particular, that a baryon is a quantized topological soliton.

The second phase was initiated by a series of influential papers by Ed Witten in the early 1980s showing how, in a certain limit, the Skyrmion picture of protons and neutrons is natural even from the perspective of quantum chromodynamics (QCD), the more fundamental theory of quarks and their interactions via gluons [251]. Witten's work inspired many others, and as a result, there was substantial progress understanding many detailed properties of protons and neutrons, including some of their excited states, in terms of Skyrmions. There was also progress understanding related particles with unit baryon number containing heavier quark types [19], and the forces between Skyrmions, leading to consequences for interacting protons and neutrons [139, 193, 233].

The third phase started (one might say) with the discovery of the Skyrmion having baryon number 2 [155, 231], and was followed by the impressive work by Braaten and collaborators in 1990, finding Skyrmions having higher baryon numbers [53]. These solutions, found through substantial numerical work, have surprisingly symmetric geometrical forms that were not expected, nor immediately understood. At this point, it was not possible to explore Skyrmions beyond baryon numbers 5 or 6. There was then further progress in understanding Skyrmion solutions, and especially those with higher baryon numbers, from a deeper mathematical perspective. The tools were those of differential geometry, rather than rigorous analysis of partial differential equations. It was realised that Skyrmions had analogies with soliton solutions of superficially quite different types of field equation, having physical applications outside nuclear physics. This was helpful for understanding the intrinsic symmetries of Skyrmions, and their dynamics. Michael Atiyah, the mathematician, contributed to these developments. One analogy was with magnetic monopoles of non-abelian gauge theory [14] – like Skyrmions, these are topological solitons in three dimensions – and another was with instanton solutions in 4-dimensional gauge theory [16]. As a result, many new static Skyrmion solutions were found, for all baryon numbers up to 22, and for several further baryon numbers up to 100 and beyond (leaving some gaps) [30, 85]. Most of these solutions are energy-minimising, given their baryon number, although some are local rather than global minima. Also a Skyrmion crystal solution, periodically filling all of space, was established [62, 158].

Another analogy between non-abelian monopoles and Skyrmions is in their classical scattering behaviour. Both for monopoles and Skyrmions, the dynamics is completely smooth. Although there are magnetic Coulomb forces between separated non-abelian monopoles, the monopoles have no point singularities in their cores. This is a basic consequence of the field dynamics being fundamental; the monopole cores are smooth and of finite size, and the point-particle description is only an approximation. Most remarkably, the monopoles scatter in a non-trivial and interesting way – in particular, they scatter at 90° in a head-on collision [14]. Skyrmion field dynamics is a little more complicated, but similar behaviour occurs.

In the fourth phase, it has been possible to explore the quantum mechanical properties of many of these Skyrmions, and compare with observed properties of nuclei. Some of this work, especially for small nuclei like the deuteron, Helium-3 and Helium-4, was done earlier [51, 61, 154, 240], but to a large extent, only the rigid-body motions of the Skyrmions were considered and therefore only the ground state and selected excited states were successfully modelled, not the full excitation spectra of these nuclei. More recent work has taken into account further, dynamical degrees of freedom – Skyrmion vibrations. The intrinsic symmetry of a Skyrmion determines the Skyrmion's permitted spin states when it is quantized as a rigidly-rotating body. The symmetry also controls the degeneracies and nature of the vibrational modes of the Skyrmion [108], and the permitted spin states when it is vibrationally excited. Some quantized vibrational motions need to be handled in a nonlinear way. This goes well beyond simply finding static Skyrmion solutions, but does not require a full study of time-dependent solutions of the classical Skyrme field equation, which remains a formidable challenge. It is now possible to model the excitation spectra of well-known nuclei like Carbon-12 [162, 203] and Oxygen-16 [116, 117] using Skyrmions. In some sense this is easier than modelling Helium-3 and Helium-4, because many more excited states are known experimentally for these larger nuclei, and there is a clearer picture of how these states are classified into vibrational and rotational bands. An improved understanding of Helium-4 has also recently emerged [204].

Here's an outline of the rest of this book. Skyrme theory is a nonlinear quantum field theory constructed from pion fields. Since a pion is itself not a fundamental particle, but a composite of a quark and antiquark, Skyrme's theory is what is called an effective field theory (EFT) [78, 167, 242]. In Chapter 2 we sketch how particles and fields are currently understood, and summarise the Standard Model, the fundamental theory of elementary particles, including quarks. Then we explain how Skyrme's theory emerges from this as an EFT. One could just discuss the theory without this justification. That was how Skyrme proceeded, because the Standard Model hardly existed in 1960 whereas pions were well known, but it is useful to put Skyrme's theory into a broader and more modern context. Both the Standard Model and Skyrme's theory are Lagrangian field theories. In Chapter 3 we introduce the Lagrangians of general finite-dimensional dynamical systems and of field theories, and in particular recall the way conserved Noether currents arise from a theory's continuous symmetries.

In Chapter 4 we present Skyrme theory in detail, deriving its field equation from the Lagrangian and explaining the theory's symmetries. We discuss how the topological baryon number $B$ arises, and describe the basic $B = 1$ Skyrmion solution, which is called a hedgehog because of the character of its pion field. This was all known to Skyrme. We also consider the forces between Skyrmions. In Chapter 5 we discuss the rigid-body quantization of the $B = 1$ Skyrmion, and explain why it should be quantized as a fermion. This leads to a model for protons and neutrons as spin $\frac{1}{2}$ particles, and the delta resonances as spin $\frac{3}{2}$ excitations of these. The initial calibration of Skyrme theory by Adkins, Nappi and Witten [7] was designed to fit these quantum states.

Chapter 6 presents in some detail the mathematical developments that led to the discovery of a large range of Skyrmion solutions with all baryon numbers up to $B = 22$, for a massless pion field. These developments occurred largely in the 1990s following the discovery of solutions with small baryon numbers having surprising geometrical shapes. A very useful tool for finding solutions and understanding their symmetries is the rational map ansatz [130], an approximate method for separating variables in the Skyrme field equation. The spatially periodic Skyrmion crystal is also described. Chapter 7 discusses the quantization of higher-$B$ Skyrmions as rigid bodies. The quantum states obtained this way are labelled by spin and parity, and also by isospin, the quantity that keeps track of the relative number of protons and neutrons in a nucleus. The rational map ansatz is helpful to ensure that the quantization of a larger nucleus is consistent with the fermionic character of its constituent protons and neutrons. However, only a limited number of observed states of nuclei can be identified as purely rotational states of a rigid body.

In Chapter 8 we look more carefully at the effect of the pion mass being non-zero (though relatively small) [5,32,33]. The Skyrmions for $B$ less than 8 are hardly changed, but for $B = 8$ and higher the effect is substantial, and the Skyrmions have a more compact structure than in the massless pion case, with different symmetries. We present Skyrmion solutions with massive pions for selected baryon numbers up to $B = 256$. Some of these are constructed with the help of a multi-layer extension of the rational map ansatz.

Chapter 9 discusses further the rigid-body quantization of Skyrmions with even baryon numbers, up to $B = 12$, and reconsiders the energy and length calibration of Skyrme theory, to avoid reliance on the highly unstable delta resonances.

In Chapter 10 we move beyond rigid-body Skyrmion dynamics, and consider the classical, harmonic vibrational spectra of Skyrmions. To a limited extent, anharmonic, larger-amplitude deformations of Skyrmions are also discussed. It is shown that the quantization of vibrational degrees of freedom together with rotational motion gives a rich nuclear spectrum closer to what is experimentally observed. This is mostly recent work, and takes into account the subtle issue of the Coriolis coupling of vibrations and rotations. Vibrational excitations of the $\alpha$-particle (Helium-4 nucleus), the isospin-doublet Beryllium-7/Lithium-7, and the Carbon-12 nucleus are analysed in detail.

In Chapters 11 and 12, these ideas are extended to the doubly-magic nuclei Oxygen-16 and Calcium-40, and it is shown that their spectra can be classified as rotational-vibrational bands of Skyrmions with intrinsic tetrahedral symmetry. The model for Oxygen-16 allows for anharmonic motion of the four $\alpha$-particle constituents, from a tetrahedral configuration through a square configuration into the dual tetrahedron. It gives perhaps the best description of a nucleus so far, based on Skyrme theory, and accounts for dozens of excited states. Unfortunately, the vibrational frequencies have not yet been calculated in detail for large Skyrmions so the modelling remains somewhat phenomenological.

Chapter 13 presents an analysis of the electromagnetic transitions between selected nuclear states. In particular, $E2$ (quadrupole) transition strengths are calculated, using Skyrme theory to find the relevant quadrupole moments.

Chapter 14 discusses variants and extensions of Skyrme theory, including models where the pion field is coupled to fields representing the heavier $\rho$ and $\omega$ mesons. In the concluding Chapter 15, we sketch the holographic Skyrme model of Sakai and Sugimoto [209]. This has its foundation in string theory, but in detail it reformulates $(3+1)$-dimensional Skyrme theory as a Yang–Mills gauge theory in $4+1$ dimensions. Here, a 3-dimensional Skyrmion appears as a holographic image of a 4-dimensional Yang–Mills instanton, and the Skyrmion is automatically coupled to a tower of vector and axial-vector mesons with completely determined couplings.

In the last ten years, there has been great interest in a variant of Skyrme theory whose solutions are Magnetic Skyrmions [86]. Magnetic Skyrmions were first observed around 2010, although the theoretical ideas go back further. The subject is now so popular and widely studied that it is these objects that are often simply called Skyrmions. This book, however, is about Skyrme's original Skyrmions and their application in nuclear physics. The Magnetic Skyrmions share a similar topological mechanism to ensure their stability, but they are different in several ways. They exist, basically, in two dimensions, in thin layers of rather exotic magnetic materials. Their field is the direction of the local magnetisation vector, not the pion field, and this has fewer components than a pion field. Physically, Magnetic Skyrmions only exist in a condensed matter context and move very slowly, whereas (nuclear) Skyrmions exist freely in 3-dimensional space and can move at arbitrary speeds less than the speed of light. Another important difference is that Magnetic Skyrmions behave classically, and there is no need to quantize them. They are sufficiently large – about 100 nm across –

that many images have been made of them by a variety of experimental techniques. There are, unfortunately, no direct images of nuclei at their characteristic length scale of about 1 fm to confirm or rule out Skyrme's original theory. However, the clear evidence for the existence of Magnetic Skyrmions has increased interest in the original Skyrmions and stimulated further research into them.

Finally, here are a few suggestions for further reading: For a survey of the work of Skyrme and of others during the first and second phases of research on Skyrmions, see the collection of articles by Brown [57], to which should be added one more of Skyrme's papers [216]. The research on Skyrmions from this period, with many interesting details about its historical development, is reviewed in the monograph of Makhankov, Rybakov and Sanyuk [171]. Some of the mathematical aspects of Skyrme theory have been reviewed by Gisiger and Paranjape [103], and multi-Skyrmion solutions but not their quantization were discussed in some detail in Chapter 9 of the monograph by Sutcliffe and the present author [184]. Weigel's monograph [243], reviewing Skyrme theory, focusses on the connection between Skyrmions and quark models, and includes a discussion of exotic baryon states, including pentaquarks. More recent work on Skyrmions, including work on Magnetic Skyrmions and the holographic approach to Skyrmions, is surveyed in the volume edited by Rho and Zahed [205], which is an expanded version of the earlier survey volume edited by Brown and Rho [58].

Chapter 2

# Fields and Particles

This chapter outlines how quantum field theory emerges as the theory describing all fundamental forms of matter, superseding the notions of point particles and classical fields. Even particles like pions, protons and neutrons, which are composites of the more fundamental quarks, can be understood using an effective quantum field theory (EFT). It is possible to construct an EFT with pion, proton and neutron fields, but Skyrme proposed a more radical type of EFT constructed from pion fields alone. Skyrme theory has topological soliton solutions – Skyrmions – and when these are quantized, they become models for the proton, neutron and larger nuclei.

## 2.1 The Classical Notions of Particles and Fields

Physicists have two fundamental ways to describe matter. One way is as particles and the other is as fields. The description as particles goes back to Galileo and Newton. Ideal particles have their mass concentrated at moving points, whose trajectories through space obey Newton's laws. Newton's equations of motion allow for quite general forces between two particles, and even more complicated forces involving several particles. The form of these forces has to be postulated. The equations of motion are then a system of ordinary differential equations which can be solved analytically in some cases, but otherwise numerically.

Newton's picture of pointlike particles is successful particularly in the case of the inverse square law force of gravity, because it can be shown rigorously that large spherical bodies like the Earth and the Sun behave as if all the mass is concentrated at the central point. More generally, though, the gravitational forces between extended bodies is less simple and has to be derived by summing or integrating the effects of all the constituent particles.

For extended classical bodies like fluids and elastic solids, we cannot usually see the constituent particles, and such bodies are described better as continua of matter. Their dynamical properties are expressed mathematically in terms of classical fields. A field $\phi(t, \mathbf{x})$ is mathematically just a function of time and space, but it describes something physical. A key example is the density of matter. In the point-particle picture, the mass density is infinite at the particle locations and zero elsewhere, but for very many particles, a smooth mass density can be defined by averaging over small regions. Another field, important for a fluid, is its velocity. This is also assumed to be a smooth function, varying in time and space.

In these examples, the fields are an effective description of matter. We do not believe in them as the fundamental description of the atomic or molecular constituents, but they are useful over a large range of scales, from microns up to 1000 km, the scale of atmospheric and oceanic flows, and beyond. There are well-established equations for the fields describing a fluid – the Navier–Stokes equations. These are partial differential equations having many known special solutions, although the general solutions, especially for turbulent flows, are hard to study even numerically. An important aspect of the equations is that they incorporate a few phenomenological parameters, including the density of the fluid, its compressibility, and its viscosity. In principle, these parameters can be derived from the underlying molecular physics, but in practice this is hardly possible except for some simple systems, like an ideal gas. Instead, they are fitted through observations. The density of uncompressed water at room temperature is a good example of such a fitted parameter.

This is typical of effective field theory more generally, There is often an underlying, more fundamental theory involving physics at a shorter length scale, but it is not practical to deduce all aspects of the effective theory from the more fundamental one. Sometimes the fundamental theory is not really known.

During the 19th century, the notion of a more fundamental classical field emerged, in the realm of electromagnetism and light. It developed through the study of electric and magnetic phenomena by Faraday and others. The space around electric charges and currents was seen to be filled with electric and magnetic fields. These fields were initially thought to be always localised close to the charges and currents, but when Maxwell perfected the equations for the fields, now unified into an electromagnetic field, it was found that there were solutions representing electromagnetic waves travelling arbitrary distances through space, matching the properties of light. In particular, electromagnetic waves all had the same speed – the speed of light as measured by Fizeau and others – and they exhibited the observed interference and polarisation effects of light. These wavelike features of light could not be explained by Newton's older corpuscular (i.e. particle) theory of light. No medium for the electromagnetic waves was required, except space itself.

The electromagnetic field is now understood to be a fundamental entity, independent of matter particles, and present throughout space for all time. Although it was not known precisely in Maxwell's time, we now recognise that sources of light do involve charges and currents, usually produced by moving matter at the atomic scale and requiring quantum theory to be understood in detail. (In the 19th century, known sources of light included flames and hot metal bodies, which were not thought to have any associated charges or currents. Now we know that they have – the excited electrons in the hot atoms.) In modern language, we say the electromagnetic field is coupled to charged particles.

We will hardly need to consider gravity in this book but for completeness, let's mention how it fits in. A classical gravitational field surrounds all massive particles according to Newton's theory. Einstein's general relativity theory has replaced this. It describes gravitational effects not in terms of fields in a fixed spacetime background, but as a continuous deformation of spacetime itself. The geometry, and in particular the spacetime metric, becomes a dynamical field, and as in Maxwell's theory there are wave solutions in empty space – gravitational waves travelling at the speed of light.

### 2.1.1  *Quantum theory*

The development of quantum theory in the early 20th century initially occurred largely independently for fields and for particles. Planck's insights, developed by Einstein, led to the notion that the electromagnetic field, and especially its wavelike solutions, should be quantized. A quantum state of the electromagnetic field consists of photons, and a photon is a kind of particle, having energy and momentum but zero mass, so it moves at the speed of light. Photons are therefore unusual, regarded as particles. They are not spatially pointlike, but manage to unify the idea of Newton's corpuscles with the wavelike interference effects observed for light. At the same time they are consistent with Einstein's special theory of relativity.

The quantum theory of the gravitational field has not been firmly established, and no graviton particle (analogous to a photon) has yet been observed, although classical gravitational waves have been recently detected at LIGO, and are observed to travel vast distances at the speed of light. So, if a graviton exists, it is likely to have zero mass like a photon.

The quantum mechanics of massive particles, initiated by Bohr and developed by de Broglie, Schrödinger and Heisenberg to understand atomic structure, looks different. This is a non-relativistic theory, although small relativistic corrections can be incorporated for particles whose speeds begin to approach that of light. These developments didn't really upset the dichotomy between particles and fields. Particles were the physical objects with positive mass that could be at rest or in motion, whereas the known fields were the electromagnetic field and the gravitational field. In particular, a massive particle could still be thought of as pointlike. Its quantum state is a delocalised wavefunction defined everywhere in space, whose squared magnitude describes the probability for finding the particle at a particular location. But a wavefunction should not be thought of as a field in the sense that we are using the term.

## 2.2 Quantum Field Theory and the Standard Model

A convergence between fields and particles occurred through the development of quantum field theory starting in the 1920s and 30s. It was discovered that a unified quantum theory of photons and the lightest charged particles with positive mass – electrons – needed a formalism which introduced an electron field, supplementing the electromagnetic field. The advantage of a quantized electron field over the standard quantum mechanics of electrons is that states can have any number of electrons, and the theory can accommodate electrons moving at very high speeds, close to the speed of light. This is because, from the start, quantum field theory has the Lorentz symmetry known from the classical theory of light, and built into the framework of special relativity. As shown by Dirac, the theory makes an inevitable prediction that there must exist the antiparticle of the electron – the positron – which has the same mass as an electron (and behaves the same way gravitationally) but the opposite electric charge. A collision of two electrons with sufficient energy can easily result in further electrons and positrons being produced. It is also possible for an electron and positron to annihilate into a pair of photons, or a larger number of photons. These particle-number changing processes can be understood straightforwardly in quantum field theory.

More broadly, although quantum field theory is formulated in terms of fields, the physical states predicted by it are interpreted as particles. This applies simultaneously to massive particles like electrons and to massless particles like photons. Quantum field theory can deal with processes where the number of particles of various types can vary, while still ensuring that conservation laws, like conservation of total electric charge and total energy, are satisfied.

Quantum field theory is usually studied using perturbation theory, where it is assumed, at least formally, that the field amplitudes are small and that distinct fields are relatively weakly coupled. Particle states arise from the quantization of the wavelike fluctuations of the fields around their constant, classical vacuum configuration. This means that in their simplest quantum states, massive particles like the electron have energy and momentum, but they are not spatially localised. Electrons, like photons, are quantized, spatially extended waves. It is possible to construct, by quantum superposition, a state where a massive particle is very likely to be in a small spatial volume, but the theory doesn't give the particle any precise intrinsic structure, either as a pointlike or extended object. This feature is not really a problem for electrons or quarks, because they have no experimentally established structure – they are the quantum analogues of classical point particles. But for protons and neutrons, which do have intrinsic structure – a measured radius, for example – this is a drawback.

Detailed calculations in the (perturbative) quantum field theory of photons and electrons – Quantum Electrodynamics, or QED – are organised using Feynman diagrams. The calculations are hard, but agree exceedingly well with all observations. The theory extends to further particles like the muon and its antiparticle. By the 1970s, quantum field theory had been extended to include many more fields, basically one field for each particle believed to be fundamental. The result is the Standard Model of particle physics, which describes the six types of quark, the leptons (electron, muon and tau, and their corresponding neutrinos), and the gauge bosons (photon, W-bosons and Z-boson, gluons). The part of the theory involving quarks and gluons is known as Quantum Chromodynamics (QCD), because of the so-called colour charges of these particles, which are the QCD analogue of the electric charges in QED. The final ingredient is the Higgs field and its corresponding particle, the Higgs boson. In this complicated, though successful quantum field theory, the most important particles for us are the two lightest (i.e. least massive) quarks, known as the up and down quarks, and denoted u and d. These are the constituents of the proton and neutron, from which all atomic nuclei are built. We shall say more about them later.

Even in this unified picture there are still two important distinctions between the particles traditionally described as matter, and the particles like photons that are traditionally associated with fields. First, the matter particles – the quarks, and the leptons including neutrinos – all have positive mass (although the neutrino masses are very small and not yet fully known). The photon and gluons are massless, and the W-bosons and Z-boson responsible for weak interactions are also in some primordial sense massless, acquiring a large mass through their interaction with the Higgs field. The second, even more clear-cut distinction is that all the matter particles have spin $\frac{1}{2}$ and are fermions, whereas the remaining particles (photon etc.) have spin 1 or spin 0 and are bosons. The spins of these particles are built into the algebraic structure of quantum field theory.

The different behaviours of fermions and bosons extends from quantum mechanics to quantum field theory. An important consequence is that the field associated with a fermion, e.g. an electron field, is not directly observable despite obeying a classical equation. This is because of the Pauli principle, which allows at most one fermionic particle to occupy a given quantum state. As a result, the quantized particles cannot combine coherently to produce a measurable classical effect. Quantum states of a bosonic particle, on the other hand, can be occupied by arbitrarily many of the particles, leading to the macroscopic effects described by a classical field like Maxwell's electromagnetic field.

Note that the physical particles, observed in experiments, are not necessarily the fundamental ones. Quarks in particular are elusive, and almost certainly permanently confined inside other particles. They leave no tracks, although their effect is seen indirectly in very high energy particle collisions, where they can be associated with jets of outgoing, visible particle tracks. In the relatively low-energy regime of nuclei, quarks are essentially invisible.

## 2.3 Pions and Nucleons

The physical particles important for us – the proton, neutron and pions – are understood to be bound states of quarks, or of quarks and their antiparticles [256]. A quark-antiquark bound state is called a meson. The three lightest mesons are the pions. They are bound states of u or d quarks and $\bar{\text{u}}$ or $\bar{\text{d}}$ antiquarks in the combinations

$$\pi^+(\text{u}\bar{\text{d}}) \ 140\,\text{MeV}, \quad \pi^0(\text{u}\bar{\text{u}} - \text{d}\bar{\text{d}}) \ 135\,\text{MeV}, \quad \pi^-(\text{d}\bar{\text{u}}) \ 140\,\text{MeV}. \quad (2.1)$$

Transitions between the u$\bar{\text{u}}$ and d$\bar{\text{d}}$ states are rapidly mediated by gluons, so the $\pi^0$ is a quantum superposition of these states.

The quarks and antiquarks each have spin $\frac{1}{2}$, and the pions are bound states of these with spin 0. States with spin 1 and higher are also possible, but have considerably higher masses (energies). Examples of these are the $\rho$ and $\omega$ mesons, which will reappear much later. The pion electric charges are, respectively, $+1, 0, -1$ in units of the proton charge, as shown by the superscripts in (2.1), and the pion masses are given in energy units (for comparison, the electron mass is 0.5 MeV, i.e. much lighter, and the proton mass is 938 MeV, i.e. considerably heavier). The $\pi^-$ is the antiparticle of the $\pi^+$, so must have the same mass. The neutral $\pi^0$ can and does have a different mass, but the near equality of the pion masses hints at a larger approximate symmetry.

At a fundamental level, this near equality is believed to result from the u and d quarks having, accidentally, nearly equal masses of just a few MeV. Most of the mass of the pions is attributed to the gluon field surrounding the quarks. In nuclear theory (and more generally, in low-energy particle theory) it is helpful to recognise the approximate symmetry between u and d quarks, and between pions, as arising from a (hypothetical) exact symmetry called *isospin symmetry*. This symmetry acts on the three pions by $SO(3)$ *isorotations*, rotating them into each other. It is very helpful to think of isorotations as acting on an internal 3-dimensional *isospace*. Using isospin symmetry, one can account for many similarities between the pion properties, not just their masses but also the way they mutually interact, and interact with protons and neutrons. The triplet of pions combine into an isovector (i.e. a vector in isospace) with isospin $I = 1$. The proton and neutron, referred to collectively as *nucleons*, combine into a doublet with isospin $I = \frac{1}{2}$ (as do the u and d quarks), and an important consequence of the symmetry is the near equality of the proton mass 938 MeV and neutron mass 940 MeV. Isospin is formulated mathematically rather like spin, which is why the $SO(3)$ isospin group can be extended to $SU(2)$ and have half-integer representations. The proton has $I_3 = \frac{1}{2}$ and the neutron $I_3 = -\frac{1}{2}$, where $I_3$ is the third component of isospin, the projection of isospin on to the 3-axis in isospace. Isospin symmetry will be vital when we formulate Skyrme theory[1].

The nucleons are the lightest of the other class of particles with quark

---

[1] In nuclear physics, the standard notations for isospin and its projection are $T$ and $-T_3$ rather than $I$ and $I_3$.

constituents – the baryons. Baryons are three-quark bound states – a consequence of the QCD colour gauge group being $SU(3)$. As quarks have spin $\frac{1}{2}$, a bound state of three quarks also has half-integer spin, and is a fermion. Using u and d quarks, there are potentially four combinations making a baryon: uuu, uud, udd, ddd. However, because of the Pauli principle, uuu and ddd states have higher energy, and the states of lowest energy are the combinations uud and udd, which are respectively a proton and a neutron, both having spin $\frac{1}{2}$. The electric charges of quarks are exotic, because for the proton and neutron charges to come out as 1 and 0, the u and d quarks have to have electric charges $\frac{2}{3}$ and $-\frac{1}{3}$ respectively. One can check that the pion charges are consistent with this. Three-antiquark states also exist – the antiproton and antineutron – but they rarely occur in nuclear physics and we will hardly mention them. Recall that the u and d quarks have masses of just a few MeV. That means that most of the nucleon mass arises from internal kinetic energy and the energy of the gluons.

Important for us later is to note that, in addition to the proton and neutron, there are four further baryons with quark constituents uuu, uud, udd and ddd, having spin $\frac{3}{2}$ and charges between 2 and $-1$. These are the higher-energy delta resonances, which form a quartet of states with isospin $I = \frac{3}{2}$. Their masses are approximately 1230 MeV. Delta resonances are not stable particles, but are observed as very short-lived quasi-bound states in collisions of pions and nucleons.

Let's say something next about the stability of the nucleons and the pions. Of all these particles, only the proton is absolutely stable. A free neutron decays to a proton and two leptons (an electron and an antineutrino) with a long lifetime of about 15 minutes. This is an example of radioactive beta decay. At the quark level, one d quark is converting to one u quark, a process mediated by a W-boson. The decay is possible because the neutron mass is slightly greater than the sum of the masses of the proton and the two leptons.

The charged pions have a moderately long life by particle physics standards – about $10^{-8}$ s – but the neutral $\pi^0$ decays much more rapidly. The decay products are again leptons, or photons in the case of the $\pi^0$. It is possible using a particle accelerator to create a beam of charged pions that travel many metres before decaying, and to aim them at a target consisting of protons and neutrons inside ordinary matter.

Although a free neutron is unstable, and numerous larger nuclei undergo beta decay as one of their constituent neutrons decay, it is important to realise that in many atomic nuclei, composed of both protons and neutrons, the neutron is completely stabilised. This is purely for energetic reasons. If one neutron in such a nucleus decayed to a proton, the nucleus would acquire higher energy (largely because of the Pauli principle applied to the protons), with some of the additional energy going into the escaping electron and antineutrino. By energy conservation, this transition is therefore forbidden in many nuclei, and they are stable to beta decay. In particular, many familiar light and medium-mass nuclei from which the everyday world is made, e.g. Carbon-12, Oxygen-16 and Iron-56[2], are completely stable.

One of the chief discoveries concerning protons, neutrons and all larger nuclei is that in every known physical process, there is overall conservation of baryon number, i.e. conservation of the net number of nucleons. Collisions knock nucleons about, and collisions at very high energy can result in the production of nucleon-antinucleon pairs; beta decay changes neutrons into protons, and in some circumstances the reverse of this happens. However, if we count each nucleon as having one unit of baryon number and each antinucleon as having minus one unit, then the total baryon number $B$ is always conserved. One of the most stringent tests of this involves watching for proton decay in a large tank of water. There are plenty of protons in the hydrogen and oxygen atoms in water, and proton decay is energetically possible if baryon number is not conserved, because a proton could decay to a positron or a $\pi^+$, together with perhaps a neutrino and/or photon, conserving electric charge and energy. But such a decay has never been seen, and it is inferred that the proton lifetime is more than $10^{34}$ years, if not infinite. Conservation of baryon number can be interpreted in quark language as saying that net quark number is conserved. For the book-keeping to work, one needs to identify the baryon number $B$ of a quark as $\frac{1}{3}$ and that of an antiquark as $-\frac{1}{3}$. Pions and other mesons, being quark-antiquark states, have zero baryon number; nucleons, being three-quark states, have unit baryon number. Conservation of baryon number is one of the key ideas motivating Skyrme theory.

In nuclear physics, the nucleons are the most important particles, and there is, as yet, no complete understanding of nuclear phenomena starting from quarks. Admittedly, some of the most sophisticated studies of proton and neutron structure are based on quark dynamics but the study of the structure of the smallest nuclei like the deuteron (a bound state of a proton

---

[2]In this notation for a nucleus, we give the chemical name, determined by the proton number, and the baryon (atomic mass) number, which is the sum of the proton and neutron numbers. So Carbon-12 has 6 protons and 6 neutrons. Also often used is the notation $^{12}$C, where C is the chemical symbol for carbon.

and neutron) and the α-particle (a bound state of two protons and two neutrons) in terms of quarks has hardly started. The theoretical calculations of, respectively, the 6- and 12-quark dynamics are too hard. So we need to consider possible effective field theories for nucleons and pions, of which Skyrme theory is an example.

From here on, for much of this book, we will ignore the leptons and even the photons. We just focus on the strong interaction (QCD) physics of pions and nucleons. This determines almost all properties of nuclei in the first approximation, including nuclear masses and binding energies, spins and parities. The spectra of excited states of nuclei, and many aspects of nuclear collision dynamics, are aspects of strong interaction physics too. Using slightly old-fashioned terminology, we are focussing on the strong nuclear force, and ignoring the weak nuclear force (mediated by the W-boson) responsible for beta decay. We also ignore for now the electromagnetic forces that distinguish protons from neutrons, and we certainly ignore gravity, which plays a negligible role in nuclear physics, except in astrophysical settings like neutron stars.

Ignoring electromagnetic effects is a bit dangerous, and we will need to consider them again later. The substantial electrostatic energy of many protons close together is the reason why larger nuclei tend to have more neutrons than protons, and why they tend to fission, limiting how large they can be. Also, the repulsive electrostatic force between smaller nuclei inhibits the fusion of these nuclei, despite the strong, attractive nuclear force at short range. Finally, emission of gamma rays (high-energy photons) is the key mechanism for an excited nucleus to decay towards the ground state. This is radioactive gamma decay, and the gamma ray energies give the most precise information about the energy spectrum of a nucleus. By ignoring such decays, we are making all the excited states below the breakup threshold for the nucleus into quantum mechanical stationary states with infinite lifetime. This is similar to what is done in atomic physics. First one works out the stationary states of electrons orbiting a nucleus, and then allows for the relatively small effect of transitions from high-energy to low-energy electron states due to photon emission. The characteristic timescale for strong interactions in nuclei is about $10^{-24}$ s, whereas excited states of nuclei have a relatively long lifetime of order $10^{-15}$ s to $10^{-9}$ s for gamma ray emission, so to first approximation, gamma decays can be ignored.

## 2.4 Nuclei and Isospin

In Fig. 2.1 we show an energy-level diagram for the nuclei with baryon number 7 [226]. This is a typical diagram for a relatively small nucleus, and there are many similar examples [59, 77]. The lower-lying levels have precisely known energies, and the spins and parities of the states are also known. The states above the ground state decay by gamma ray emission. The higher-lying states are more unstable and decay predominantly by nuclear breakup, which is a strong interaction process. These higher states are therefore resonances, so their energies are more uncertain – the energy levels have a significant width. The nuclei with baryon number 7 include Beryllium-7, with 4 protons and 3 neutrons, and Lithium-7, with 3 protons and 4 neutrons; also Boron-7 with 5 protons and 2 neutrons, and Helium-7 with 2 protons and 5 neutrons. Lithium-7 in its ground state is stable, and Beryllium-7 is almost stable, with a lifetime of 53 days. In fact, an isolated Beryllium-7 nucleus is stable, but in a neutral atom it can capture an electron, which converts one proton to one neutron and produces a Lithium-7 nucleus, with the emission of a neutrino. This is a weak interaction process typical for proton-rich nuclei. Boron-7 and Helium-7 are much less stable.

It is clear that the energies and spin/parities of Beryllium-7/Lithium-7 states match up rather well. These nuclei form an isospin doublet with $I = \frac{1}{2}$, similar to the proton/neutron doublet. The third component of isospin is assigned the values $I_3 = \frac{1}{2}$ for Beryllium-7 and $I_3 = -\frac{1}{2}$ for Lithium-7. $I_3$ is related to the electric charge $Q$ and baryon number $B$ through the formula $Q = \frac{1}{2}B + I_3$ (the Gell-Mann–Nishijima formula with no heavy quarks present). Equivalently, $2I_3$ is the excess of protons over neutrons. The Boron-7 states have $I_3 = \frac{3}{2}$ and the Helium-7 states $I_3 = -\frac{3}{2}$. The ground states of these nuclei form part of an $I = \frac{3}{2}$ multiplet, and for isospin symmetry to work, it is essential that it is completed into a quartet of states, with $I_3 = \frac{3}{2}, \frac{1}{2}, -\frac{1}{2}, -\frac{3}{2}$. That requires some of the higher-lying states of Beryllium-7/Lithium-7 to have similar properties to Boron-7/Helium-7, in particular their energy and spin/parity. The spectrum in Fig. 2.1 shows the relevant states, and they are joined up with a dotted line to indicate their status as a quartet of $I = \frac{3}{2}$ states. It is not easy to directly assign isospin $I = \frac{1}{2}$ or $I = \frac{3}{2}$ to excited states of Beryllium-7/Lithium-7. This is done by looking at the complete spectrum of nuclei with baryon number 7.

Energy increases with $I$, so $I$ takes its minimum value compatible with the $I_3$ values. In almost all cases, a nucleus in its ground state has $I = |I_3|$,

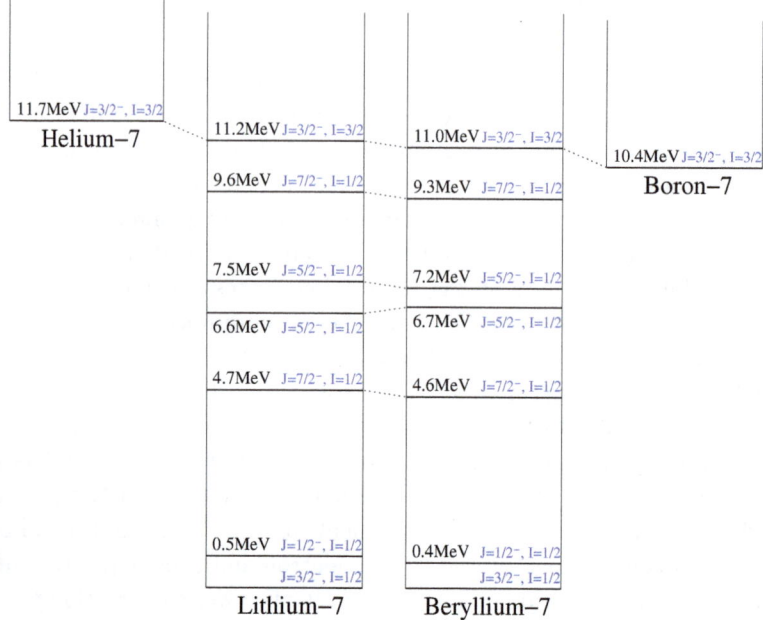

Fig. 2.1: Energy spectrum for nuclei with baryon number 7

but excited states may have a higher value of $I$. Within conventional nuclear physics, where protons and neutrons are treated as approximately independent particles, this increase of energy is interpreted as being a consequence of the Pauli principle, which favours the proton and neutron numbers being equal or nearly equal for a given baryon number. We will see that in Skyrme theory, isospin arises by quantizing the internal rotational motion of the Skyrme field configuration in isospace, which explains the increase of energy with isospin in a different way. In the Bethe–Weizsäcker mass formula for nuclear masses (ground state energies), one term is the asymmetry energy [39, 244]. This is quadratic in $I_3$, which is equivalent to being quadratic in $I$ for $I = |I_3|$.

Carbon-12 is a typical nucleus with $B$ even, whose ground state and most excited states have isospin 0. But a few states of Carbon-12 with excitation energy above 15 MeV can be identified as part of a triplet of states with isospin 1, where the other states in the triplet are low-lying states of Nitrogen-12 and Boron-12. Some even higher states can be assigned to an isospin 2 quintet that includes the ground states of Oxygen-12 and Beryllium-12.

Many isospin multiplets have been identified, and they are tabulated in ref.[170]. Some are incomplete, but this is because of experimental difficulties rather than for any fundamental theoretical reason. As nuclei get larger, proton-rich nuclei are harder to produce and less stable than neutron-rich nuclei, so states with high $I_3$ in a multiplet may be missing or have unconfirmed spin/parity. Also, for a nucleus whose ground state has isospin 0 or $\frac{1}{2}$, it is difficult to pick out the high-energy states with higher isospin, because there are so many states with similar energies, and some are not precisely classified.

Isospin symmetry is not an exact symmetry. It is broken, and increasingly so for larger nuclei. The main reason is the increasing importance of the electrostatic Coulomb energy as the proton number increases. So stable nuclei become neutron-rich. Although $I_3$ can always be defined, the total isospin quantum number $I$ makes no sense for baryon numbers beyond about 60. An obvious transition occurs above Calcium-40. Doubly-magic Calcium-40 is stable[3]; it has 20 protons and 20 neutrons, and can be assigned isospin 0. Calcium-48 is also doubly-magic and stable, having 20 protons and 28 neutrons, but assigning it as having isospin 4 is not helpful. However, well beyond Calcium-40 there is still interest in so-called self-conjugate nuclei, which have equal numbers of protons and neutrons, and in pairs of mirror nuclei, where the numbers of protons and neutrons are exchanged. The largest self-conjugate nucleus that has been identified is Xenon-108, with 54 protons and 54 neutrons, but it is very short-lived. Tin-100 is a self-conjugate doubly-magic nucleus.

An interesting phenomenon occurs for the pair Scandium-41/Calcium-41, which are mirror nuclei. Their low-lying states match up rather well in spin/parity. However, the Scandium-41 states (a Calcium-40 core with an extra proton) have energy about 6.5 MeV higher than the Calcium-41 states (Calcium-40 core with an extra neutron). This energy gap is larger than the gaps between the first few states of each of these nuclei separately, with their various spins and parities, implying that the isospin breaking is quite large. But it seems that these nuclei can still be interpreted as an isospin doublet, with the isospin breaking treated using first-order perturbation theory. That means that the states and their spin/parity properties are as if isospin symmetry were exact, and the isospin-breaking part of the

---

[3]In the shell model of nuclei [45, 221], the *magic numbers* for closed shells are 2, 8, 20, 28, 50, 82, 126, ..., and a doubly-magic nucleus is one where both the proton and neutron numbers are magic. Such nuclei are particularly stable and require considerable energy to excite.

Hamiltonian just leads to an overall upward shift of the energy of Scandium-41 states relative to those of Calcium-41.

While investigating the quantization of Skyrmions, we will work with an isospin-symmetric theory for most of the time, and quantized states will be classified by their isospin quantum numbers. In Skyrme theory, the isospin $I$ arises through the quantization of the internal orientational degrees of freedom of a Skyrmion, so there is a quantum number $K_3$ that is the projection of isospin on to the third body-fixed axis of the Skyrmion in isospace, complementing the more familiar and measurable $I_3$, the projection on to the third space-fixed axis in isospace. The interplay of Skyrmion symmetries with these quantum numbers will be a key topic in later chapters.

## 2.5 Effective Field Theory

How do fields representing nucleons and pions come about? Let's do a thought experiment. The current paradigm is to regard QCD as fundamental. The fields are those corresponding to u and d quarks and gluons. The pions and nucleons (and antinucleons) are bound states of quarks and antiquarks. But let's consider the possibility that quarks and gluons are not fundamental after all, but have substructure. Indeed, several attempts have been made to construct a theory where they have; the motivation is that a deeper theory may have fewer fundamental fields and fewer parameters, and may be able to predict the mass ratios of all six quark types, currently a mystery. One of the unattractive aspects of the Standard Model is that it has very many independent fields (difficult to count, because each field has several components) and there are at least 20 free parameters, mainly quark and lepton masses, but also coupling constants. It is a long-standing dream to have a simpler theory, explaining the broad range of masses and the field coupling parameters that determine how strongly the particles interact and how quickly they decay.

Suppose we had such a theory. What good would the Standard Model and its QCD subtheory be? Well, still very good. Most likely it would explain many phenomena just as well as it does now, although it wouldn't incorporate the possibility of creating the fundamental particles of the deeper theory. But those particles would probably be very massive, and hard to produce without a very high energy particle accelerator, otherwise we would have detected them by now. From the perspective of the new theory, the old Standard Model would become an *effective field theory (EFT)* with an impressive but limited range of validity. It would still accurately describe particles of various spins and masses, even though they were no longer fundamental. Only beyond some very high energy scale would the Standard Model break down.

This thought experiment provides the justification for trying to construct a field theory with fundamental fields for the pions and nucleons. It would be an EFT valid below the typical energy scale of QCD, which is about 1 GeV. Pions are particles of spin 0 so they are described by scalar fields. Nucleons have spin $\frac{1}{2}$, so are described by Dirac fields, like the electron and quarks. The field theory should be Lorentz invariant, to be compatible with the special theory of relativity. This is because some phenomena in nuclear physics involve quite high energies, around 100 MeV, leading to the possibility of pion particle production. (Nucleon pair production is energetically out of reach in mainstream nuclear physics.) It should also be a quantum field theory rather than a classical field theory, because quantum mechanics is vital for a consistent treatment of almost all of nuclear physics. For example, quantum mechanical tunnelling phenomena were originally discovered in the context of nuclear $\alpha$-particle emission – radioactive alpha decay, where a heavy nucleus emits a Helium-4 nucleus.

As an aside, we should mention that there is a reasonable, alternative point of view. Much of nuclear physics occurs at energies of a few MeV (this refers to binding energies per nucleon and excitation energies, leaving out the rest masses of the nucleons), where nucleons are not moving anywhere approaching the speed of light, and pion production is energetically out-of-reach. Here, nuclei can be modelled using non-relativistic quantum mechanics, with a Schrödinger equation controlling the many-nucleon dynamics. The role of the pion field is then similar to the role of electric and magnetic fields in an atom. It generates the forces and the potentials that appear in the quantum mechanical Hamiltonian, and hence in the Schrödinger equation. However, the forces between protons and neutrons are not at all simple like the electrostatic Coulomb attraction between an electron and an atomic nucleus. Within a nucleus are complicated forces between nucleons that depend on the nucleon separations, their spins and their momenta. Adding to the complication, it is not enough to know the potential between each pair of nucleons, because there are independent 3-body and 4-body forces.

So this brings us back to an EFT involving pions. Such a theory can bring some order to the chaos of inter-nucleon forces. Pion fields are known to describe the small forces between nucleons when they are well separated. A (nonlinear) field theory can determine how to extrapolate these forces to shorter separations, where they are much stronger. Therefore, an EFT of both pions and nucleons is now a focus for much theoretical work in nuclear physics [78]. In EFT, isospin symmetry acts by (internal) $SO(3)$ rotations on the three pion fields, and on the nucleon fields as an isospin doublet. One of the merits of EFT is that one does not have to input the spatial dependence of the forces – how they decay with increasing distance. Instead, the EFT only has some field coupling constants to adjust.

In QED, the Coulomb forces between electrically charged particles can be derived from the simplest Feynman diagram that induces particle scattering. This diagram involves one-photon exchange between two charged particles. Higher-order diagrams can systematically account for corrections to the Coulomb force at small separations, arising from the fact that particles in quantum mechanics are not truly pointlike, but are surrounded by a cloud of virtual particle-antiparticle pairs. But these are small effects in the case of electrons and positrons.

Similarly, in the EFT of pions and nucleons, the dominant force between protons and neutrons when they are well separated is derived from the Feynman diagrams where there is an exchange of one of the three types of pion. But the corrections due to higher-order diagrams are more substantial and have to be treated less rigorously than in QED, by introducing a cutoff at short length scales. This EFT has taken rather long to perfect, but its status has recently become firmer. The theory has few parameters – the pion and nucleon masses, the basic coupling strength between a pion and a nucleon, some pion-pion nonlinear interaction strengths, and some parameters to do with the cutoff scale. These parameters can be fitted using observed nucleon-nucleon scattering data. The theory then predicts to some extent the strength and nature of 3-body and 4-body forces. Finally these forces, or rather their potentials, can be used as input in a Schrödinger equation to calculate properties of nuclei with several nucleons. The spin/parity and binding energy of several small nuclei can be calculated, as can the spin/parities and energies of excited states, including resonances, see e.g. refs.[191, 235].

This is impressive, but the calculations are numerically very intensive when applied to a medium-sized nucleus like Carbon-12, and require the resources of some of the world's largest supercomputers. An EFT of both pions and nucleons also has some problems. Conceptually, it is not completely satisfactory that the EFT is used only to find the inter-nucleon potentials, and not directly to find nuclear states and their properties. Also, the potentials at very short separations are not properly determined, because of the cutoff. The output of a calculation is vast, as it gives the complete quantum state for all the constituent nucleons, but it is hard to get a picture of what is going on. Spatial correlations of particles can be quantified, but not heuristically understood or visualised.

In summary, there are strengths and weaknesses of an EFT of pions and nucleons. Such a theory is relativistic, and it has quite a small number of parameters. Physical calculations rely on Feynman diagrams, and summing these up systematically is the essence of perturbation theory. However, this runs into difficulties because the basic pion-nucleon coupling is not very small, and higher-order diagrams make a greater contribution than in QED. The coupling becomes larger as the particle momenta increase, and higher-momentum effects are important at short distances (because of the quantum mechanical uncertainty relation connecting momentum and position). These have to be handled through an *ad hoc* short-distance cutoff.

## 2.6 Skyrme's Effective Field Theory

The Skyrme theory is an effective field theory of pions, but its key difference from the pion-nucleon EFT discussed above is that it has no fundamental nucleon fields. Otherwise, it has similar pion-pion interactions, and similar symmetries. The nucleons now emerge from the theory as quantized Skyrmions – quantized topological soliton solutions of the nonlinear pion field equation – and because these are smooth and have a finite size, there is no need to introduce an explicit cutoff. In fact, the quantum mechanics of the Skyrmion degrees of freedom has no short-distance singularities, and can be thought of as naturally evading the need for a cutoff in the field theory. From the basic Skyrmion solution, one can extract a pion-nucleon coupling strength. This coupling strength is recognised as not being small, so perturbation theory and its associated Feynman diagrams are best avoided, but this is again natural in Skyrme theory.

Skyrme theory is almost entirely about nucleons, pions, and their dynamics. We need hardly know about quarks and the gluons that mediate their interactions. Quarks have never been directly detected, and no observed state has either fractional baryon number or fractional electric charge. Historically, Skyrme theory was proposed by Skyrme around 1960, whereas the idea of quarks was only proposed by Gell-Mann and Zweig in 1964. Skyrme never mentioned quarks, although those interested in Skyrmions since 1964 have always had the quark picture in the background.

In Chapter 4 we will describe the Skyrme theory in detail. The simplest version of the theory has fewer parameters than pion-nucleon EFT, at the expense of less accuracy. There is an energy scale and a length scale, and a pion mass parameter. The energy and length scale need to be calibrated against nuclear mass and length data, and there are various ways of doing this. There is far more emphasis on exact solutions of the nonlinear, classical pion field equation – the Skyrme field equation – than in a pion-nucleon EFT. Heuristically, in Skyrme theory, the pion field in a nucleus becomes a classical pion condensate with a topological twist that stabilises the field, even though individual pion particles can decay. The Skyrmion solutions provide a unified picture of nucleon structure and of nucleon-nucleon interactions, and this is one of the theory's most attractive features. The strength of the forces between Skyrmions can be calculated within the theory, and do not need to be separately postulated.

A Skyrmion with baryon number $B$ greater than 1 can be interpreted as a collection of nucleons that are highly correlated and tightly bound – rather too much so. It corresponds to a configuration of nucleons that are as close together as they can be in a nucleus, at the lowest point of their potential energy landscape, where they partially merge. The Skyrmions need to be softened by allowing them to deform and vibrate, as well as rotate in space and isospace. Some of the key vibrational degrees of freedom have a tendency to split a Skyrmion into clusters with smaller baryon numbers, a process that involves relatively small energy. Quantizing the dynamics of such suitably chosen Skyrmion degrees of freedom provides a heuristically compelling picture of the collective motions of nucleons in nuclei. We will see that some nuclear spectra can be almost entirely interpreted in terms of quantized collective motions of a Skyrmion, both rotational and vibrational. This works particularly well for doubly-magic nuclei like Oxygen-16, or those at some distance from shell closures, like Carbon-12. So far, Skyrme theory has had difficulty with less-correlated nuclei like Helium-5, where a single nucleon orbits outside a closed shell [107]. However, recent work on the interaction of Skyrmions has clarified how the nuclear spin-orbit coupling arises in Skyrme theory [122], and how it contributes to the interaction between two nucleons [121].

## Chapter 3

# Lagrangians and Symmetries

Skyrme theory is a Lagrangian field theory whose fundamental *Skyrme field* is a matrix expression that combines the three pion fields. The action is the time-integral of the Lagrangian – itself the integral over space of a Lagrangian density constructed from the Skyrme field and its first time- and space-derivatives. In Chapter 4 we will derive the field equation of Skyrme theory, the Euler–Lagrange equation, using the principle of least action.

The Skyrme field has an infinite number of dynamical degrees of freedom, but when we later consider the dynamics of Skyrmions, we will not allow all these to be excited. Attention will be restricted to a finite number of collective degrees of freedom, for example, the position of the Skyrmion's centre of mass and the angles determining its orientation, supplemented by the amplitudes of selected vibrational modes. This more restricted system also has a Lagrangian, whose classical and quantized dynamics will need to be considered.

So in this chapter we recall the general structure of Lagrangian dynamics, first for a system with finitely many degrees of freedom, and then for field theory [184]. Lagrangian dynamics is well adapted to incorporate symmetries of a system. According to Noether's theorem, for each generator of a continuous (Lie) symmetry group there is a corresponding conservation law. Symmetries are important for Skyrmions. The conserved quantities that arise are energy and momentum, spin and isospin, all of which have quantum analogues.

## 3.1 Finite-Dimensional Systems

In Lagrangian dynamics, the configuration space is a smooth manifold $M$, of dimension $D$, say, and the system moves along a trajectory through $M$.

Let $\mathbf{q} = (q^1, \ldots, q^D)$ denote local coordinates on $M$. The trajectory can be expressed as $\mathbf{q}(t)$, or in component form as $q^\alpha(t)$, $1 \leq \alpha \leq D$.

The Lagrangian combines a potential energy – a scalar function $V(\mathbf{q})$ on $M$ – and a kinetic energy. The kinetic energy is quadratic in the coordinate velocities $\dot{q}^\alpha$, and the coefficient matrix of this quadratic expression defines a physical inertia tensor. Mathematically, the inertia tensor is a Riemannian metric on $M$, given locally by a symmetric, positive-definite matrix $g_{\alpha\beta}(\mathbf{q})$. We denote the inverse of the metric by $g^{\alpha\beta}$, so $g^{\alpha\beta} g_{\beta\gamma} = \delta^\alpha_\gamma$ with $\delta^\alpha_\gamma$ the Kronecker delta symbol ($\delta^\alpha_\gamma = 1$ if $\alpha = \gamma$, 0 otherwise). Here and below, the summation convention is used; if an index is repeated, it is summed over.

A general trajectory $\mathbf{q}(t)$ has the associated Lagrangian

$$L(t) = \frac{1}{2} g_{\alpha\beta}(\mathbf{q}) \dot{q}^\alpha \dot{q}^\beta - V(\mathbf{q}). \tag{3.1}$$

This depends on the instantaneous position $\mathbf{q}$ and velocity $\dot{\mathbf{q}} = \frac{d\mathbf{q}}{dt}$. Although the Lagrangian is defined in terms of some local coordinate system, it is actually coordinate-invariant provided $g_{\alpha\beta}$ transforms in the appropriate way under coordinate changes (i.e. as a covariant 2-tensor on $M$).

The principle determining physically allowed trajectories $\mathbf{q}(t)$ is the *principle of least action*. Consider all trajectories that begin at $\mathbf{q}^{(1)}$ at $t = t_1$ and end at $\mathbf{q}^{(2)}$ at $t = t_2$ ($t_2 > t_1$). (The velocities at $t_1$ and $t_2$ do not need to be specified.) The action is defined as

$$S = \int_{t_1}^{t_2} L(t) \, dt, \tag{3.2}$$

for any trajectory satisfying the initial and final condition. The actual, physical trajectory is one for which $S$ is stationary (usually it is minimal).

A standard result in the calculus of variations implies that a physical trajectory must satisfy the Euler–Lagrange equation

$$\frac{d}{dt}\left(\frac{\partial L}{\partial \dot{q}^\alpha}\right) - \frac{\partial L}{\partial q^\alpha} = 0, \tag{3.3}$$

which here takes the form

$$\frac{d}{dt}\left(g_{\alpha\beta} \dot{q}^\beta\right) - \frac{\partial}{\partial q^\alpha}\left(\frac{1}{2} g_{\beta\gamma} \dot{q}^\beta \dot{q}^\gamma - V\right) = 0. \tag{3.4}$$

This equation of motion can be reexpressed as

$$\ddot{q}^\beta + \Gamma^\beta_{\gamma\delta} \dot{q}^\gamma \dot{q}^\delta + g^{\beta\alpha} \partial_\alpha V = 0, \tag{3.5}$$

where $\partial_\alpha$ denotes $\frac{\partial}{\partial q^\alpha}$ and

$$\Gamma^\beta_{\gamma\delta} = \frac{1}{2} g^{\alpha\beta} (\partial_\gamma g_{\delta\alpha} + \partial_\delta g_{\gamma\alpha} - \partial_\alpha g_{\gamma\delta}) \tag{3.6}$$

is the Levi-Civita connection on $M$, determined from the metric $g_{\alpha\beta}$. Note that the equation for the coordinate acceleration $\ddot{q}^\beta$ has a contribution quadratic in velocity (a geometric effect) and a contribution from the gradient of $V$ (a classical force).

The law of conservation of energy is obtained by multiplying (3.5) by $\dot{q}^\alpha$ and summing over $\alpha$. One finds that

$$\frac{d}{dt}\left(\frac{1}{2} g_{\alpha\beta} \dot{q}^\alpha \dot{q}^\beta + V\right) = 0 \tag{3.7}$$

and therefore

$$\frac{1}{2} g_{\alpha\beta} \dot{q}^\alpha \dot{q}^\beta + V = E, \tag{3.8}$$

where the constant $E$ is the conserved total energy. $\dot{\mathbf{q}}$ is the coordinate velocity, but on a Riemannian manifold the geometrical speed $v$ – the distance moved per unit time – depends on the metric, and is $v = \sqrt{g_{\alpha\beta} \dot{q}^\alpha \dot{q}^\beta}$. Therefore the first term in $E$ is $\frac{1}{2} v^2$.

In a constant, uniform potential $V$, the Euler–Lagrange equation simplifies to

$$\ddot{q}^\beta + \Gamma^\beta_{\gamma\delta} \dot{q}^\gamma \dot{q}^\delta = 0, \tag{3.9}$$

the equation for motion along a geodesic of $M$. Conservation of energy implies that the motion is at constant speed.

A special case of the above Lagrangian formalism is the Newtonian dynamics of one or more particles. For one particle of mass $m$ moving in $d$ dimensions, $M$ is $\mathbb{R}^d$ with coordinates $x^i$, $1 \le i \le d$ and the Euclidean metric $g_{ij} = m\delta_{ij}$. The Lagrangian is

$$L = \frac{1}{2} m \dot{x}^i \dot{x}^i - V(\mathbf{x}), \tag{3.10}$$

leading to the equation of motion

$$m \ddot{x}^i = -\partial_i V, \tag{3.11}$$

so the force on the particle is minus the gradient of $V$.

## 3.2 Symmetries and Conservation Laws

Suppose a Lie group $G$ of symmetries acts on the configuration space $M$, leaving invariant the potential $V$ and metric $g_{\alpha\beta}$. There is then a set of conservation laws for solutions of the equation of motion (3.5), one for each generator of the group.

Let $\mathbf{q} \mapsto \mathbf{q} + \varepsilon\boldsymbol{\xi}(\mathbf{q})$ be the effect of one of the symmetry generators of $G$, with $\varepsilon$ infinitesimal. The invariance conditions are that the Lie derivatives of the potential and metric in the direction of the vector field $\boldsymbol{\xi}$ vanish. Explicitly, these conditions are

$$\xi^\gamma \partial_\gamma V = 0,$$
$$\xi^\gamma \partial_\gamma g_{\alpha\beta} + \partial_\alpha \xi^\gamma g_{\gamma\beta} + \partial_\beta \xi^\gamma g_{\alpha\gamma} = 0. \tag{3.12}$$

If the trajectory $\mathbf{q}(t)$ shifts to $\mathbf{q}(t) + \varepsilon\boldsymbol{\xi}(\mathbf{q}(t))$, the velocity components $\dot{q}^\alpha$ change to $\dot{q}^\alpha + \varepsilon \partial_\gamma \xi^\alpha \dot{q}^\gamma$, and it is straightforward to show, using the invariance conditions above, that the shift of $L$ vanishes at order $\varepsilon$. Noether's theorem then asserts that

$$Q = \xi^\gamma g_{\gamma\beta} \dot{q}^\beta \tag{3.13}$$

is a conserved quantity [136]. Let us verify this directly. Taking the time derivative of $Q$ and using the equation of motion (3.5), we find

$$\frac{dQ}{dt} = \xi^\gamma \frac{d}{dt}(g_{\gamma\beta}\dot{q}^\beta) + \partial_\alpha \xi^\gamma g_{\gamma\beta} \dot{q}^\alpha \dot{q}^\beta$$
$$= \xi^\gamma \left(\frac{1}{2} \partial_\gamma g_{\alpha\beta} \dot{q}^\alpha \dot{q}^\beta + \partial_\gamma V\right) + \partial_\alpha \xi^\gamma g_{\gamma\beta} \dot{q}^\alpha \dot{q}^\beta. \tag{3.14}$$

By combining terms and symmetrising over indices this becomes

$$\frac{dQ}{dt} = \frac{1}{2}(\xi^\gamma \partial_\gamma g_{\alpha\beta} + \partial_\alpha \xi^\gamma g_{\gamma\beta} + \partial_\beta \xi^\gamma g_{\alpha\gamma})\dot{q}^\alpha \dot{q}^\beta + \xi^\gamma \partial_\gamma V. \tag{3.15}$$

Both terms on the right hand side vanish, by the invariance conditions (3.12), so $Q$ is conserved. The same result also follows from a variant of a rather more general argument we shall use in the context of field theory.

An example is the conserved angular momentum for a Newtonian particle moving in three dimensions in a spherically symmetric potential $V(|\mathbf{x}|)$. An infinitesimal rotation around the axis $\mathbf{n}$ (the notation $\mathbf{n}$ will always denote a unit vector) shifts a trajectory $\mathbf{x}(t)$ to $\mathbf{x}(t) + \varepsilon\mathbf{n} \times \mathbf{x}(t)$ but leaves the Lagrangian invariant. The conserved quantity (one component for each Cartesian direction of $\mathbf{n}$) is the angular momentum vector

$$\mathbf{L} = m\mathbf{x} \times \dot{\mathbf{x}}. \tag{3.16}$$

## 3.3 Lagrangian Field Theory

Classical Lagrangian field theory is concerned with the dynamics of one or more fields defined throughout space and evolving in time. Let us suppose space is $\mathbb{R}^d$, so spacetime is $\mathbb{R} \times \mathbb{R}^d$. Spacetime coordinates are written in combined form as $x = (t, \mathbf{x})$, and we shall often identify $x^0 = t$. A scalar field $\phi(x)$ is a single function on $\mathbb{R} \times \mathbb{R}^d$. The field represents an infinite number of dynamical degrees of freedom because, formally, the value of $\phi$ at each spatial point is one degree of freedom, evolving in time. The field values at distinct points are locally coupled together, because the Lagrangian depends not just on $\phi$ and its time derivative $\partial_0 \phi$, but also on its space derivatives $\partial_i \phi$, which are combined into its gradient $\nabla \phi$.

The simplest type of Lagrangian for the field $\phi$ is

$$L = \int \left( \frac{1}{2}(\partial_0 \phi)^2 - \frac{1}{2}\nabla\phi \cdot \nabla\phi - u(\phi) \right) d^d x. \tag{3.17}$$

The Lagrangian density $\mathcal{L}$, the integrand here, is a local quantity. $u$ is some function of $\phi$ (not explicitly dependent on $x$), often taken to be a polynomial. There is a natural splitting of this Lagrangian into kinetic energy and potential energy terms, $L = T - V$. The kinetic energy of the field is

$$T = \int \frac{1}{2}(\partial_0 \phi)^2 \, d^d x, \tag{3.18}$$

and

$$V = \int \left( \frac{1}{2}\nabla\phi \cdot \nabla\phi + u(\phi) \right) d^d x \tag{3.19}$$

is the potential energy. Note that the gradient of the field contributes to $V$. $u$ should be bounded below, otherwise the dynamics is liable to produce singular fields. It is generally arranged that the minimal value of $u$ is zero.

For a field theory defined on $\mathbb{R} \times \mathbb{R}^d$ we shall require Euclidean symmetry and time-translation symmetry. Euclidean symmetry combines spatial translations and rotations. Translational symmetry is ensured by having no explicit dependence on $\mathbf{x}$ in the Lagrangian, and integrating over $\mathbb{R}^d$ using the standard measure. Rotational symmetry requires combining the gradient terms into a scalar, as in (3.17). Acceptable generalisations of (3.17) could involve a further kinetic term

$$\int (\nabla\phi \cdot \nabla\phi)(\partial_0 \phi)^2 \, d^d x, \tag{3.20}$$

and the potential energy could include a further term like

$$\int (\boldsymbol{\nabla}\phi \cdot \boldsymbol{\nabla}\phi)^2 \, d^d x \,. \tag{3.21}$$

Terms similar to these occur in Skyrme theory.

The field theory with Lagrangian (3.17) possesses more symmetry. Because the time and space derivatives of $\phi$ both occur quadratically and with a relative minus sign, the theory is Lorentz invariant (the speed of light is unity). This is vital for a theory like Skyrme theory, aiming to describe elementary particles. In Lorentz-invariant theories we use the condensed notation $\partial_\mu \phi \, \partial^\mu \phi$ to denote $(\partial_0 \phi)^2 - \boldsymbol{\nabla}\phi \cdot \boldsymbol{\nabla}\phi$. Generally, greek indices run from 0 to $d$ in $(d+1)$-dimensional spacetime, and are raised or lowered using the Minkowski metric $\eta^{\mu\nu}$, with signature $(1, -1, \ldots, -1)$, so $\partial_\mu \phi \, \partial^\mu \phi \equiv \eta^{\mu\nu} \partial_\mu \phi \, \partial_\nu \phi$.

The action associated with a Lagrangian density $\mathcal{L}(\partial_\mu \phi, \phi)$ is

$$S = \int_{t_1}^{t_2} L \, dt = \int_{t_1}^{t_2} \int \mathcal{L}(\partial_\mu \phi, \phi) \, d^d x \, dt \,. \tag{3.22}$$

The *principle of least action* is that $S$ should be stationary under infinitesimal field variations, for given initial and final data on $\mathbb{R}^d$: $\phi(t_1, \mathbf{x}) = \phi^{(1)}(\mathbf{x})$, $\phi(t_2, \mathbf{x}) = \phi^{(2)}(\mathbf{x})$. So consider an infinitesimal variation of the field trajectory $\phi(t, \mathbf{x})$ to $\phi(t, \mathbf{x}) + \delta\phi(t, \mathbf{x})$ where $\delta\phi \to 0$ as $|\mathbf{x}| \to \infty$, and $\delta\phi = 0$ at times $t_1$ and $t_2$. The first-order variation of $S$ is

$$\delta S = \int_{t_1}^{t_2} \int \left( \frac{\partial \mathcal{L}}{\partial(\partial_\mu \phi)} \partial_\mu \delta\phi + \frac{\partial \mathcal{L}}{\partial \phi} \delta\phi \right) d^d x \, dt \,. \tag{3.23}$$

Integrating by parts, this becomes

$$\delta S = \int_{t_1}^{t_2} \int \left\{ \left( -\partial_\mu \frac{\partial \mathcal{L}}{\partial(\partial_\mu \phi)} + \frac{\partial \mathcal{L}}{\partial \phi} \right) \delta\phi \right\} d^d x \, dt \,. \tag{3.24}$$

$\delta S$ vanishes for all $\delta\phi$ provided

$$\partial_\mu \frac{\partial \mathcal{L}}{\partial(\partial_\mu \phi)} - \frac{\partial \mathcal{L}}{\partial \phi} = 0 \,, \tag{3.25}$$

and this is the field equation, the Euler–Lagrange equation satisfied by the physical field.

For the basic Lagrangian (3.17), the field equation is the nonlinear wave equation

$$\partial_\mu \partial^\mu \phi + \frac{du}{d\phi} = 0 \,. \tag{3.26}$$

If the field is static, this simplifies to the nonlinear Laplace equation

$$\nabla^2 \phi = \frac{du}{d\phi}, \tag{3.27}$$

which is the condition for $\phi$ to be a stationary point of the field potential energy $V$. A solution $\phi(\mathbf{x})$ is stable if it is a minimum of $V$.

The Lagrangian formalism can easily be extended to a theory of $n$ scalar fields $\phi = (\phi_1, \ldots, \phi_n)$. The Lagrangian depends on all $n$ component fields and their derivatives, and the action needs to be stationary with respect to independent variations of each field. The field equations for a Lorentz-invariant theory are therefore

$$\partial_\mu \frac{\partial \mathcal{L}}{\partial(\partial_\mu \phi_l)} - \frac{\partial \mathcal{L}}{\partial \phi_l} = 0, \quad 1 \leq l \leq n. \tag{3.28}$$

Because the field takes values in the linear space $\mathbb{R}^n$, this type of theory is called a linear scalar field theory, even though the field equations (3.28) are generally nonlinear.

An important phenomenon in a theory with an $n$-component field is the possibility of internal symmetries, unrelated to the symmetries of space or spacetime. For example, for

$$\mathcal{L} = \frac{1}{2} \partial_\mu \phi_l \, \partial^\mu \phi_l - u(\phi_l \phi_l), \tag{3.29}$$

the repeated index $l$ is summed over from 1 to $n$, so $\mathcal{L}$ is invariant under internal rotations

$$\phi_l(x) \mapsto R_{lm} \phi_m(x) \tag{3.30}$$

with $R_{lm}$ an $SO(n)$ matrix. This symmetry leads to conservation laws. Isospin symmetry in Skyrme theory is an example of this type of symmetry, involving $SO(3)$ isospace rotations of the three pion fields.

Yet another variant of field theory is where the (multi-component) field $\phi$ takes values in a non-trivial manifold $Y$. Skyrme theory is also an example of this. Such a theory is called a nonlinear sigma model. Sometimes it is formulated as a linear theory, with the field subject to a nonlinear constraint[1]. This is convenient if $Y$ is a simple algebraic submanifold of a Euclidean space. For example, in Skyrme theory $Y$ is a round, unit 3-sphere and this sits conveniently in a Euclidean space of one higher dimension.

Some field theories have a large internal symmetry group, but choosing the vacuum boundary condition at spatial infinity reduces the symmetry.

---

[1] Historically, in this formulation, one field component was often denoted $\sigma$, i.e. sigma.

This is called spontaneous symmetry breaking. The spontaneous breaking of a continuous internal symmetry group $G$ has important consequences for the field dynamics. Let $H$ be the subgroup of $G$ that leaves the chosen vacuum field configuration invariant. Small amplitude oscillations around this vacuum can be decomposed into the directions tangent and orthogonal to the orbit $G/H$. The tangent directions are "flat" directions, since the function $u$ is unchanging in these directions, by symmetry. $u$ is generally not flat in the orthogonal directions, but increases quadratically. In the flat directions, the oscillating field components $\psi$ satisfy the wave equation

$$\partial_0 \partial_0 \psi - \nabla^2 \psi = 0 \tag{3.31}$$

to linear order in $\psi$. The plane wave solutions are $\psi = \psi_0 e^{-i(\mathbf{k}\cdot\mathbf{x}-\omega t)}$, with the relation between frequency $\omega$ and wave-vector $\mathbf{k}$

$$\omega = |\mathbf{k}|, \tag{3.32}$$

so $\omega$ is arbitrarily close to zero. In the quantized theory there are massless, elementary scalar particles associated with such waves. The number of distinct particles is $\dim G - \dim H$, the dimension of the orbit $G/H$. This is Goldstone's theorem [105], whose proof does not assume the small amplitude, linear approximation that we have just made, so it is not completely straightforward. There are long-range interactions mediated by the exchange of these massless Goldstone particles. By contrast, in a theory of scalar fields with a symmetry group $G$ that is not spontaneously broken, all particles are generally massive and interactions are short-range, i.e. exponentially decaying with distance.

In the example (3.29), suppose that $u$ has its minimum at some non-zero field value. This field value, constant through space and time, is a possible vacuum; however, because of the internal symmetry group $SO(n)$, the vacuum is not unique. Nevertheless, one vacuum configuration has to be chosen. Its orbit is $SO(n)/SO(n-1)$, which is the $(n-1)$-sphere, and the number of Goldstone particles is $n-1$.

We end this section with a remark about the relationship between field theory and the finite-dimensional dynamical systems we considered earlier. Let us consider the example with the $n$-component field $\phi = (\phi_1, \ldots, \phi_n)$. By a field configuration $\phi(\mathbf{x})$, we mean a multiplet of smooth functions $(\phi_1(\mathbf{x}), \ldots, \phi_n(\mathbf{x}))$ defined throughout space at a given time, satisfying boundary conditions so that the energy is finite. A configuration is not necessarily a static solution of the field equations, but it could be the instantaneous form of a dynamical field. The configuration space $\mathcal{C}$ consisting of all these field configurations $\phi(\mathbf{x})$ is the infinite-dimensional analogue of the manifold $M$ with points $\mathbf{q}$ – the configuration space of the finite-dimensional dynamical system.

The field potential energy, including the gradient term, depends only on the configuration, so it is a scalar function on $\mathcal{C}$. The kinetic energy depends quadratically on the time derivative of the field configuration, so it depends quadratically on the velocity through $\mathcal{C}$, and is positive definite. It therefore defines an infinite-dimensional Riemannian metric on $\mathcal{C}$.

## 3.4 Noether's Theorem in Field Theory

If a Lagrangian field theory has an infinitesimal symmetry, then there is an associated current $J^\mu(x)$ that is conserved, meaning that

$$\partial_\mu J^\mu \equiv \partial_0 J^0 + \boldsymbol{\nabla} \cdot \mathbf{J} = 0. \tag{3.33}$$

Both spacetime symmetries and internal symmetries lead to such conservation laws. Let us recall how $J^\mu$ is found, and give some examples.

Suppose the theory is for a scalar field $\phi$. Consider a general infinitesimal variation of $\phi$,

$$\phi(x) \mapsto \phi(x) + \varepsilon \Delta \phi(x). \tag{3.34}$$

This variation is a symmetry if one can show, without using the field equation, that the corresponding variation of the Lagrangian density $\mathcal{L}$ is a total divergence,

$$\mathcal{L}(x) \mapsto \mathcal{L}(x) + \varepsilon \partial_\mu K^\mu(x). \tag{3.35}$$

The action then varies only by a surface term. Sometimes $K^\mu$ will be zero, and the action strictly invariant.

Now let us calculate the change in $\mathcal{L}$ more explicitly:

$$\mathcal{L}(x) \mapsto \mathcal{L}(x) + \varepsilon \frac{\partial \mathcal{L}}{\partial(\partial_\mu \phi)} \partial_\mu(\Delta\phi) + \varepsilon \frac{\partial \mathcal{L}}{\partial \phi} \Delta\phi \tag{3.36}$$

$$= \mathcal{L}(x) + \varepsilon \partial_\mu \left( \frac{\partial \mathcal{L}}{\partial(\partial_\mu \phi)} \Delta\phi \right) + \varepsilon \left( \frac{\partial \mathcal{L}}{\partial \phi} - \partial_\mu \frac{\partial \mathcal{L}}{\partial(\partial_\mu \phi)} \right) \Delta\phi.$$

Using the field equation, the last bracket vanishes. Then, by identifying (3.35) and (3.36), we see that the current

$$J^\mu \equiv \frac{\partial \mathcal{L}}{\partial(\partial_\mu \phi)} \Delta\phi - K^\mu \tag{3.37}$$

is conserved, i.e. $\partial_\mu J^\mu = 0$ if $\phi$ satisfies its field equation. This is Noether's theorem. For a multi-component field, the first term in $J^\mu$ becomes a sum over the field components.

Current conservation, $\partial_\mu J^\mu = 0$, implies the conservation of the Noether charge

$$Q = \int J^0 \, d^d x. \tag{3.38}$$

$Q$ is time-independent because

$$\frac{dQ}{dt} = \int \partial_0 J^0 \, d^d x = -\int \nabla \cdot \mathbf{J} \, d^d x = 0. \tag{3.39}$$

Here we have used the divergence theorem, and assumed that $\mathbf{J} \to \mathbf{0}$ as $|\mathbf{x}| \to \infty$.

As an example, consider complex Klein–Gordon theory. This theory has a single, complex scalar field $\phi(x)$, and Lagrangian density

$$\mathcal{L} = \frac{1}{2} \partial_\mu \bar\phi \, \partial^\mu \phi - \frac{1}{2} m^2 \bar\phi \phi. \tag{3.40}$$

The phase rotation $\phi \mapsto e^{i\alpha}\phi$, $\bar\phi \mapsto e^{-i\alpha}\bar\phi$ is an internal $U(1)$ symmetry, leaving $\mathcal{L}$ invariant. Infinitesimally, setting $\alpha = \varepsilon$, we find $\Delta\phi = i\phi$, $\Delta\bar\phi = -i\bar\phi$. Here $K^\mu = 0$, so the conserved current is

$$J^\mu = -\frac{i}{2}(\bar\phi \partial^\mu \phi - \phi \partial^\mu \bar\phi). \tag{3.41}$$

One can verify that $\partial_\mu J^\mu = 0$ directly, using the Klein–Gordon field equation

$$\partial_\mu \partial^\mu \phi + m^2 \phi = 0 \tag{3.42}$$

and its complex conjugate. The nonlinear Klein–Gordon theory, where the term in $\mathcal{L}$ proportional to $m^2$ is replaced by a general function $u(\bar\phi\phi)$, has the same symmetry and the same conserved current.

As a second, rather general example, consider translations in spacetime

$$x^\nu \mapsto x^\nu + \varepsilon^\nu \tag{3.43}$$

with $\varepsilon^\nu$ infinitesmal, for an arbitrary field theory not depending explicitly on the spacetime coordinates. The effect on a field $\phi$ is

$$\phi(x) \mapsto \phi(x+\varepsilon) = \phi(x) + \varepsilon^\nu \partial_\nu \phi(x), \tag{3.44}$$

and similarly for derivatives of $\phi$. The effect on the Lagrangian density, no matter what the details of its structure, is

$$\mathcal{L} \mapsto \mathcal{L} + \varepsilon^\nu \partial_\nu \mathcal{L} = \mathcal{L} + \varepsilon^\nu \partial_\mu (\delta^\mu_\nu \mathcal{L}). \tag{3.45}$$

This variation is of the form (3.35), so we can derive a conservation law. Since the parameter $\varepsilon^\nu$ is a spacetime vector, and $\Delta\phi = \partial_\nu \phi$, the conserved current is a tensor

$$T^\mu_\nu = \frac{\partial \mathcal{L}}{\partial(\partial_\mu \phi)} \partial_\nu \phi - \delta^\mu_\nu \mathcal{L}. \tag{3.46}$$

$T^\mu_\nu$ is the energy-momentum tensor, and it satisfies $\partial_\mu T^\mu_\nu = 0$.

The conserved charge associated with time-translation symmetry is the field energy

$$E = \int T^0_0 \, d^d x = \int \left( \frac{\partial \mathcal{L}}{\partial(\partial_0 \phi)} \partial_0 \phi - \mathcal{L} \right) d^d x, \qquad (3.47)$$

and the conserved charge associated with spatial translations is the field momentum vector

$$P_i = -\int T^0_i \, d^d x = -\int \frac{\partial \mathcal{L}}{\partial(\partial_0 \phi)} \partial_i \phi \, d^d x. \qquad (3.48)$$

Another general conservation law, the conservation of angular momentum, arises from the symmetry under spatial rotations.

Skyrme theory has spacetime symmetries and also internal symmetries, so there are several conservation laws. In the simplest version of Skyrme theory, with massless pions, the three pions are Goldstone bosons of a spontaneously broken, internal $SO(4)$ symmetry, with unbroken symmetry group $SO(3)$. Goldstone's theorem implies that the masslessness of the pions doesn't depend on the details of the Skyrme Lagrangian.

As $SO(4)$ is a 6-dimensional Lie group, Skyrme theory with massless pions has six independent, conserved internal Noether currents. These conservation laws are valid, even though the $SO(4)$ symmetry is spontaneously broken. In the theory with massive pions, the $SO(4)$ symmetry is explicitly broken and only the unbroken $SO(3)$ symmetry survives, so there are three internal Noether currents. These are the components of the isospin current. The form of these various currents will be clarified in the next chapter.

It is remarkable that the conservation of baryon number in Skyrme theory is not associated with any symmetry, but instead arises for a topological reason. There is a conserved baryon current, but it is not a Noether current.

# Chapter 4

# Skyrme Theory

## 4.1 $SU(2)$

The Skyrme field is an $SU(2)$ matrix field, so we start by briefly describing $SU(2)$, the *special unitary group* of $2 \times 2$ matrices. $SU(2)$ is a Lie group, a group which is also a manifold. Although one may think of $SU(2)$ simply as a group of matrices depending on three parameters, more geometrically it has the form of a 3-sphere.

The elements of $SU(2)$ are $2 \times 2$ complex matrices $U$ satisfying

$$UU^\dagger = 1_2 \quad \text{and} \quad \det U = 1, \tag{4.1}$$

where $U^\dagger$ is the hermitian conjugate (complex conjugate of the transpose) of $U$ and $1_2$ is the unit $2 \times 2$ matrix. The first condition says that $U$ is unitary and the second that the determinant of $U$ is 1 (the conditions for $U$ to be special unitary). Note that the first condition implies that $U^{-1}$, the inverse of $U$, is $U^\dagger$. The general matrix that lies in $SU(2)$ can be written as

$$U = \begin{pmatrix} a_0 + ia_3 & ia_1 + a_2 \\ ia_1 - a_2 & a_0 - ia_3 \end{pmatrix}, \tag{4.2}$$

where $(a_0, a_1, a_2, a_3)$ are real parameters subject to the constraint

$$a_0^2 + a_1^2 + a_2^2 + a_3^2 = 1. \tag{4.3}$$

It is straightforward to check that the conditions (4.1) are satisfied for $U$ in this form. It can also be checked that the group axioms are satisfied by these matrices.

The constraint (4.3) says that the four parameters $(a_0, a_1, a_2, a_3)$ lie on a unit sphere in a 4-dimensional space, i.e. they lie on a 3-sphere, $S^3$. One can regard $(a_1, a_2, a_3)$ as three independent parameters, constrained by $\mathbf{a} \cdot \mathbf{a} \leq 1$.

Then $a_0$ is determined up to a sign. A trigonometric parametrisation of $(a_0, a_1, a_2, a_3)$ in terms of three Euler angles is also possible.

It is useful now to introduce the three Pauli matrices (familiar from the quantum mechanics of spin $\frac{1}{2}$ particles, although that is not the context here),

$$\tau_1 = \begin{pmatrix} 0 & 1 \\ 1 & 0 \end{pmatrix}, \quad \tau_2 = \begin{pmatrix} 0 & -i \\ i & 0 \end{pmatrix}, \quad \tau_3 = \begin{pmatrix} 1 & 0 \\ 0 & -1 \end{pmatrix}. \tag{4.4}$$

The Pauli matrices are hermitian and traceless, and obey the product relations $\tau_i \tau_j = \delta_{ij} 1_2 + i\epsilon_{ijk} \tau_k$. These relations imply that $\text{Tr}(\tau_i \tau_j) = 2\delta_{ij}$, where Tr denotes the trace of a matrix, the sum of the diagonal elements. In terms of Pauli matrices, the general $SU(2)$ matrix (4.2) can be expressed elegantly as

$$U = a_0 + i\mathbf{a} \cdot \boldsymbol{\tau}, \tag{4.5}$$

where $a_0^2 + \mathbf{a} \cdot \mathbf{a} = 1$. Here we have combined the parameters $(a_1, a_2, a_3)$ into a real vector $\mathbf{a}$ and similarly the three Pauli matrices into a vector $\boldsymbol{\tau}$ of matrices. The parameter $a_0$ is really multiplying the unit matrix $1_2$ but it is convenient to suppress this. From now on, any number representing a matrix means that number times the unit matrix, so we will, for example, write $UU^\dagger = 1$ to indicate that $U$ is unitary. Note that for $U$ expressed this way, $U^{-1} = U^\dagger = a_0 - i\mathbf{a} \cdot \boldsymbol{\tau}$.

Elements of $SU(2)$ infinitesimally close to the identity are of the form

$$U = 1 + \epsilon X, \tag{4.6}$$

where $\epsilon$ is infinitesimal and real, and $X$ a matrix of order 1. To order $\epsilon$, the condition $UU^\dagger = 1$ reduces to $X + X^\dagger = 0$, so $X$ is antihermitian, and writing $U = \begin{pmatrix} 1 + \epsilon X_{11} & \epsilon X_{12} \\ \epsilon X_{21} & 1 + \epsilon X_{22} \end{pmatrix}$, we see that the condition $\det U = 1$ reduces to $\text{Tr} X \equiv X_{11} + X_{22} = 0$. $X$ is therefore antihermitian and traceless. More explicitly then, such a $U$ can be expressed in terms of the hermitian, traceless Pauli matrices as

$$U = 1 + i\epsilon \mathbf{a} \cdot \boldsymbol{\tau}, \tag{4.7}$$

where $\mathbf{a}$ is an arbitrary real vector. The tangent space of a Lie group at the identity element is, by definition, the Lie algebra of the Lie group. For $SU(2)$, we see that the Lie algebra, denoted $su(2)$, is the space of antihermitian traceless $2 \times 2$ matrices $X$, which have the explicit form

$X = i\mathbf{a}\cdot\boldsymbol{\tau}$. Elements of $su(2)$ can be added and subtracted, and they also close under the commutator bracket, as

$$\begin{aligned}[i\mathbf{a}\cdot\boldsymbol{\tau}, i\mathbf{b}\cdot\boldsymbol{\tau}] &= -a_i b_j [\tau_i, \tau_j] \\ &= -2i\epsilon_{ijk} a_i b_j \tau_k \\ &= i\mathbf{c}\cdot\boldsymbol{\tau},\end{aligned} \quad (4.8)$$

where $\mathbf{c}$ is the real vector $-2\mathbf{a}\times\mathbf{b}$.

The Lie algebra $su(2)$ plays an important role in formulating the derivative of an $SU(2)$-valued field. Suppose $U(x)$ is an $SU(2)$ matrix depending differentiably on a variable $x$. The derivative $\frac{dU}{dx}$ is the $2\times 2$ matrix obtained by differentiating all the matrix entries of $U$. In general this lies neither in $SU(2)$ nor in $su(2)$; however, the matrix $\frac{dU}{dx}U^{-1}$ does lie in $su(2)$. We show this as follows: For $\delta x$ small, the matrix $U(x+\delta x)$ is close to $U(x)$, so it is the *product* of $U(x)$ with an element of $SU(2)$ close to the identity. Therefore, for some real $\mathbf{a}(x)$,

$$\begin{aligned}U(x+\delta x) &= (1 + i\delta x\, \mathbf{a}(x)\cdot\boldsymbol{\tau})U(x) \\ &= U(x) + i\delta x\, \mathbf{a}(x)\cdot\boldsymbol{\tau}\, U(x),\end{aligned} \quad (4.9)$$

so

$$\left\{\frac{1}{\delta x}\bigl(U(x+\delta x) - U(x)\bigr)\right\} U^{-1}(x) = i\mathbf{a}(x)\cdot\boldsymbol{\tau}. \quad (4.10)$$

As $\delta x \to 0$, the left hand side becomes $\frac{dU}{dx}U^{-1}$, and we see from the right hand side that this is in $su(2)$. Similarly, $U^{-1}\frac{dU}{dx}$ is in $su(2)$. In general $U^{-1}\frac{dU}{dx} \neq \frac{dU}{dx}U^{-1}$, but these quantities are related by conjugation by $U$.

This argument generalises to an $SU(2)$ field $U(x)$ that depends on several spacetime variables $x^\mu$. The quantities $\partial_\mu U U^{-1}$ and $U^{-1}\partial_\mu U$ both lie in the Lie algebra $su(2)$. Because of their vector character in spacetime, they are called currents.

There's an important relationship between the Lie groups $SU(2)$ and $SO(3)$, where $SO(3)$ is the group of real, orthogonal $3\times 3$ rotation matrices having determinant 1. $SU(2)$ double covers $SO(3)$, as there is a map from $SU(2)$ to $SO(3)$ where the pair of $SU(2)$ matrices $U$ and $-U$ map to a single $SO(3)$ matrix. This map is a group homomorphism, preserving the group multiplication. For an $SU(2)$ matrix $U$, the explicit $SO(3)$ matrix $M$ corresponding to it is

$$M_{ij} = \frac{1}{2}\mathrm{Tr}(\tau_i U \tau_j U^{-1}), \quad (4.11)$$

and clearly $M$ is unchanged if $U$ changes sign.

Geometrically, as $SU(2)$ is a 3-sphere and $U$ and $-U$ are antipodal on the 3-sphere, the group $SO(3)$ can be regarded as a 3-sphere with antipodal points identified. For us, the most important difference between these groups is their topology. All closed loops on $SU(2)$ are contractible to a point, but on $SO(3)$ there is one class of non-contractible closed loops. To see this, note that any loop on $SO(3)$ can be lifted to a continuous path in $SU(2)$. One needs to select one of the $SU(2)$ matrices $U$ or $-U$ corresponding to each $SO(3)$ matrix along the loop, choosing the sign to ensure the path is continuous. The loop on $SO(3)$ is non-contractible if the lift of the loop connects elements of $SU(2)$ with opposite signs. The best-known example is a loop of rotations about a fixed axis $\mathbf{n}$, where the rotation angle $\theta$ runs from 0 to $2\pi$. This is a closed loop in $SO(3)$, but lifts to a path in $SU(2)$ running from 1 to $-1$, so the loop is non-contractible in $SO(3)$. Explicitly, the path in $SU(2)$ consists of the matrices

$$\cos\frac{\theta}{2} + i\sin\frac{\theta}{2}\mathbf{n}\cdot\boldsymbol{\tau}, \tag{4.12}$$

as can be verified using (4.11).

The existence of non-contractible loops in $SO(3)$ is key to allowing Skyrmions to be quantized as fermions.

## 4.2 The Skyrme Lagrangian and Field Equation

Skyrme theory [215, 216] is a nonlinear theory of dynamical pion fields in $(3+1)$-dimensional spacetime, with the Skyrme field $U(x)$ being an $SU(2)$-valued field that is also a Lorentz scalar. Recall that $x$ is shorthand for the spacetime coordinates $x^\mu = (t, \mathbf{x})$. We impose the boundary condition $U(\mathbf{x}) \to 1$ as $|\mathbf{x}| \to \infty$ ($U(\infty) = 1$, for short). Remarkably, and this was Skyrme's main motivation, the theory has topological soliton solutions, called Skyrmions, that can be interpreted as baryons.

In order to make explicit the nonlinear pion theory, we use the expression for a general $SU(2)$ matrix in terms of Pauli matrices (4.5), and write the Skyrme field as

$$U(x) = \sigma(x) + i\boldsymbol{\pi}(x)\cdot\boldsymbol{\tau}, \tag{4.13}$$

where $\boldsymbol{\pi} = (\pi_1, \pi_2, \pi_3)$ is the triplet of pion fields and $\sigma$, called the sigma field, is determined through the constraint $\sigma^2 + \boldsymbol{\pi}\cdot\boldsymbol{\pi} = 1$, ensuring that $U \in SU(2)$. Not only the magnitude, but also the sign of $\sigma$ is determined by imposing field continuity and the boundary conditions $\sigma(\infty) = 1$, $\boldsymbol{\pi}(\infty) = \mathbf{0}$. When the pions are treated as massless, the theory has some additional, attractive mathematical features, and some extra symmetry. So we start with this version of the theory, and add a pion mass term later.

The theory with massless pions is defined by the Lagrangian

$$L = \int \left\{ -\frac{F_\pi^2}{16} \text{Tr}(\partial_\mu U U^{-1} \partial^\mu U U^{-1}) \right.$$
$$\left. + \frac{1}{32e^2} \text{Tr}([\partial_\mu U U^{-1}, \partial_\nu U U^{-1}][\partial^\mu U U^{-1}, \partial^\nu U U^{-1}]) \right\} d^3x$$
$$= \int \left\{ -\frac{F_\pi^2}{16} \text{Tr}(R_\mu R^\mu) + \frac{1}{32e^2} \text{Tr}([R_\mu, R_\nu][R^\mu, R^\nu]) \right\} d^3x. \quad (4.14)$$

In the second expression, we have introduced the convenient shorthand for the current, $R_\mu = \partial_\mu U U^{-1}$. $F_\pi$ and $e$ are parameters whose values are calibrated using experimental data. They can be scaled away by using energy and length units of $F_\pi/4e$ and $2/eF_\pi$, respectively, which we adopt from now on. In terms of these standard units, called *Skyrme units*, the Skyrme Lagrangian becomes

$$L = \int \left\{ -\frac{1}{2} \text{Tr}(R_\mu R^\mu) + \frac{1}{16} \text{Tr}([R_\mu, R_\nu][R^\mu, R^\nu]) \right\} d^3x, \quad (4.15)$$

and its integral over time is the action of Skyrme theory. The current $R_\mu$ and its commutators $[R_\mu, R_\nu]$ take their values in the Lie algebra $su(2)$. Spacetime indices are raised using the Minkowski metric, which reverses the sign of the current's spatial components. Note that the commutator $[R_\mu, R_\nu]$ vanishes if $\mu = \nu$. $L$ therefore has terms which are simultaneously quadratic in the time and space derivatives of $U$, but has no term quartic in the time derivative.

Finding the Skyrme field equation, the Euler–Lagrange equation, is a little tricky. Variations should maintain $U$ inside the group $SU(2)$. So we vary $U$ by *multiplying* by an element of $SU(2)$ close to the identity,

$$U(x) \mapsto (1 + X(x))U(x), \quad (4.16)$$

where $X$ is small and in the Lie algebra $su(2)$. The field variation is therefore $\delta U = XU$. As $U^{-1}(x) \mapsto U^{-1}(x)(1 - X(x))$, the current varies to

$$R_\mu \mapsto \partial_\mu((1+X)U)) U^{-1}(1-X)$$
$$= (\partial_\mu U + \partial_\mu X U + X \partial_\mu U)(U^{-1} - U^{-1}X)$$
$$= R_\mu + \partial_\mu X + X R_\mu - R_\mu X$$
$$= R_\mu + \partial_\mu X + [X, R_\mu], \quad (4.17)$$

to first order in $X$. Therefore $\delta R_\mu = \partial_\mu X + [X, R_\mu]$. In the resulting variation of the action, the commutator term involving $X$ drops out, because of the trace identity $\mathrm{Tr}([A,B]C) = \mathrm{Tr}(A[B,C])$. Using this trace identity again, integrating by parts and dropping surface terms to isolate $X$, one concludes that the variation of the action is zero for any $X$ provided

$$\partial_\mu \left( R^\mu + \frac{1}{4}[R_\nu, [R^\nu, R^\mu]] \right) = 0, \qquad (4.18)$$

and this is the Skyrme field equation, a nonlinear wave equation for $U(x)$. An interesting feature of (4.18) is that it is in the form of a current conservation equation $\partial_\mu \tilde{R}^\mu = 0$, where $\tilde{R}^\mu = R^\mu + \frac{1}{4}[R_\nu, [R^\nu, R^\mu]]$. This is something that occurs generally when the Lagrangian density does not depend directly on the fields but only on their derivatives.

Any constant, uniform field is a solution of the field equation, but the only such solution satisfying the boundary condition is $U = 1$ for all **x** and $t$. This solution is the classical vacuum solution of Skyrme theory, having minimal energy.

So far we have expressed the Lagrangian and field equation in terms of the (right) current $R_\mu = \partial_\mu U U^{-1}$, but there is an alternative formulation using the left current $L_\mu = U^{-1}\partial_\mu U$, which is related by a conjugation. Replacing $R_\mu$ by $L_\mu$ in the Lagrangian has no effect, because of the traces, and the field equation becomes $\partial_\mu \tilde{L}^\mu = 0$, where $\tilde{L}^\mu = L^\mu + \frac{1}{4}[L_\nu, [L^\nu, L^\mu]]$.

The Skyrme Lagrangian has an internal $(SU(2) \times SU(2))/\mathbb{Z}_2 \cong SO(4)$ symmetry arising from the transformations $U(x) \mapsto \mathcal{O}_1 U(x) \mathcal{O}_2$, where $\mathcal{O}_1$ and $\mathcal{O}_2$ are constant elements of $SU(2)$; the transformation is unchanged if $\mathcal{O}_1$ and $\mathcal{O}_2$ both change sign, which is why we quotient by $\mathbb{Z}_2$. It is called *chiral symmetry*, because there are independent transformations acting on the left and on the right of $U$. However, the boundary condition $U(\infty) = 1$ spontaneously breaks this chiral symmetry to $SO(3)$ isospin symmetry given by the conjugation

$$U(x) \mapsto \mathcal{O} U(x) \mathcal{O}^{-1}, \qquad \mathcal{O} \in SU(2). \qquad (4.19)$$

The isospin symmetry group is $SO(3)$ rather than $SU(2)$, because conjugation by $\mathcal{O}$ has the same effect as conjugation by $-\mathcal{O}$.

There are conserved Noether currents associated with chiral symmetry. The infinitesimal version of left multiplication by $\mathcal{O}_1$ is the variation $U(x) \mapsto (1 + \varepsilon Y)U(x)$, where $Y$ is an arbitrary constant element of $su(2)$, so $\Delta U = YU$. This is similar to the field variation used to derive the Skyrme field equation, but with $X = \varepsilon Y$ now a constant. The conserved (left) Noether

current is
$$J_L^\mu = \text{Tr}(Y\tilde{R}^\mu), \tag{4.20}$$
so the conservation law $\partial_\mu J_L^\mu = 0$ is very similar to the field equation. In fact, since $Y$ is arbitrary and since $\tilde{R}^\mu$ like $Y$ is in $su(2)$, the conservation of this current is completely equivalent to the field equation. The conserved (right) Noether current associated with right multiplication by $\mathcal{O}_2$ is
$$J_R^\mu = \text{Tr}(Y\tilde{L}^\mu). \tag{4.21}$$
Again, the conservation law $\partial_\mu J_R^\mu = 0$ is completely equivalent to the field equation.

It is useful to identify the isospin current. This is associated with the isorotations (4.19), where $\mathcal{O}_1 = \mathcal{O}$ and $\mathcal{O}_2 = \mathcal{O}^{-1}$. The infinitesimal symmetry transformation of the field is now
$$U(x) \mapsto (1+\varepsilon Y)U(x)(1-\varepsilon Y)$$
$$= U(x) + \varepsilon[Y, U(x)], \tag{4.22}$$
and the corresponding Noether current is the combination of chiral currents
$$J^\mu = \text{Tr}(Y(\tilde{R}^\mu - \tilde{L}^\mu)). \tag{4.23}$$
As $Y$ is arbitrary, the $su(2)$-valued isospin current, defined as
$$J^\mu_{\text{isospin}} = \tilde{R}^\mu - \tilde{L}^\mu, \tag{4.24}$$
satisfies the conservation law
$$\partial_\mu J^\mu_{\text{isospin}} = \partial_\mu(\tilde{R}^\mu - \tilde{L}^\mu) = 0. \tag{4.25}$$

A chiral transformation generally mixes the sigma and pion fields, violating the boundary conditions, but the isorotation (4.19) is simpler and preserves the boundary conditions. The latter transformation is $\sigma \mapsto \sigma$ and $\boldsymbol{\pi} \mapsto M\boldsymbol{\pi}$, where $M$ is the $SO(3)$ matrix (4.11) corresponding to the $SU(2)$ matrix $\mathcal{O}$.

## 4.3 Skyrme Field Topology

A field configuration $U(\mathbf{x})$ in Skyrme theory, at a fixed time, is a map from $\mathbb{R}^3$ to the 3-sphere $S^3$, the underlying manifold of the group $SU(2)$. The boundary condition $U(\infty) = 1$ means that we can identify all points at spatial infinity, regarding them as a single point, because the field has a single value there. The topological effect is to close $\mathbb{R}^3$ off, effectively making space into a 3-sphere of infinite radius.

Topologically, the Skyrme field configuration $U(\mathbf{x})$ is now a map from $S^3$ of infinite radius to $S^3$ (the group $SU(2)$) of unit radius. Such a map is classified by its homotopy class – an integer. This integer is denoted by $B$ and identified as the *baryon number* of the configuration. Conveniently, for a field configuration that is differentiable almost everywhere (as we expect a Skyrmion of finite energy to be), the baryon number has a simpler definition as the *topological degree* of the map $U$. This has an integral expression involving the derivatives of $U$,

$$B = -\frac{1}{24\pi^2} \int_{\mathbb{R}^3} \epsilon_{ijk} \operatorname{Tr}(R_i R_j R_k) \, d^3x, \qquad (4.26)$$

where $R_i = \partial_i U U^{-1}$ are the spatial components of the current $R_\mu$. The baryon number $B$, being an integer, is conserved under any continuous deformation of the field, including time-evolution, so $B$ is the principal attribute of a Skyrmion. It is conserved for topological reasons, not as a Noether charge. We denote by $\mathcal{C}_B$ the infinite-dimensional space of all Skyrme field configurations with baryon number $B$.

### 4.3.1 Topological degree of a map

It is important to understand the formula (4.26), so we will pause to discuss the topological degree of a map more broadly [49], and include as a basic example the degree of a map from $S^1$ to $S^1$, i.e. from a circle to a circle, where the degree coincides with the winding number. Sometimes the degree of a map between higher-dimensional manifolds is called a winding number, but we shall avoid this.

The topological degree is defined for a map $\Psi$ between two closed differential manifolds of the same dimension, $\Psi : M \mapsto N$. Let $\dim M = \dim N = n$. Both $M$ and $N$ must be oriented, and the map should be differentiable, with continuous derivatives, although this last requirement can be relaxed a little. For simplicity, let us assume $M$ is connected. We may then assume that $N$ is connected too, since the image of $M$ will always lie in one of the connected components of $N$.

We need a normalised volume form $\Omega$ to be defined on $N$. Locally, this is an $n$-form on $N$ satisfying the normalisation condition

$$\int_N \Omega = 1. \qquad (4.27)$$

If $N$ is a Riemannian manifold, then there is a natural volume form associated with the metric, and a normalised volume form is obtained from this by multiplying by a positive constant.

Now consider $\Psi^*(\Omega)$, the volume form on $N$ pulled-back to $M$ using the map $\Psi$. In terms of local coordinates $\mathbf{x}$ on $M$ and $\mathbf{y}$ on $N$, if $\Omega = \omega(\mathbf{y})\,dy^1 \wedge dy^2 \wedge \cdots \wedge dy^n$ and $\Psi$ is represented by functions $\mathbf{y}(\mathbf{x})$, then the pull-back is

$$\Psi^*(\Omega) = \omega(\mathbf{y}(\mathbf{x}))\frac{\partial y^1}{\partial x^j}dx^j \wedge \frac{\partial y^2}{\partial x^k}dx^k \wedge \cdots \wedge \frac{\partial y^n}{\partial x^l}dx^l$$

$$= \omega(\mathbf{y}(\mathbf{x}))\det\left(\frac{\partial y^i}{\partial x^j}\right) dx^1 \wedge dx^2 \wedge \cdots \wedge dx^n, \qquad (4.28)$$

where $J(\mathbf{x}) \equiv \det(\partial y^i/\partial x^j)$ is the Jacobian of the map at $\mathbf{x}$.

The topological degree of $\Psi$ is now defined as the integral over $M$ of this pulled-back version of $\Omega$,

$$\deg\Psi = \int_M \Psi^*(\Omega). \qquad (4.29)$$

We shall show below that the degree is an integer, and therefore a topological invariant of $\Psi$. The degree is also independent of the choice of $\Omega$, because the difference of two normalised volume forms on $N$ is an $n$-form whose integral is zero, and hence an exact form. The pull-back of the difference is therefore also exact on $M$, and integrates to zero – this key result comes from cohomology theory, the topological theory of differential forms on manifolds.

The simplest example is for a map $\Psi : S^1 \mapsto S^1$. Let us fix coordinates $\theta$ and $\chi$ on the domain and target, both having range $[0, 2\pi]$. The map $\Psi$ is given by a function $\chi(\theta)$ that must satisfy $\chi(2\pi) = \chi(0)$ mod $2\pi$ to give a geometrically continuous map. If we choose the obvious normalised volume form $\frac{1}{2\pi}d\chi$ on the target $S^1$, then the formula (4.29) reduces to

$$\deg\Psi = \frac{1}{2\pi}\int_0^{2\pi} \frac{d\chi}{d\theta} d\theta = \frac{1}{2\pi}(\chi(2\pi) - \chi(0)). \qquad (4.30)$$

The topological degree is therefore an integer; it is the winding number of the map.

For Skyrmions, we require the degree of a map from $\mathbb{R}^3 \simeq S^3$ to $SU(2) = S^3$, and can make use of the Lie group structure of the target. Recall that elements of $SU(2)$ can be written as

$$U = a_0 + i\mathbf{a}\cdot\boldsymbol{\tau}, \qquad (4.31)$$

where $\tau_1, \tau_2, \tau_3$ are the Pauli matrices, and where $a_0^2 + \mathbf{a}\cdot\mathbf{a} = 1$. The standard normalised volume form on $SU(2)$ can be expressed as

$$\Omega = \frac{1}{24\pi^2}\operatorname{Tr}\left(dUU^{-1} \wedge dUU^{-1} \wedge dUU^{-1}\right). \qquad (4.32)$$

Here, $dUU^{-1}$ is an $su(2)$ matrix of 1-forms, and the multiple wedge product means doing the usual matrix multiplication and at the same time taking the wedge product of the matrix entries. After taking the trace, the result is an ordinary 3-form on $SU(2)$. $\Omega$ is invariant under left and right multiplication by fixed elements of $SU(2)$, $U \mapsto \mathcal{O}_1 U \mathcal{O}_2$, and since $(SU(2) \times SU(2))/\mathbb{Z}_2 = SO(4)$, $\Omega$ is invariant under the full rotation group of the target 3-sphere. To understand the normalisation factor, consider $U$ close to the identity, where $a_0$ is close to 1 and essentially constant. Then (4.32) simplifies to

$$\Omega = \frac{-6i}{24\pi^2} \operatorname{Tr}(\tau_1 \tau_2 \tau_3)\, da_1 \wedge da_2 \wedge da_3 = \frac{1}{2\pi^2}\, da_1 \wedge da_2 \wedge da_3\,. \quad (4.33)$$

This is the correct normalisation, because the unit 3-sphere has total volume $2\pi^2$ and has volume element $da_1 \wedge da_2 \wedge da_3$ near $a_0 = 1$.

If $\Psi$ is a map from $\mathbb{R}^3$ to $SU(2)$, represented by a function (a Skyrme field) $U(\mathbf{x})$, then

$$\deg \Psi = \frac{1}{24\pi^2} \int_{\mathbb{R}^3} \operatorname{Tr}\left( dUU^{-1} \wedge dUU^{-1} \wedge dUU^{-1} \right), \quad (4.34)$$

where $dUU^{-1}$ now denotes $\partial_i U(\mathbf{x})\, U^{-1}(\mathbf{x})\, dx^i$. This is the integral of the pull-back of $\Omega$ to $\mathbb{R}^3$. If we set $R_i = \partial_i U U^{-1}$, we recover the formula (4.26) used to define the baryon number of a Skyrme field configuration.

A very useful feature of the topological degree of a map $\Psi : M \mapsto N$, for any manifolds $M$ and $N$, is that there is a second, apparently independent way to compute it. For each point $\mathbf{y}$ on $N$, the set of points on $M$ mapped by $\Psi$ to $\mathbf{y}$ is called the set of preimages of $\mathbf{y}$. Choose a point $\mathbf{y}$ such that its preimages form a set (possibly empty) of isolated points $\{\mathbf{x}^{(1)}, \ldots, \mathbf{x}^{(Q)}\}$, and such that at each of the points $\mathbf{x}^{(q)}$ the Jacobian of the map is non-zero. Such points $\mathbf{y}$ are generic, occurring almost everywhere on $N$. Let

$$\widetilde{\deg}\,\Psi = \sum_{q=1}^{Q} \operatorname{sign}\left( J(\mathbf{x}^{(q)}) \right), \quad (4.35)$$

where $\operatorname{sign}(J(\mathbf{x}^{(q)}))$ is the sign of the Jacobian at $\mathbf{x}^{(q)}$. We say that $\widetilde{\deg}\,\Psi$ *counts the preimages* of $\mathbf{y}$ with their multiplicity, which is 1 or $-1$, depending on whether $\Psi$ is locally orientation preserving or orientation reversing. Clearly $\widetilde{\deg}\,\Psi$ is an integer. It is a theorem that $\widetilde{\deg}\,\Psi = \deg\,\Psi$, as we will show in a moment. This theorem tells us firstly, that because $\widetilde{\deg}\,\Psi$ is an integer, the integral $\deg\,\Psi$ is an integer, and secondly, that because the integral $\deg\,\Psi$ is clearly independent of $\mathbf{y}$, the preimage count $\widetilde{\deg}\,\Psi$ is independent of the choice of $\mathbf{y}$.

To prove that $\widetilde{\deg \Psi} = \deg \Psi$, we deform the volume form $\Omega$ on $N$ so that it is concentrated on a small neighbourhood of the point $\mathbf{y}$, and still normalised. $\deg \Psi$ is unaffected, because, as we argued earlier, it doesn't depend on the choice of $\Omega$. The pulled-back volume form $\Psi^*(\Omega)$ is now concentrated on small neighbourhoods of each of the preimages $\{\mathbf{x}^{(1)}, \ldots, \mathbf{x}^{(Q)}\}$. Moreover, the integral of $\Psi^*(\Omega)$ over one of these preimage neighbourhoods is simply $\pm 1$, which can be understood by a naive local change of coordinates from $\{x^i\}$ back to $\{y^i\}$, which introduces an inverse Jacobian factor $|J(\mathbf{x})|^{-1}$, and reproduces the volume form $\Omega$ up to an orientation preserving/reversing sign. This local integral is therefore unity, up to a sign, by the normalisation condition. Summing over the preimages, we see that the formula (4.29) for $\deg \Psi$ reduces to the expression (4.35) for $\widetilde{\deg \Psi}$.

We will see later that preimage counting is often the easiest way of determining the topological degree of a map.

## 4.4 Skyrme Field Energy

If one restricts to static fields $U(\mathbf{x})$ in $\mathbb{R}^3$, then the Skyrme energy functional derived from the Lagrangian (4.15) is

$$E = \int \left\{ -\frac{1}{2}\mathrm{Tr}(R_i R_i) - \frac{1}{16}\mathrm{Tr}([R_i, R_j][R_i, R_j]) \right\} d^3x. \tag{4.36}$$

Static solutions of the Skyrme field equation (4.18) are critical points (either minima or saddle points) of this energy. The solution that globally minimises the energy for given baryon number $B$ is called a Skyrmion, and its energy is denoted by $E_B$. We also sometimes refer to local minima and saddle points with close-to-minimal energy as Skyrmions.

The energy (4.36) for any field configuration, provided it is finite, has the following property under spatial rescaling [71]. First note that it decomposes into two positive parts, $E = E_2 + E_4$, corresponding to the quadratic and quartic terms in the current $R_i$ and hence in the field's spatial derivatives. Therefore, under the rescaling $\mathbf{x} \mapsto \mu \mathbf{x}$, the energy becomes

$$E(\mu) = \frac{1}{\mu} E_2 + \mu E_4. \tag{4.37}$$

The two parts scale in opposite ways, leading to a minimal value of $E(\mu)$ for some positive $\mu$. This implies that any Skyrmion will have a well-defined scale and will neither expand to cover all of space nor contract to be localised at a single point. If we start from the Skyrmion itself and rescale by $\mu$, then $E(\mu)$ must take its minimal value when $\mu = 1$, so $E_2$ and $E_4$

are equal for a Skyrmion, each being half the total energy $E_B$. It is now clear why the Lagrangian consisting of only the first term in (4.15) does not support stable Skyrmions. This problem is cured by the addition of the second term in (4.15), the *Skyrme term*. Another term which is quartic or higher order in the space and time derivatives could do equally well in this respect, but the Skyrme term is the unique quartic expression that is Lorentz invariant and still only quadratic in the time derivative of the field.

Faddeev established a lower bound on the energy of a Skyrmion with (positive) baryon number $B$ [82],

$$E_B \geq 12\pi^2 B. \tag{4.38}$$

This is known as the Faddeev–Bogomolny bound, because it is analogous to energy bounds established by Bogomolny [44] for several other types of topological soliton. It is proved by completing the square and rewriting the Skyrme energy function (4.36) as

$$E = \int -\frac{1}{2}\text{Tr}\left(R_i - \frac{1}{4}\epsilon_{ijk}[R_j, R_k]\right)\left(R_i - \frac{1}{4}\epsilon_{ilm}[R_l, R_m]\right) d^3x$$
$$- \frac{1}{2}\int \epsilon_{ijk} \text{Tr}(R_i R_j R_k) \, d^3x, \tag{4.39}$$

and then noting that the first term is non-negative, and the second is $12\pi^2 B$.

The bound would be saturated (i.e. satisfied with equality) if one could find a Skyrmion satisfying the Faddeev–Bogomolny equation

$$R_i - \frac{1}{4}\epsilon_{ijk}[R_j, R_k] = 0. \tag{4.40}$$

However this equation has no solution, so Skyrmions have energy strictly greater than the lower bound. This is easiest to understand using a reinterpretation of the Skyrme energy, which we discuss next.

### 4.4.1  *Elastic strain formulation*

There is a more geometrical formulation of the static Skyrme energy (4.36), which is often useful and illuminating [176]. As in nonlinear elasticity theory, the energy density of a Skyrme field depends on the local strain associated with the map $U : \mathbb{R}^3 \mapsto S^3$. The elastic strain tensor $D_{ij}$ is defined at each point $\mathbf{x} \in \mathbb{R}^3$ by

$$D_{ij} = -\frac{1}{2}\text{Tr}(R_i R_j). \tag{4.41}$$

This symmetric, positive definite $3 \times 3$ matrix quantifies the local deformation of the metric geometry induced by the map $U$. The image under $U$ of an infinitesimal spatial ball of radius $\epsilon$ and centre $\mathbf{x}$ in $\mathbb{R}^3$ is, to leading order in $\epsilon$, an infinitesimal (solid) ellipsoid on the target 3-sphere with principal axes $\epsilon\lambda_1, \epsilon\lambda_2, \epsilon\lambda_3$, where $\lambda_1^2, \lambda_2^2, \lambda_3^2$ are the three eigenvalues of the matrix $D_{ij}$. The signs of $\lambda_1, \lambda_2$ and $\lambda_3$ are chosen so that $\lambda_1\lambda_2\lambda_3$ is positive (negative) if $U$ is locally orientation preserving (reversing). In terms of these eigenvalues, the static energy $E$ and the baryon number $B$ are integrals over $\mathbb{R}^3$ of the corresponding densities $\mathcal{E}$ and $\mathcal{B}$ given by

$$\mathcal{E} = \lambda_1^2 + \lambda_2^2 + \lambda_3^2 + \lambda_1^2\lambda_2^2 + \lambda_2^2\lambda_3^2 + \lambda_3^2\lambda_1^2, \tag{4.42}$$

$$\mathcal{B} = \frac{1}{2\pi^2}\lambda_1\lambda_2\lambda_3. \tag{4.43}$$

Here, $\lambda_i^2$ can be interpreted as the squared strain of length along the $i$th principal axis, and $\lambda_i^2\lambda_j^2$ can be interpreted as the squared strain of area in the plane spanned by the $i$th and $j$th principal axes. $|\lambda_1\lambda_2\lambda_3|$ is the volume strain. The combination of terms in (4.42) matches the combination of quadratic and quartic terms in (4.36) and is very natural for a theory in three spatial dimensions.

Using the simple inequality

$$(\lambda_1 \pm \lambda_2\lambda_3)^2 + (\lambda_2 \pm \lambda_3\lambda_1)^2 + (\lambda_3 \pm \lambda_1\lambda_2)^2 \geq 0, \tag{4.44}$$

and expanding out, it follows using the formulae (4.42) and (4.43) that $\mathcal{E} \geq 12\pi^2|\mathcal{B}|$, and therefore the total Skyrme energy satisfies the Faddeev–Bogomolny bound (4.38).

However, the bound cannot be attained with equality for any finite-energy field configuration with $B \neq 0$. This is because equality requires the left hand side of (4.44) to be zero everywhere, implying that the three strain eigenvalues are all 0, or all $\pm 1$. In the region where the eigenvalues are 0, the map $U$ takes a constant value, which doesn't contribute to $B$. In the remaining region of space, where the map is not constant, it is required to be an isometry, preserving all lengths. This is obviously not possible since no finite region of (flat) $\mathbb{R}^3$ is isometric to a finite region of the (curved) 3-sphere of unit radius.

From the viewpoint of nonlinear elasticity, a Skyrmion can be regarded as compressing all of $\mathbb{R}^3$ onto a unit 3-sphere, which obviously generates non-zero elastic energy. Note however that the Faddeev–Bogomolny bound can be attained if space itself is assumed to be a 3-sphere of unit radius [176, 182]. The identity map from a unit 3-sphere to itself is trivially an isometry, so it is the $B = 1$ Skyrmion on the unit 3-sphere. The bound is also attained by the vacuum solution, for which the strain eigenvalues are 0 everywhere, and $E = B = 0$.

## 4.5 Hedgehog Skyrmions

The spherically-symmetric $B = 1$ Skyrmion was described in the original work of Skyrme and takes the *hedgehog* form

$$U(\mathbf{x}) = \cos f(r) + i \sin f(r)\, \hat{\mathbf{x}} \cdot \boldsymbol{\tau}, \tag{4.45}$$

where $\hat{\mathbf{x}}$ is the unit, outward-pointing radial vector, and $f$ is a real radial profile function. In terms of $\sigma$ and $\boldsymbol{\pi}$ fields,

$$\sigma = \cos f(r), \quad \boldsymbol{\pi} = \sin f(r)\, \hat{\mathbf{x}}. \tag{4.46}$$

The profile function must satisfy the boundary conditions $f(0) = \pi$ and $f(\infty) = 0$. The latter condition ensures that $U(\infty) = 1$, the vacuum value, while the former ensures that $U(\mathbf{0})$ is unambiguous and that $B = 1$. In fact $U(\mathbf{0}) = -1$, and we call this field value at the origin the antivacuum. The name hedgehog derives from the fact that the pion field $\boldsymbol{\pi}$ everywhere points radially outwards. The value of $B$ is confirmed by substituting the hedgehog ansatz into the expression (4.26) for the baryon number, giving

$$B = -\frac{2}{\pi} \int_0^\infty f' \sin^2 f \, dr = \frac{1}{\pi} f(0) = 1. \tag{4.47}$$

Alternatively, it is easy to verify that if $f'$ is negative for all $r$, then each point of the target space $SU(2)$ (except $U = 1$) has exactly one preimage in $\mathbb{R}^3$, with positive Jacobian.

Here, spherically-symmetric does not mean that the Skyrme field $U$ is just a function of the radial coordinate $r$, since such a field must have zero baryon number. When we refer to a spatial symmetry of a Skyrmion, we mean that the field has the property that the effect of a spatial rotation can be compensated by, or is equivalent to, an isorotation (4.19). This implies that both the energy density $\mathcal{E}$ and baryon density $\mathcal{B}$ are strictly symmetric.

For a hedgehog field configuration, the radial and angular strain eigenvalues are

$$\lambda_1 = -f', \quad \lambda_2 = \lambda_3 = \frac{\sin f}{r}. \tag{4.48}$$

The energy, in its nonlinear elasticity version, therefore reduces to

$$E = 4\pi \int_0^\infty \left\{ r^2 f'^2 + 2\sin^2 f\, (1 + f'^2) + \frac{\sin^4 f}{r^2} \right\} dr. \tag{4.49}$$

The variational equation for $f(r)$ is the second-order, nonlinear ordinary differential equation

$$(r^2 + 2\sin^2 f) f'' + 2r f' + \sin 2f \left( f'^2 - 1 - \frac{\sin^2 f}{r^2} \right) = 0, \tag{4.50}$$

and the same equation is obtained by substituting the hedgehog ansatz (4.45) into the static Skyrme field equation. A solution of (4.50) satisfying the boundary conditions was rigorously proved to exist by Kapitanski and Ladyzenskaia [148]. The solution cannot be obtained in closed form but it is a simple task to compute it numerically using a shooting method. The numerical solution is presented in Fig. 4.1. Its calculated energy is $E = 1.232 \times 12\pi^2$, to three decimal places, exceeding the Faddeev–Bogomolny bound $E = 12\pi^2$ by approximately 23%.

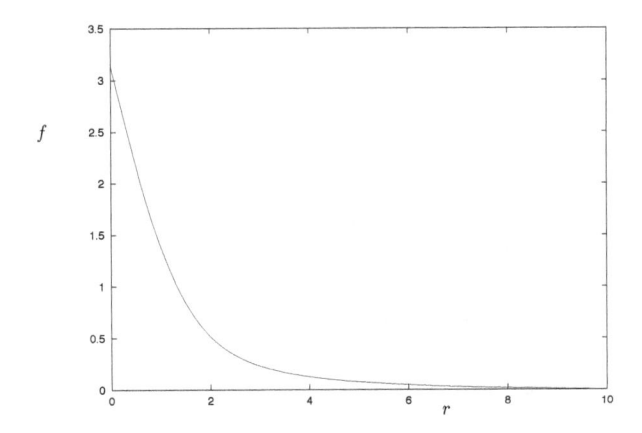

Fig. 4.1: Profile function $f(r)$ of the $B = 1$ Skyrmion

Linearisation of the radial equation (4.50) reveals that the large $r$ asymptotic behaviour of the profile function is $f(r) \sim C/r^2$, where the coefficient $C$ is numerically found to be $C = 2.16$. The leading-order asymptotic fields are

$$\sigma = 1, \quad \boldsymbol{\pi} = \frac{C}{r^2}\hat{\mathbf{x}}, \qquad (4.51)$$

so from far away, a hedgehog Skyrmion resembles a triplet of orthogonal pion dipoles, each with dipole strength $4\pi C$. In Section 4.7 we will discuss the asymptotic interaction of well-separated Skyrmions in terms of dipole-dipole forces.

The hedgehog Skyrmion (4.45) can be centred at any point in space and given any orientation by acting with the translation and rotation groups of $\mathbb{R}^3$. The space of static $B = 1$ Skyrmion solutions, known as the *moduli space*, is therefore 6-dimensional. In general, one expects the moduli space of a Skyrmion to be 9-dimensional, since in addition to translations and rotations there are independent isorotations. However, for a hedgehog an isorotation is equivalent to a spatial rotation, which is why the Skyrmion is spherically symmetric and why three moduli are lost.

Esteban [81] has proved more generally, without making any symmetry assumption, the rigorous existence of a $B = 1$ Skyrmion – a global minimiser of the energy functional (4.36) with unit baryon number. It has also been shown by a spherical averaging argument that this is indeed the $B = 1$ hedgehog Skyrmion [208]. Numerical evidence from a study of linearised vibrations also confirms that the hedgehog Skyrmion is at least a local minimum of the energy, and stable [108].

There are further solutions involving the hedgehog ansatz (4.45). $U$ is well defined provided $f(0) = n\pi$, where $n$ is any positive integer, and eq.(4.47) shows that in this case, $B = n$. The pion field still points radially, but alternately inwards or outwards as $r$ increases. The equation for the profile function $f$ appears to have solutions for all $n$ [140, 216], and they have been constructed numerically for several values of $n$.

However, for $n > 1$, the hedgehog solution does not represent the minimal-energy Skyrmion with $B = n$, and in fact these solutions are not even bound against breakup into $n$ well-separated $B = 1$ Skyrmions. For example, the $B = 2$ hedgehog has energy $E = 3.67 \times 12\pi^2$, about three times the energy of the $B = 1$ hedgehog, and it has six unstable modes [236,250]. For all $B > 1$, the hedgehog solutions are, almost certainly, unstable saddle points of the energy. Stable, minimal-energy Skyrmions with $B > 1$ have less symmetry than a hedgehog.

The $n = -1$ solution is the $B = -1$ antiSkyrmion, whose profile function is obtained from that of the $B = 1$ Skyrmion by the reflection $f \mapsto -f$. For a general Skyrmion with no particular symmetry, a corresponding anti-Skyrmion can always be obtained by reversing the sign of either one or all three of the pion fields (a reflection or inversion in isospace). AntiSkyrmions have the same energy density as Skyrmions, but opposite baryon density.

## 4.6 Visualising Skyrmions

To visualise Skyrmions, and the $B = 1$ hedgehog Skyrmion in particular, it is useful to plot a surface of constant baryon density and colour it using P. O. Runge's colour sphere. The colours indicate the value of the normalised, unit pion field $\hat{\boldsymbol{\pi}} = \boldsymbol{\pi}/|\boldsymbol{\pi}|$ on the surface. No attempt at colouring is made at points where $\sigma = \pm 1$ and $\boldsymbol{\pi} = 0$, but these are absent from the surfaces we show. The equator of the colour sphere corresponds to $\hat{\pi}_3 = 0$. Here, the primary colours red, green and blue show where the field $\hat{\pi}_1 + i\hat{\pi}_2$ takes the values $1, e^{\frac{i2\pi}{3}}$ and $e^{\frac{i4\pi}{3}}$, respectively, and the intermediate colours yellow, cyan and magenta show the values $e^{\frac{i\pi}{3}}, -1$ and $e^{\frac{i5\pi}{3}}$. The $\hat{\pi}_3$-value is assigned to the "lightness", so that white and black, at the poles on the colour sphere, show where $\hat{\pi}_3 = \pm 1$, respectively. The hedgehog form of the $B = 1$ Skyrmion means that any sphere centred at the origin is a surface of constant baryon density, and the colouring of the Skyrmion reproduces the colour sphere itself (see Fig. 4.2). Many Skyrmions having baryon numbers greater than 1 and less symmetry will appear later, coloured using the Runge colour sphere.

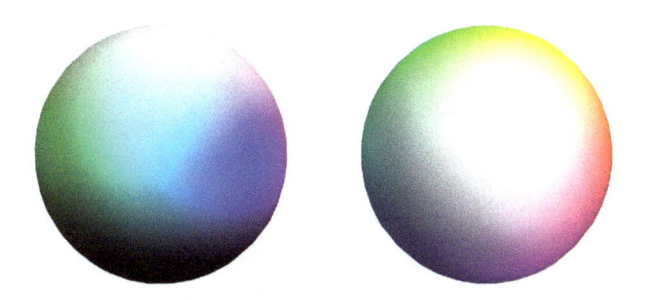

Fig. 4.2: $B = 1$ Skyrmion (two different orientations)

Because the pion fields are scalar fields, charges (sources) of equal sign attract. In this aspect, the Skyrme field is similar to a gravitational field where bodies of equal mass attract, and unlike an electric field where bodies of equal charge repel; the Newtonian gravitational potential, like the Skyrme field, is a scalar, whereas the electrostatic potential of an electric charge is the time-component of a 4-vector. As a consequence, parts of Skyrmions with the same colour tend to attract, and low-energy configurations can be constructed by gluing Skyrmions together with nearby colours matching. This is what makes the colouring so useful. However for $B > 1$ there is some frustration, i.e. non-matching colours, as we will see. If there were no frustration, as occurs with a suitably oriented Skyrmion-antiSkyrmion pair, for example, then the Skyrmions could annihilate.

## 4.7 Asymptotic Interactions of Hedgehogs

As noted above, the asymptotic field of a hedgehog Skyrmion is that of a triplet of orthogonal pion dipoles, so one can find the asymptotic force between two well-separated $B = 1$ Skyrmions by computing the interaction energy between the dipole triplets. It is convenient to rewrite the asymptotic field (4.51) in the form

$$\pi_j = \frac{C}{r^2}\hat{x}^j = \frac{\mathbf{p}_j \cdot \mathbf{x}}{4\pi r^3}, \qquad (4.52)$$

where we have introduced the three orthogonal dipole moments

$$\mathbf{p}_j = 4\pi C \mathbf{e}_j, \qquad (4.53)$$

with $\{\mathbf{e}_j\}$ being the standard basis vectors of $\mathbb{R}^3$. A rotation of a Skyrmion rotates this frame of dipoles, but the magnitudes are unchanged. The interaction energy of two individual dipoles with moments $\mathbf{p}$ and $\mathbf{q}$, and separation vector $\mathbf{X}$, is

$$E_{\text{dip}} = \frac{1}{2\pi}(\mathbf{p} \cdot \tilde{\partial})(\mathbf{q} \cdot \tilde{\partial})\frac{1}{|\mathbf{X}|}, \qquad (4.54)$$

where $\tilde{\partial}_i = \frac{\partial}{\partial X^i}$. This is similar to the interaction energy of two electric dipoles, but has the opposite sign, because the pion field is a scalar, so like charges attract.

We can use the translation and isorotational symmetries to position the first Skyrmion at the origin in standard orientation, and position the second at $\mathbf{X}$, with $|\mathbf{X}| \gg 1$. The dipole moments of the second Skyrmion are $\mathbf{q}_j = M\mathbf{p}_j$, where $M$ is some $SO(3)$ rotation matrix. There is a dipole interaction between $\mathbf{p}_j$ and $\mathbf{q}_k$ only if $j = k$, so summing the interactions of the three pairs and using (4.54) we obtain the total interaction energy

$$E_{\text{int}} = 8\pi C^2(\tilde{\partial} \cdot M\tilde{\partial})\frac{1}{|\mathbf{X}|}. \qquad (4.55)$$

To understand this better, let us express the matrix $M$ in terms of a rotation by $\psi$ about an axis $\mathbf{n}$,

$$M_{ij} = \cos\psi\, \delta_{ij} + (1 - \cos\psi)n_i n_j + \sin\psi\, \epsilon_{ijk} n_k. \qquad (4.56)$$

The interaction energy then takes the form

$$E_{\text{int}} = -8\pi C^2(1 - \cos\psi)\frac{1 - 3(\hat{\mathbf{X}} \cdot \mathbf{n})^2}{|\mathbf{X}|^3}. \qquad (4.57)$$

This result was originally obtained by Skyrme [216], and verified by Jackson et al. [139] and Vinh Mau et al. [233]. Clearly, by a suitable choice of the axis

**n**, the two Skyrmions can be made to either repel or attract, corresponding to a positive or negative interaction energy. The attraction is maximal (i.e. $E_{\text{int}}$ is maximally negative) if $\hat{\mathbf{X}} \cdot \mathbf{n} = 0$ and $\psi = \pi$, i.e. if one Skyrmion is rotated relative to the other by 180° about a direction perpendicular to the line joining them. This relative orientation is known as the *attractive channel*.

The dipole calculation described above is not a rigorous derivation of the asymptotic interaction energy since it assumes that a Skyrmion generating an asymptotic dipole triplet field reacts to an external field like a dipole triplet. However, a more formal calculation of the interaction energy confirms the result [210].

The formal calculation exploits an important idea for constructing a field configuration that superposes two well-separated Skyrmions, $U_1(\mathbf{x})$ and $U_2(\mathbf{x})$. This is the *product ansatz* [216],
$$U(\mathbf{x}) = U_1(\mathbf{x})U_2(\mathbf{x}), \qquad (4.58)$$
which can be used for any pair of well-separated Skyrmions, not just hedgehog Skyrmions with $B = 1$. Because $SU(2)$ is a group, the product $U(\mathbf{x})$ is definitely a Skyrme field configuration. If $U_1$ and $U_2$ satisfy vacuum boundary conditions, then so does $U$. Since $U_2$ is close to 1 well away from its core, the deformation of $U_1$ is small near its own core; similarly $U_2$ suffers a small deformation near its core, due to the tail of $U_1$.

If $U_1$ has baryon number $B_1$ and $U_2$ has baryon number $B_2$, then $U$ has baryon number $B = B_1 + B_2$. This can be verified by using the integral formula for the baryon number (4.26), but can be understood more simply as follows. By a small continuous deformation, the configuration $U_1$ can be strictly localised to a finite ball, with $U_1 = 1$ outside. $U_2$ can be similarly localised. Then, provided the separation is large enough that the balls have no overlap, the product field has undistorted fields $U_1$ and $U_2$ in these balls, and the total baryon number is clearly $B_1 + B_2$, from the integral formula. A continuous deformation does not change the baryon number, so it remains the same if we revert to the product with a small overlap of the Skyrmion tails. Further, we can now reduce the separation if we wish. This is also a continuous change, so the baryon number is $B_1 + B_2$ even as the core separation goes to zero.

The product ansatz has proved to be a useful way to combine well-separated Skyrmions, and is often used to construct initial data prior to a numerical search for Skyrmion solutions with higher baryon numbers. However, it is not a good way to approximate the resulting Skyrmion solutions, as these do not consist of well-separated subclusters. The product of

$U_1$ and $U_2$ usually fails to capture the symmetry of a true Skyrmion. The worst aspect of the product ansatz is that the products $U_1 U_2$ and $U_2 U_1$ are almost always different, but there is nothing to choose between them, and their difference is substantial if the separation is not large. The limitations of the product ansatz became clear many years ago, in the search for the $B = 2$ Skyrmion. The product of two $B = 1$ hedgehogs in the attractive channel, at any separation, fails to exhibit the correct symmetry of the exact solution, and has too large an energy.

## 4.8 Adding a Pion Mass Term

An additional term

$$L_{\text{mass}} = m^2 \int \text{Tr}(U - 1)\, d^3x, \qquad (4.59)$$

or a variant of this, can be included in the Lagrangian (4.15) of Skyrme theory to give the pions a mass $m$ (in Skyrme units). This term does not depend on derivatives of $U$, and it contributes $m^2 \text{Tr}(1 - U)$ to the field potential energy density. In the derivation of the Skyrme field equation, the additional variation of the Lagrangian density resulting from the variation $\delta U = XU$ is $m^2 \text{Tr}(XU)$. Since $X$ is traceless, only the traceless part of $U$, i.e. $U - \frac{1}{2}\text{Tr}(U)$ where the $\frac{1}{2}$ cancels against the trace of the suppressed unit matrix, contributes here. This is the part of $U$ involving the pion fields, with the sigma field set to zero. The Skyrme field equation in the case of massive pions is therefore

$$\partial_\mu \tilde{R}^\mu + m^2 \left(U - \frac{1}{2}\text{Tr}(U)\right) = 0. \qquad (4.60)$$

The pion mass term breaks chiral symmetry explicitly, and the left and right currents are no longer conserved. However, an isorotation – conjugation of $U$ by $\mathcal{O}$ – is still a symmetry, so the isospin current (4.24) is conserved. This conservation law can be verified directly from the field equation (4.60) and its analogue involving $\tilde{L}^\mu$, since these have the same pion mass terms.

The only constant, uniform solutions of the field equation are now the vacuum $U = 1$, and the antivacuum $U = -1$. The vacuum has vanishing energy and is stable, but the antivacuum has a uniform energy density $4m^2$ and is unstable. An antivacuum solution is in any case excluded by the boundary condition $U(\infty) = 1$, but it can occur approximately in limited regions of space.

The equation for the $B = 1$ hedgehog Skyrmion profile is modified when $m$ is positive and takes the form

$$(r^2+2\sin^2 f)f''+2rf'+\sin 2f\left(f'^2-1-\frac{\sin^2 f}{r^2}\right)-m^2 r^2 \sin f = 0. \quad (4.61)$$

The solution approaches the vacuum more rapidly than before as $r$ increases, so the hedgehog Skyrmion is more compact, and also has higher energy.

In Chapter 8 we will discuss in much more detail the modifications to Skyrmion solutions that result from the pion mass term.

Chapter 5

# Quantization of Skyrmions

## 5.1 Quantization of Skyrme Fields

Quantization is a vital issue for Skyrmions, because Skyrmions are supposed to model physical nucleons and nuclei, and a single nucleon is a spin $\frac{1}{2}$ fermion. In this chapter we introduce the general idea of Skyrme field and Skyrmion quantization, and then discuss in some detail the quantization of the $B = 1$ hedgehog Skyrmion. The quantization of Skyrmions whose baryon numbers are greater than 1 will be considered in later chapters, after we have better understood the relevant classical Skyrmion solutions.

It is not viable to systematically quantize Skyrme theory as a quantum field theory. The theory is highly nonlinear and the Skyrmions depend in an essential way on the nonlinearity. They even differ topologically from the vacuum or any small perturbation of the vacuum. The usual practical approach to quantum field theory is perturbation theory, aided by Feynman diagrams. The basic fields of Skyrme theory are three pion fields, and three pion particles arise from the quantization of small fluctuations of these fields around the vacuum $\sigma = 1$, $\boldsymbol{\pi} = \boldsymbol{0}$. In the massless theory, only derivatives of the pion fields occur in the Lagrangian, and the pions are the Goldstone bosons of the spontaneously broken chiral symmetry. The interaction terms in the Lagrangian, beyond the quadratic terms, lead to pion interactions, but there are technical difficulties with these because Skyrme field theory is non-renormalisable, so loop diagrams require an explicit high-momentum cutoff. In any case, pion physics is of limited interest to us, and pion-pion interactions are not easily studied experimentally, except indirectly. Pion beams have been produced for decades, but not pion targets, nor collisions between pion beams.

In chiral EFTs, the pion interaction terms are supplemented by couplings to explicit nucleon fields [78]. This allows a perturbative treatment of pion-nucleon interactions, and more importantly, of nucleon-nucleon forces mediated by pion exchange. Again, an explicit cutoff is needed to control divergences in loop diagrams. From the nucleon-nucleon scattering amplitudes one can reconstruct a nucleon-nucleon potential. This is an input to a Schrödinger equation describing the interactions of several nucleons. 3-body and 4-body potentials can also be derived. This approach to nuclear physics is rather successful, but requires powerful and expensive computational resources.

In Skyrme theory there are no explicit nucleon fields, but instead there are the classical Skyrmion solutions representing nucleons and nuclei. It is possible to study the interactions of pions with Skyrmions perturbatively, and interpret the results in terms of pion-nucleon interactions when $B = 1$ [75]. The basic method goes back to the 1970s, and to studies of quantized kink solitons in one dimension [201]. The classical kink, like a Skyrmion, can be at rest, and linearised perturbations around it are wavelike mesons that scatter off the kink. The field dynamics can be quantized perturbatively, and the resulting states are meson particles scattering off the kink. At lowest order the kink does not back-react, but this violates momentum conservation. Pions scattering off a Skyrmion similarly do not conserve momentum or isospin if one works at lowest order. At higher order, interactions between the meson fields and the centre of mass of the kink soliton automatically occur. This allows for momentum transfer between the mesons and the kink, with overall conservation of momentum. Similarly, in Skyrme theory, the conservation laws associated with spacetime symmetries and with isospin symmetry can be maintained in a systematic perturbative expansion, although this is technically quite elaborate.

This approach to Skyrme theory gives a consistent and satisfactory model for quantized pion-nucleon interactions, and could be extended to give models for pions scattering off larger nuclei, although there are no serious studies of this. Pion scattering off a nucleus can change the isospin and electric charge of the nucleus. One experimental application is to use pion scattering off a nucleus with low isospin to create higher-isospin, exotic nuclei. For example, pion scattering off Hydrogen-3 or Helium-3 can be used explore possible three-neutron bound states and resonances [164].

Instead, our main interest is in the energy spectra of nuclei in their ground and excited states. There are no incoming or outgoing pion particles, so we can ignore the pion waves. It is sufficient to construct quantized models for Skyrmion dynamics using finitely many degrees of freedom –

often referred to as *collective coordinates*. This is analogous to the multi-nucleon quantum mechanics derived using chiral EFT. The degrees of freedom are rigid motions of Skyrmions, whose quantization leads to states with momentum, spin and isospin, supplemented by vibrational degrees of freedom, whose excitations lead to a Skyrmion partially separating into smaller clusters, and possibly breaking up as the amplitudes become large. Pion field dynamics is responsible for all these dynamical phenomena, but no pion waves or real pion particles are involved.

The collective coordinates of Skyrmions do not map simply onto the positional coordinates of nucleons. This makes Skyrme theory genuinely different from most other nuclear physics models. One reason is that the simple $B = 1$ Skyrmions modelling single nucleons partially merge into larger Skyrmions, and this radically changes the collective coordinate geometry. They are not like the hard balls illustrated in most pictures of atomic nuclei. One cannot directly see the nucleons in a larger Skyrmion, and one certainly cannot pin down the nucleon positions, nor say which are protons and which are neutrons. The other reason is that Skyrmions have orientational degrees of freedom in a way that nucleons do not. The $B = 1$ Skyrmions are not restricted from the start to be spin $\frac{1}{2}$ particles, and they have the possibility of being excited to spin $\frac{3}{2}$ when interacting strongly with other Skyrmions.

Some aspects of formal quantum field theory do persist in Skyrme theory. One well-known approach to quantum field theory is the Schrödinger wavefunctional picture, where there is formally a wavefunction on a function space of infinite dimensions (the classical field configuration space). The quantum Hamiltonian combines a Laplacian on this space and a potential, just as in ordinary quantum mechanics. This Schrödinger picture is intuitively attractive but fiendishly difficult to implement in field theories, because of analytical problems. The field configuration space is often curved – it is in Skyrme theory – which leads to further difficulties. Nevertheless, some topological aspects of quantization are conveniently considered in the Schrödinger picture. The topology of the field configuration space of Skyrme theory is understood, and important constraints on the wavefunction can be deduced. Their consequences can then be worked out in the context of the finite-dimensional quantum mechanics of Skyrmion collective coordinates. We discuss these topological ideas next, and the consequences will be discussed in depth later.

## 5.2 Topology and Quantization

The configuration space of Skyrme fields with baryon number $B$ is $\mathcal{C}_B = \text{Maps}_B(\mathbb{R}^3 \mapsto SU(2))$. This is the infinite-dimensional space of all finite-energy field configurations at a given time with baryon number $B$, satisfying the vacuum boundary condition. For all $B$ these spaces are topologically equivalent to the space $\mathcal{C}_0 = \text{Maps}_0(\mathbb{R}^3 \mapsto SU(2))$, the space of field configurations with baryon number 0. This is because the field $U$ is valued in a group. If we fix a single configuration $U_B \in \mathcal{C}_B$, then multiplication by $U_B$ maps invertibly between $\text{Maps}_0$ and $\text{Maps}_B$. Since we imposed the vacuum boundary condition, these spaces are also topologically the same as $\text{Maps}_0(S^3 \mapsto S^3)$ with the base point condition $U(\infty) = 1$.

The space $\text{Maps}_0(S^3 \mapsto S^3)$ is connected (i.e. has one connected piece) but *not simply connected*. It has non-contractible loops. Topologically, there is just one type of non-contractible loop in the space, and going round such a loop twice makes it contractible. Technically, this is because the first homotopy group of the space is [49, 184]

$$\pi_1(\text{Maps}_0(S^3 \mapsto S^3)) = \pi_4(S^3) = \mathbb{Z}_2. \tag{5.1}$$

Thus, $\mathcal{C}_0$ has a universal double cover, and so does $\mathcal{C}_B$.

It is a general principle of quantization that wavefunctions do not have to be single-valued on the classical configuration space, but only on its universal cover. In Skyrme theory there are therefore two possible quantizations. In one, the wavefunction is single-valued on $\mathcal{C}_B$; in the other, it is only single-valued on the double cover, and changes sign when one goes round a non-contractible loop [87].

It is now understood that the second of these quantizations is correct, i.e. the wavefunction $\Psi$ in a sector with any baryon number $B$ changes sign after going round a non-contractible loop. This is not just a choice, but required in Skyrme theory or any similar low-energy effective field theory of QCD, when the number of quark colours is odd. We expand on the reasons for this below.

A small subspace of field configurations in $\mathcal{C}_1$ consists of hedgehog Skyrmions centred at the origin, in all possible orientations. This subspace is a copy of $SO(3)$, which, as we explained earlier, is not simply connected. Its universal double cover is the group $SU(2)$. A loop in $SO(3)$ consisting of rotations around some axis by an angle running from 0 to $2\pi$ is non-contractible, and it can be shown that the copy of this loop, consisting of a $B = 1$ Skyrmion rotated by an angle running from 0 to $2\pi$, remains non-contractible in the full configuration space $\mathcal{C}_1$ [248].

This has the following consequences:

(1) If we rotate a $B = 1$ Skyrmion by $2\pi$, then its wavefunction changes sign. The quantum state of a $B = 1$ Skyrmion therefore represents a nucleon with half-integer spin. Even in the presence of other $B = 1$ Skyrmions far away, a rotation of just one of them by $2\pi$ is still a non-contractible loop, and this single Skyrmion's spin should remain a half-integer, so the quantization sign choice needs to be the same for all $B$.

(2) In the sector of any baryon number $B$, the wavefunction changes sign if the locations of two $B = 1$ Skyrmions are exchanged (without rotating them in the process). This is because the exchange is also a non-contractible loop, as was shown by Finkelstein and Rubinstein [87].

(3) In general, if $B$ is odd then a rigid rotation of a complete field configuration by $2\pi$ is a non-contractible loop, while if $B$ is even, it is contractible [104]. Thus, the spin of a quantized Skyrmion with baryon number $B$ is half-integer if $B$ is odd, and integer if $B$ is even.

(4) For a $B = 1$ hedgehog Skyrmion, an isorotation by $2\pi$ is equivalent to a spatial rotation by $2\pi$. Thus, in any $B = 1$ quantum state, the isospin is half-integer. More generally, the isospin of any quantized Skyrmion is half-integer if $B$ is odd, and integer if $B$ is even.

Consequences 1 and 2 link spin with statistics. The quantized $B = 1$ Skyrmion has half-integer spin and is a fermion. If we had quantized the theory the other way, then the wavefunction would be unchanged under both a $2\pi$ rotation and under Skyrmion exchange. The quantized $B = 1$ Skyrmion would have been a boson with integer spin. Remarkably therefore, in Skyrme theory the spin-statistics theorem is *derived from topology*.

## 5.3 Rigid-Body Quantization

In the rigid-body approximation to quantization, a Skyrmion of any baryon number $B$ is allowed to translate, rotate and isorotate rigidly, but is not allowed to deform. This is a low-energy approximation to quantized dynamics that reduces the infinite-dimensional space of field configurations to a finite-dimensional space, an orbit of the symmetry group of Skyrme theory. The symmetry group consists of

$$\text{(translations)} \times \text{(rotations)} \times \text{(isorotations)}. \qquad (5.2)$$

The translation part is rather trivial. The Skyrmion centre of mass **X** is free to move, and the Skyrmion can gain momentum, either classically or quantum mechanically. The wavefunction acquires a plane wave factor $e^{i\mathbf{k}\cdot\mathbf{X}}$ if the momentum $\mathbf{p} = \hbar\mathbf{k}$ has a definite value. From now on we usually ignore translations. The remaining group acting is $SO(3) \times SO(3)$. Naively, a rigid-body wavefunction is a function on this group, but to allow for non-contractible loops, wavefunctions need to be defined on the covering space $SU(2) \times SU(2)$. However, the combination of $2\pi$ rotations in space and in isospace is always contractible, so one may quotient by the diagonal $\mathbb{Z}_2$ factor. This makes the symmetry group $(SU(2) \times SU(2))/\mathbb{Z}_2 = SO(4)$, but we shall not exploit this explicitly. The rigid-body wavefunctions presented below will be functions on $SU(2) \times SU(2)$, with spin and isospin both half-integers if $B$ is odd, and integers if $B$ is even.

For any particular classical Skyrmion solution, its orbit under the action of the symmetry group $SO(3) \times SO(3)$ is usually smaller than the group itself. This is because the Skyrmion is usually invariant under some symmetry subgroup, which we call the *intrinsic symmetry group* of the Skyrmion (sometimes omitting "intrinsic"). The rigid-body wavefunction needs to be invariant under each intrinsic symmetry element up to a sign.

On the copy of $SU(2)$ that is the double cover of the group of rotations, there is a basis of rigid-body wavefunction $|J, L_3, J_3\rangle$, where $J$ is the total spin quantum number (spin, for short), $L_3$ the third component of spin relative to body-fixed axes, and $J_3$ the third component of spin relative to space-fixed axes (both in units of $\hbar$). Since we also have an isorotational copy of $SU(2)$, a basis for rigid-body wavefunctions can be written as

$$|J, L_3, J_3\rangle \otimes |I, K_3, I_3\rangle. \tag{5.3}$$

$I$ is the total isospin quantum number and $I_3$ is the third component of isospin as usually defined in either particle or nuclear physics. The electric charge of a multi-baryon quantum state (in units of the proton charge) is $Q = \frac{1}{2}B + I_3$. Physically, $B$ represents the total number of protons and neutrons, whereas $I_3$ is half the difference between the numbers of protons and neutrons. So $Q$ is the proton number.

$J$ and $I$ take definite values for rigid-body energy eigenstates, because the spin and isospin operators commute with the Hamiltonian. The values of $J_3$ and $I_3$ can also have definite values, as these operators mutually commute, and also commute with the total spin and isospin operators. $J_3$ and $I_3$ are not constrained, i.e. they take all the standard $(2J+1)(2I+1)$ allowed values for given $J$ and $I$, and the energy is independent of them. Thus, we are going to suppress these labels and write basis states as

$$|J, L_3\rangle \otimes |I, K_3\rangle. \tag{5.4}$$

$K_3$, the third component of isospin relative to body-fixed axes, is a rather specific property of quantized Skyrmions, and occurs because the pion field structure gives a classical Skyrmion an orientation in isospace. Crucially, the intrinsic symmetry of a Skyrmion places constraints on the body-fixed spin and isospin projections $L_3$ and $K_3$, and the allowed states are generally superpositions of the basis states (5.4), combining different values of $L_3$ and $K_3$. The precise form of the constraints depends on a choice of the underlying Skyrmion's standard orientation in both space and isospace. Moreover, because of these constraints, not all values of $J$ and $I$ are allowed. This is the most interesting aspect of rigid-body Skyrmion quantization, because it allows a comparison with what is seen in the spectra of nuclei, where some spins are allowed and some forbidden. If the match is good, we can conclude that the nucleus has an underlying intrinsic structure with the same symmetry as the Skyrmion. (This remark is a simplification, because vibrational excitations also need to be considered for all but the smallest nuclei, and these extend the range of allowed spins.) In Chapter 7 we will give examples of the constraints on spin/isospin wavefunctions and how to solve them for Skyrmions with various baryon numbers greater than 1. But for now, we focus on the quantization of the $B = 1$ Skyrmion.

## 5.4 Quantized Hedgehog Skyrmion – Proton and Neutron

The rigid-body quantization of the $B = 1$ hedgehog Skyrmion has been discussed in detail by Adkins, Nappi and Witten [7]. This Skyrmion has a large intrinsic symmetry group with continuous parameters. A rotation about any axis **n** by any angle $\alpha$ is equivalent to an isorotation about the same axis by the same angle. Acting with both the rotation and the isorotation on a quantum state of the Skyrmion, the wavefunction $|\Psi\rangle$ is unchanged, that is,

$$e^{i\alpha \mathbf{n} \cdot \mathbf{L}} e^{i\alpha \mathbf{n} \cdot \mathbf{K}} |\Psi\rangle = |\Psi\rangle \,. \tag{5.5}$$

It follows, by considering $\alpha$ infinitesimal and all **n**, that

$$(\mathbf{L} + \mathbf{K})|\Psi\rangle = 0 \,, \tag{5.6}$$

so the "grand spin" $\mathbf{L} + \mathbf{K}$ must vanish[1].

---

[1]The grand spin operators obey the usual angular momentum commutation relations. This shows that the signs in eq.(5.5) are right. One might suspect a factor $e^{-i\alpha \mathbf{n} \cdot \mathbf{K}}$ is required, but the operators $\mathbf{L} - \mathbf{K}$ do not close under commutation.

By applying the familiar Clebsch–Gordon rules for adding angular momenta, we know that it is possible to combine spin $J$ with isospin $I$ into a state with total grand spin 0 only when $J = I$. The resulting state is unique. More precisely, in terms of body-fixed spin and isospin operators, we require $\mathbf{L} \cdot \mathbf{L} = \mathbf{K} \cdot \mathbf{K}$. But space-fixed and body-fixed total angular momentum operators are the same, i.e. $\mathbf{J} \cdot \mathbf{J} = \mathbf{L} \cdot \mathbf{L}$ and $\mathbf{I} \cdot \mathbf{I} = \mathbf{K} \cdot \mathbf{K}$, so $\mathbf{J} \cdot \mathbf{J} = \mathbf{I} \cdot \mathbf{I}$ and therefore $J = I$; the rigid-body quantum states of the $B = 1$ Skyrmion have equal spin and isospin.

Since the baryon number $B = 1$ is odd, for any axis $\mathbf{n}$ we also require

$$e^{i2\pi \mathbf{n} \cdot \mathbf{L}}|\Psi\rangle = -|\Psi\rangle, \tag{5.7}$$

so $J$ and hence $I$ must be half-integer. The allowed states therefore have spin and isospin

$$J = I = \frac{1}{2}, \frac{3}{2}, \ldots, \tag{5.8}$$

and the energy of such states in Skyrme units is

$$E = \mathcal{M}_1 + \frac{1}{2\lambda}J(J+1), \tag{5.9}$$

where $\mathcal{M}_1$ is the classical mass (energy) of the $B = 1$ Skyrmion and $\lambda$ is the Skyrmion's moment of inertia. The Skyrmion has just a single moment of inertia, because of its large symmetry. We will present formulae for the moment of inertia tensors of general Skyrmions later, and an integral formula for $\lambda$.

The $J = I = \frac{1}{2}$ states correspond to the nucleons with spin $\frac{1}{2}$, i.e. the proton and neutron, p and n. The proton has $I_3 = \frac{1}{2}$ and the neutron $I_3 = -\frac{1}{2}$, reproducing their respective electric charges 1 and 0. The $J = I = \frac{3}{2}$ states correspond to the delta resonances $\Delta^{++}, \Delta^+, \Delta^0$ and $\Delta^-$, with spin $\frac{3}{2}$. Such states are produced in pion-nucleon collisions, and in other processes that excite nucleons to a sufficiently high energy. The delta resonances are broad. They have an energy considerably greater than the combined masses of a nucleon and pion, and decay rapidly. However, like nucleons, delta resonances can be interpreted as made of a combination of three up and down quarks. States with $J = I = \frac{5}{2}$ or higher have such high energy that they radiate pions immediately and are experimentally undetectable, so we ignore them. They would not be allowed in a naive quark model, where all baryons are three-quark states. It is a success of the Skyrme theory that, at relatively low energy, there are no predicted quantum states with $J$ and $I$ unequal.

Adkins, Nappi and Witten gave explicit, geometrically illuminating formulae for the rigid-body wavefunctions of a $B = 1$ Skyrmion, as functions on $SU(2)$. These wavefunctions are simpler to write down than the equivalent wavefunctions constructed from the basis states (5.3). The space of orientations of the Skyrmion is a single copy of $SO(3)$, as rotations are equivalent to isorotations, but because of the half-integer quantization, the wavefunctions are single-valued only on the double cover, $SU(2)$. $SU(2)$ matrices $A$ have the parametrisation

$$A = a_0 + i\mathbf{a} \cdot \boldsymbol{\tau}, \quad (5.10)$$

as in eq.(4.5), with the Cartesian coordinates $(a_0, \mathbf{a})$ restricted to a unit 3-sphere. Any wavefunction on this 3-sphere automatically has grand spin 0.

The (space-fixed) rotation group and isorotation group act independently on $B = 1$ Skyrmion wavefunctions. Their generators form independent $su(2)$ Lie algebras that together generate the $SO(4)$ that acts on functions defined on the 3-sphere. The spin operators, i.e. the rotation group generators, are

$$J_i = \frac{i}{2}\left\{-a_0\frac{\partial}{\partial a_i} + a_i\frac{\partial}{\partial a_0} - \varepsilon_{ijk}\left(a_j\frac{\partial}{\partial a_k} - a_k\frac{\partial}{\partial a_j}\right)\right\}, \quad (5.11)$$

and the isospin operators are

$$I_i = \frac{i}{2}\left\{a_0\frac{\partial}{\partial a_i} - a_i\frac{\partial}{\partial a_0} - \varepsilon_{ijk}\left(a_j\frac{\partial}{\partial a_k} - a_k\frac{\partial}{\partial a_j}\right)\right\}. \quad (5.12)$$

These operators have standard $su(2)$ commutation relations and they mutually commute. Note how their structure is not very different from the usual orbital angular momentum operators that generate $SO(3)$ rotations in $\mathbb{R}^3$.

Wavefunctions can be conveniently expressed in terms of $(a_0, \mathbf{a})$. Such wavefunctions will have an irrelevant radial dependence, i.e. a dependence on $a_0^2 + \mathbf{a} \cdot \mathbf{a}$, but the operators above are purely angular and do not detect this. (The operators and wavefunctions can be expressed in terms of Euler angles on the 3-sphere if desired.)

There are four independent states with spin and isospin $\frac{1}{2}$. These are all linear in the Cartesian coordinates. The normalised wavefunctions representing a proton or neutron ($I_3 = \pm\frac{1}{2}$), with spin up or down ($J_3 = \pm\frac{1}{2}$), are

$$p^\uparrow = \frac{1}{\pi}(a_1 + ia_2), \quad p^\downarrow = -\frac{i}{\pi}(a_0 - ia_3),$$

$$n^\uparrow = \frac{1}{\pi}(a_0 + ia_3), \quad n^\downarrow = -\frac{1}{\pi}(a_1 - ia_2). \quad (5.13)$$

Using the $J_3$ operator (5.11) one can check, for example, that $J_3\,p^\uparrow = \frac{i}{2}(a_3 - ia_0) = \frac{1}{2}p^\uparrow$. Note that these linear functions all change sign when the signs of all four coordinates are reversed, so they are double-valued on $SO(3)$, as required.

The delta resonances with spin and isospin $\frac{3}{2}$ are represented by sixteen wavefunctions that are cubic in the coordinates. The $J_3$ and $I_3$ eigenvalues take each of the four values $\frac{3}{2}, \frac{1}{2}, -\frac{1}{2}, -\frac{3}{2}$. For example, the wavefunction representing the $\Delta^{++}$ state with spin fully up is

$$\Delta^{++} = \frac{\sqrt{2}}{\pi}(a_1 + ia_2)^3, \tag{5.14}$$

and has $J_3 = I_3 = \frac{3}{2}$. These wavefunctions are again double-valued on $SO(3)$. Not all cubic functions are allowed, as it is necessary to project out those that are products of a linear function and the squared radius $a_0^2 + \mathbf{a}\cdot\mathbf{a}$.

Adkins, Nappi and Witten established the first effective calibration of Skyrme theory using these nucleon (N) and delta resonance ($\Delta$) states. Their masses in Skyrme units are given by the formula (5.9), with $J(J+1) = \frac{3}{4}$ for the nucleons and $J(J+1) = \frac{15}{4}$ for the deltas:

$$E_N = \mathcal{M}_1 + \frac{3}{4}\frac{1}{2\lambda},$$
$$E_\Delta = \mathcal{M}_1 + \frac{15}{4}\frac{1}{2\lambda}. \tag{5.15}$$

To convert $\mathcal{M}_1$ to a mass in physical units we use the conversion factor $F_\pi/4e$. Since $\lambda$ has dimensions mass $\times$ length$^2$ the conversion factor for the spin contribution is $(F_\pi/4e)^{-1}(2/eF_\pi)^{-2} = e^3 F_\pi$. Therefore the physical nucleon and delta masses are

$$E_N = \mathcal{M}_1 \frac{F_\pi}{4e} + \frac{3}{4}\frac{1}{2\lambda}e^3 F_\pi,$$
$$E_\Delta = \mathcal{M}_1 \frac{F_\pi}{4e} + \frac{15}{4}\frac{1}{2\lambda}e^3 F_\pi. \tag{5.16}$$

Numerically, it is found that $\mathcal{M}_1 = 1.232 \times 12\pi^2 = 145.9$ and $\lambda = 107$ for the $B = 1$ Skyrmion with massless pions. The required calibration factors are therefore

$$e = 5.45 \quad \text{and} \quad F_\pi = 129\,\text{MeV} \tag{5.17}$$

in order to fit the average mass 939 MeV for the nucleons and 1232 MeV for the deltas. Even without finding these calibration values, one can see from the formulae (5.16) that the nucleon mass exceeds the classical Skyrmion mass by a quarter of the delta-nucleon mass difference, i.e. by about

73 MeV, with this excess arising from the quantum mechanical spin energy of the Skyrmion.

We should clarify here that $e$ is regarded as dimensionless, so the length conversion factor $2/eF_\pi$ gives a physical result in inverse MeV. This is standard in particle and nuclear physics. A conversion to real length units is achieved using Planck's constant, which has the physical value $\hbar = 197.3$ MeV fm. It is therefore consistent to leave out the factor $\hbar^2$ in the contribution of the spin energy to the total mass (which would normally be $\hbar^2 J(J+1)/2\lambda$). Alternatively, it is possible to include the $\hbar^2$ factor explicitly and then use the conversion factor $F_\pi/4e$. So in Skyrme theory

$$e^3 F_\pi = \hbar^2 \frac{F_\pi}{4e}, \qquad (5.18)$$

and therefore $\hbar = 2e^2$ in Skyrme units. The numerical value is $\hbar = 59.3$ in the Adkins–Nappi–Witten calibration.

Having calibrated Skyrme theory using nucleon and delta masses, Adkins, Nappi and Witten proceeded to calculate further physical properties of the nucleons and deltas, like the charge densities of the nucleons and their magnetic moments. Delta to nucleon decay amplitudes, through either pion or photon emission, were also calculated. The results agree with experimental measurements to within about 30%.

Adkins and Nappi improved the calibration by varying the pion mass parameter $m$, and adjusting the values of $e$ and $F_\pi$ so that the pion mass together with the nucleon and delta masses are all fitted correctly [5]. The physical pion mass is $m_\pi = meF_\pi/2 = 138$ MeV (this has the correct units). The calibration factors are now

$$e = 4.84, \quad F_\pi = 108 \text{ MeV} \quad \text{and} \quad m = 0.526. \qquad (5.19)$$

For this value of $m$ the Skyrmion mass and moment of inertia in Skyrme units are $\mathcal{M}_1 = 1.308 \times 12\pi^2$ and $\lambda = 62.9$, requiring a decrease in $e$ and $F_\pi$ compared with their values for $m = 0$. $\hbar = 46.9$ in Skyrme units in the Adkins–Nappi calibration.

In Chapter 4 we presented the classical potential energy of two well-separated Skyrmions, as a function of their separation and relative orientation. The potential is that of a pair of interacting triplets of pion dipoles, and the result can be extended to the case of massive pions. Using this potential, the nucleon-nucleon interaction can be calculated quantum mechanically, by treating the nucleons as Skyrmions with spin $\frac{1}{2}$. The Skyrmions' separation distance is held fixed, but the Skyrmion orientations and the direction of the separation vector are quantized. The potential is small at large separation so a perturbative calculation is justified. The calculation at first-order in the potential gives a satisfactory tensor force between nucleons [193], coupling the spins to the separation vector, but some other parts of the interaction are unsatisfactory at this order. In particular, there is no central attraction (an attraction independent of spin and isospin). Recently, Halcrow and Harland have calculated to second-order and have included some non-trivial kinetic energy effects [121]. That means that the nucleon wavefunctions are perturbed by the orientation-dependence of the potential, and include some admixture of spin $\frac{3}{2}$ states. The overall interaction is presented as a sum of terms, as in the phenomenological Paris model [159]. For states with total isospin 0, there is a central interaction, an exchange interaction, a spin-spin interaction and a spin-orbit coupling (a term coupling spin to orbital angular momentum), and there are four similar terms for states with isospin 1. Seven of the eight terms generated from Skyrme theory satisfactorily match the Paris potentials. Particularly interesting is the progress in understanding the spin-orbit coupling [115], whose large observed strength is not otherwise well understood in nuclear theory.

## 5.5 Classical Interpretation of Quantized Skyrmion States

The wavefunctions given above for the nucleon and delta resonance states can be given an approximate interpretation in terms of a classically spinning $B = 1$ Skyrmion [90, 103, 178]. This classical approximation is interesting and surprising for states with spin $\frac{1}{2}$ or $\frac{3}{2}$. Although quantum mechanical orbital angular momentum has a classical limit in terms of classical particle motion, it is usually assumed that a particle's spin, especially for spin $\frac{1}{2}$, has a purely algebraic character. The classical picture of Skyrmion spin is possible because the hedgehog Skyrmion has an orientation defined by its frame of pion dipoles, and this orientation can be time-varying.

The wavefunctions of p↓ and n↑, proportional to $a_0 \mp ia_3$, have maximal magnitude on the circle $a_0^2 + a_3^2 = 1$. This circle is the $U(1)$ subgroup of $SU(2)$ parametrising rotations about the $x^3$-axis. So the Skyrmion has maximal probability to have an orientation related to its standard orientation by a rotation about the $x^3$-axis. The opposite phase variations of the two wavefunctions imply that the spin projections along the $x^3$-axis have opposite signs. In the classical approximation, the hedgehog is spinning about the positive $x^3$-axis, anticlockwise in the n↑ state and clockwise in the p↓ state. The angular velocity, in Skyrme units, is $\omega = 0.28$. This is calculated by identifying the quantized angular momentum $\frac{1}{2}\hbar$ with the classical angular momentum $\lambda\omega$, and using $\hbar = 59.3$ and $\lambda = 107$ (the Adkins–Nappi–Witten calibration). Alternatively $\omega = 0.37$ using the Adkins–Nappi calibration.

The wavefunctions of p↑ and n↓, proportional to $a_1 \pm ia_2$, have maximal magnitude on the circle $a_1^2 + a_2^2 = 1$. This circle is not a $U(1)$ subgroup, but consists of the $SU(2)$ elements that are obtained by multiplying the fixed element $A = i\tau_1$ by elements in the $U(1)$ subgroup defined earlier. The maximal probabilities are therefore for those orientations obtained by turning the hedgehog Skyrmion upside down (a rotation by $\pi$ about the $x^1$-axis) and then letting it spin about the $x^3$-axis. The spin is anticlockwise about the positive $x^3$-axis in the p↑ state and clockwise in the n↓ state. The angular velocity is the same as before. The classical interpretations of all these nucleon states are illustrated in Fig. 5.1.

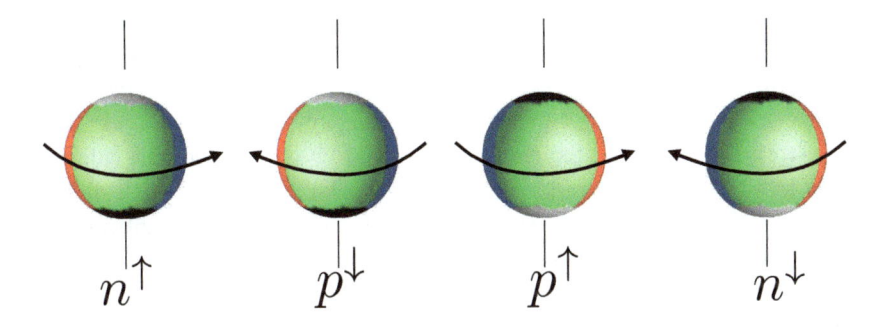

Fig. 5.1: Proton and neutron modelled as classically spinning Skyrmions

The nucleon wavefunctions are not very strongly peaked at the classical orientations we have identified. The orientation can deviate, but the wavefunctions vanish if the orientation is completely reversed. For example, the wavefunction of p↓ vanishes where $a_1^2 + a_2^2 = 1$.

Note that with respect to body-fixed axes of the hedgehog Skyrmion, a neutron state is always of the same kind. The neutron spins right-handedly around the positive body-fixed 3-axis (the directed black-to-white axis). Similarly, a proton spins left-handedly. So we can visualise a proton or neutron whose spin points in any direction, since a spin $\frac{1}{2}$ state is always spin-up relative to some spatial direction.

The deltas have higher spin than the nucleons, and the classical approximation is more convincing here. The wavefunction of the fully spin-up $\Delta^{++}$, with $J_3 = \frac{3}{2}$, illustrates this. The wavefunction is proportional to $(a_1+ia_2)^3$ and has maximal magnitude on the circle $a_1^2 + a_2^2 = 1$. It is more strongly peaked here than the corresponding p$^\uparrow$ state proportional to $a_1 + ia_2$, and the spin projection along the $x^3$-axis is three times greater, so the inferred classical angular velocity is close to 1. This angular velocity is large in the sense that the outer part of the Skyrmion, at an approximate radius of 1 in Skyrme units, is moving close to the speed of light, and the Skyrmion is probably distorted significantly by its spin. For the same reason, the $\Delta^{++}$ emits pion radiation strongly.

Battye, Krusch and Sutcliffe have attempted to construct quasi-exact, classically spinning $B = 1$ Skyrmion solutions numerically, with no constraint on the Skyrmion shape [27]. There is an angular velocity cutoff above which no solution exists, because of strong pion radiation. This seems to rule out a classical model for a delta resonance. For this reason, and also because a delta resonance rapidly decays to a nucleon and a pion, the Adkins–Nappi calibration of Skyrme theory (and also the Adkins–Nappi–Witten calibration), based on the nucleon and delta masses, is not really trustworthy. A delta cannot be treated as an excited stationary state of a rigidly-rotating hedgehog Skyrmion. We will show that a different calibration is more successful for modelling larger nuclei. Skyrmions with higher baryon numbers have far larger moments of inertia, so their quantum mechanical spin excitation energies are far smaller for modest spins, and their inferred classical angular velocities are far smaller too. Nuclear excitation energies are typically between 1 and 20 MeV, too small for quantum mechanical pion emission, so the decays of the low-lying excited states are principally electromagnetic, as in the decays of excited atoms. The spin and isospin excitations can therefore more accurately be regarded as quantum mechanical stationary states.

The classical picture of nucleon states makes it possible to simulate nucleon-nucleon scattering using purely classical Skyrme field dynamics. One needs to give the Skyrmions an initial velocity and spin, and an impact parameter – the separation of the initial straight lines along which they move. The Skyrmion scattering angles can be calculated and plotted [89, 90] (see also the videos linked to these papers). It has not yet been possible to calculate the physically important differential or total cross sections, as this would require data from many different impact parameters, and an averaging over different initial orientations too. Results that have been obtained so far lead to certain predictions for what happens to the classical spin in a collision. For example, the simulations show that in a head-on collision of longitudinally polarised nucleons (in the centre of mass frame), the outgoing nucleons emerge at right angles and they are also longitudinally polarised. Right-angle scattering in a head-on collision is a smooth process that is possible for topological solitons like Skyrmions – it was observed first in collisions of non-abelian monopoles [14].

Such results could be extrapolated to a prediction for what may occur in a fully polarised nucleon-nucleon scattering experiment, where the initial nucleon polarisations are fixed and the final polarisations are measured. A fully polarised experiment is in principle possible, although the collision rate that can be achieved typically reduces by a factor of $10^3$ for each polarisation state that is either fixed for the incoming particles, or measured for the outgoing particles. The impact parameter cannot be fixed experimentally at all, but perhaps it can be inferred from the observed scattering angle – the impact parameter is probably small if the scattering angle is large.

## 5.6 The Need for Fermionic Quantization

It was Skyrme who suggested that the $B = 1$ Skyrmion could be quantized as a spin $\frac{1}{2}$ fermion. His argument rested on the orientational configuration space of the Skyrmion being $SO(3)$, which is not simply connected, so wavefunctions need only be single-valued on the double cover. This was a remarkable suggestion. The basic fields of Skyrme theory are pion fields, and pion particles are bosons, in agreement with their spin being zero. The idea that a quantized topological soliton in an effective field theory of pions could be a fermion, and model a spin $\frac{1}{2}$ proton or neutron, was a bold step.

The consistency of a fermionic quantization was confirmed by an investigation of the complete configuration spaces $\mathcal{C}_B$ of the theory, for each baryon number, as we have discussed above. However, there is no argument from Skyrme theory alone requiring the fermionic rather than bosonic quantization.

The need for fermionic quantization can be extracted from the more fundamental gauge theory of quarks and gluons, QCD, where the physics requires the gauge group to be (colour) $SU(3)$, so that baryons are made of three spin $\frac{1}{2}$ quarks and are therefore fermions. Hypothetically, one can extend the gauge group to $SU(N_c)$, where $N_c$ is the number of quark colours; in this case a baryon contains $N_c$ spin $\frac{1}{2}$ quarks and is still a fermion if $N_c$ is odd. A connection between QCD and the Skyrme EFT was pioneered by Witten in the early 1980s [251], focussing on the large $N_c$ limit. This work was very influential in reviving interest in Skyrme theory more than 20 years after it had first been proposed.

Witten needed to assume that the EFT modelled a low-energy version of QCD in which three or more quark flavours are close to massless. This is a somewhat unphysical version, because the u and d quarks are undoubtedly close to massless, but the strange quark s (the next quark in the mass hierarchy) is considerably heavier. Treating it as light leads to a phenomenology with an $SU(3)$ flavour symmetry. Here, the proton and neutron isospin doublet extends to an $SU(3)$ flavour octet of baryons, including the $\Lambda$, $\Sigma$ and $\Xi$ baryons which have non-zero strangeness.

Witten's analysis is therefore for a version of QCD with $N_f$ flavours of light quarks, and an $SU(N_f)$ flavour symmetry, where $N_f \geq 3$. This leads to a Skyrme theory with its field $U$ taking values in the group $SU(N_f)$. This goes beyond the theory we have discussed, where $U \in SU(2)$. The various terms in the Lagrangian and action that we have considered so far are not fundamentally different in this extended theory, although for $N_f = 3$ there are eight meson fields, not just three pion fields, and the meson mass terms can be more complicated. The key difference, noted by Witten, is that it is now possible to include a Wess–Zumino term [245].

The Wess–Zumino term is an additional contribution to the action of $SU(N_f)$ Skyrme theory, generated from the underlying gauge theory. It is given by

$$S_{\text{WZ}} = -\frac{iN_c}{240\pi^2} \int \epsilon_{\mu\nu\alpha\beta\gamma} \text{Tr}(R_\mu R_\nu R_\alpha R_\beta R_\gamma) \, d^5x \,. \tag{5.20}$$

Here, $R_\mu = \partial_\mu U U^{-1}$ as usual but it takes its value in $su(N_f)$. The integration is performed over a 5-dimensional region whose boundary is 4-dimensional spacetime. One needs to imagine extending the field values $U(x)$ in spacetime in some smooth way into the interior of the 5-dimensional region, although the precise way this is done has no effect. The Wess–Zumino term does not contribute to the classical energy, but it plays an important role in the quantum theory. Its introduction breaks the time reversal and space inversion (parity) symmetries of the theory down to the combined symmetry operation

$$t \mapsto -t, \quad \mathbf{x} \mapsto -\mathbf{x}, \quad U \mapsto U^{-1} \,, \tag{5.21}$$

which appears to be realised in nature, unlike these individual symmetry operations. A topological argument shows that the coefficient $N_c$ must be an integer, and Witten argued that $N_c$ should be identified with the number of quark colours in the gauge theory, based on considerations of flavour anomalies in the quark (QCD) and Skyrme theories.

To determine whether a Skyrmion should be quantized as a fermion, one compares the amplitudes for the processes in which a Skyrmion remains at rest for some long time $T$, and in which the Skyrmion is slowly rotated through an angle $2\pi$ during this time. The usual quadratic and quartic terms in the action do not distinguish between these two processes since they are quadratic in time derivatives, but the Wess–Zumino term is only linear in the time derivative of the field $U$ and so can distinguish them. It results, in fact, in the amplitudes for these two processes differing by a factor $(-1)^{N_c}$, which shows that the Skyrmion should be quantized as a fermion when $N_c$ is odd, and in particular in the physical case $N_c = 3$. Note that this argument breaks down if one tries to apply it to the original $SU(2)$ Skyrme theory. This is because the five current terms appearing inside the trace in (5.20) are then each valued in the 3-dimensional Lie algebra $su(2)$ and cannot all be linearly independent. The antisymmetrisation with the $\epsilon$-tensor therefore gives zero.

It was regarded for many years as theoretically unsatisfactory that in the flavour-$SU(2)$ Skyrme theory, there was no direct argument requiring the fermionic quantization of Skyrmions. It was necessary to rely on the flavour-$SU(3)$ extension. The matter was resolved by Freed [91], who developed a deeper understanding of anomalies in QCD and how they propagate into a low-energy EFT. Freed's argument uses a generalised cohomology theory, unlike Witten's argument that relies on real differential cohomology as captured by the Wess–Zumino term. Freed's result is that in the $SU(2)$ Skyrme theory derived from QCD with two flavours of light quarks and an odd number $N_c$ of colours, quantized Skyrmions must again be fermions.

Chapter 6

# Skyrmions with Higher $B$ – Massless Pions

Skyrmions with baryon numbers $B$ higher than 1 have been constructed by a variety of methods. Initially, in the 1980s, they were hard to find. The product ansatz was tried. This is easy to implement but doesn't get close to true Skyrmion solutions. By numerically exploring fields with two $B = 1$ Skyrmions in the attractive channel having certain reflection symmetries, the stable $B = 2$ Skyrmion was eventually found [155, 231]. It has toroidal symmetry, which was a surprise at first [175]. Subsequently, solutions up to baryon number $B = 6$ were constructed numerically by Braaten, Townsend and Carson [53], although it turned out that only their solutions up to $B = 5$ were correct. There are always numerical difficulties as $B$ increases, because the Skyrmions get larger and closer to the boundary of the numerically available box. It was particularly striking that the $B = 3$ solution has tetrahedral symmetry, and the $B = 4$ solution has octahedral (cubic) symmetry. Naively one would expect that if three $B = 1$ Skyrmions coalesced, then there would be at most the symmetry of an equilateral triangle, and if four coalesced there would be at most tetrahedral symmetry. The enhanced symmetries as $B = 1$ Skyrmions merge were unexpected.

It was then noted, by those familiar with the theory of non-abelian Yang–Mills–Higgs monopoles, that these Skyrmions have similar shapes and symmetries to the most compact monopoles. Earlier, there had been developed a rather deep understanding of monopoles [14], and in particular, a close connection was found between monopoles and rational maps from a 2-sphere to a 2-sphere [141]. This suggested that Skyrmion solutions and their symmetries could be better understood, and new ones more easily found, if one could exploit rational maps to construct approximate Skyrmions [130]. This approach – the rational map ansatz – has been very fruitful, and will be explained in detail below. It has led, among other things, to a classification of all Skyrmions up to baryon number $B = 22$ for massless pions [30]. For larger baryon numbers, the solutions are rather different and less systematically understood. Some of them are similar to parts of a solution of the Skyrme field equation with a spatially periodic, crystalline form. This Skyrmion crystal solution is rather precisely known, and has remarkably low energy per baryon.

The Skyrmions constructed using the rational map ansatz have a polyhedral form, and are hollow inside. Many of these, for $B > 8$, are not stable when the pion mass parameter has a more realistic value of $m \approx 1$. This is fortunate, as nuclei are not modelled by hollow shells of matter. Nevertheless, in this chapter we discuss the solutions with massless pions, especially for $B \leq 8$, and leave for a later chapter a discussion of Skyrmions with massive pions. There, other methods are needed to find Skyrmion solutions. One is to glue $B = 4$ solutions together – infinitely many glued together produces a variant of the Skyrmion crystal. Another powerful method is to use a multi-layer version of the rational map ansatz.

It needs to be realised that the numerical search for Skyrmion solutions, aided by analytical ideas like the rational map ansatz, remains a challenge both for massless and massive pions, and stable solutions are not known for all $B$. Even if one finds a stable solution, it is difficult to know if it is the global energy minimiser. The landscape of global and metastable minima becomes complicated as $B$ increases, and some solutions are saddle points of the energy.

Later in this chapter, we will discuss briefly what is known rigorously about the mathematical existence of Skyrmion solutions for $B > 1$; it is surprisingly little.

## 6.1 Skyrmions with Baryon Numbers $B \leq 8$

Here we discuss the minimal-energy Skyrmions with baryon numbers $1 \leq B \leq 8$ for massless pions, constructed using numerical methods. Details of the numerical codes used to compute these solutions can be found in refs.[30]. Each of these solutions is presented in a convenient orientation in space and isospace, and centred at the origin, but can be acted upon by translations, rotations and isorotations.

All the Skyrmions have some non-trivial intrinsic symmetry, a symmetry group $K \subset O(3)$. This means that the solution is invariant under various pairs of combined (proper or improper) rotations and isorotations. The rotation-isorotation pairs $(k, \mu_k)$, for $k \in K$, have to satisfy the group laws of $K$, so $\mu_{kk'} = \mu_k \mu_{k'}$. The map $k \mapsto \mu_k$ is therefore a homomorphism, and if it has a non-trivial kernel then the Skyrmion is invariant under the pure rotations in this kernel. Pure rotational symmetries are rather uncommon but do occur, examples being the $D_2$ subgroup of symmetries of the $B = 4$ Skyrmion. However, the reverse is not possible. One might think that there could be a homomorphism mapping the isorotation element to the rotation element, and if this had a non-trivial kernel, then the Skyrmion would be invariant under selected pure isorotations. However, this leads to a contradiction for Skyrmions with non-zero baryon number. For if a Skyrme field is invariant under a pure isorotation around some axis in isospace, then everywhere in space the pion field value is restricted to that axis; the target $S^3$ of the Skyrme field is therefore only partially covered, and the baryon number is zero (by preimage counting).

As we have already noted, for $B > 1$ the minimal-energy Skyrmion is not a spherically-symmetric hedgehog. For $B = 2$, the Skyrmion is rotationally symmetric around a single axis, and its full symmetry group is the infinite dihedral group $D_{\infty h} \equiv O(2)$. Its energy density has an axially-symmetric, toroidal structure. There are further axially-symmetric solutions of the Skyrme equation for $B > 2$ [155] but these are not the minimal-energy solutions, and in fact for $B > 4$ they are not even sufficiently bound to prevent breakup into $B$ single Skyrmions, so they correspond to saddle points. Instead, Skyrmions with baryon numbers $B > 2$ have at most a discrete intrinsic symmetry group, $K$.

In displaying Skyrmions it is conventional to plot a surface of constant baryon density $\mathcal{B}$ (a baryon density isosurface), where $\mathcal{B}$ is the integrand in eq.(4.26), although energy density isosurfaces are qualitatively very similar. In Fig. 6.1 we display baryon density isosurfaces for the minimal-energy Skyrmions having baryon numbers $1 \leq B \leq 8$, and give the symmetry group $K$.

The $B = 3$ Skyrmion has tetrahedral symmetry $T_d$, and the $B = 4$ Skyrmion has octahedral symmetry $O_h$. The baryon density is concentrated around the edges and vertices of, respectively, a tetrahedron and a cube. A major motivation for the proposal of the rational map ansatz was to better understand the symmetries and shapes of these Skyrmions.

The $B = 5$ Skyrmion has the relatively small symmetry $D_{2d}$. The polyhedron along whose edges the baryon density is concentrated consists of four squares and four pentagons, the top and bottom being related by a relative rotation of $90°$. In case the reader is not familiar with extended dihedral symmetries, we briefly recount them here. The dihedral group $D_n$ is

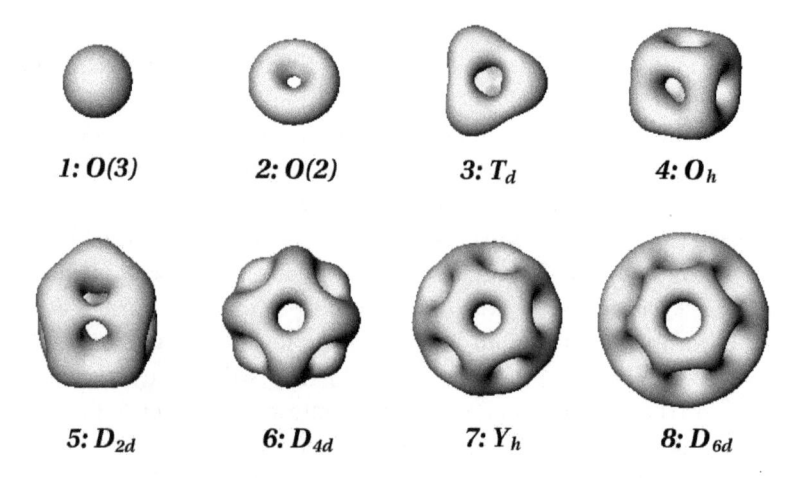

Fig. 6.1: Skyrmions from $B = 1$ to $B = 8$ (for $m = 0$)

obtained from $C_n$, the cyclic group of order $n$, by the addition of a $C_2$ axis perpendicular to the main $C_n$ symmetry axis. In total there are $n$ such $C_2$ axes, with angular separation $\pi/n$. The group $D_n$ can then be extended by the addition of a reflection symmetry in two ways: by including a reflection in the plane perpendicular to the main $C_n$ axis, which produces the group $D_{nh}$ or, alternatively, a reflection in a plane containing the main $C_n$ axis and bisecting a neighbouring pair of the $C_2$ axes, which produces the group $D_{nd}$. It is a little curious that the $B = 5$ Skyrmion has relatively little symmetry. In fact, as we mention later, there is an octahedrally-symmetric $B = 5$ solution, but it is a higher-energy saddle point.

The $B = 6$ and $B = 8$ Skyrmions have symmetries $D_{4d}$ and $D_{6d}$ respectively. The polyhedron associated with the $B = 6$ Skyrmion consists of two halves, each formed from a square with pentagons hanging down from all four sides. To join these halves the two squares must be parallel, with one rotated by 45° relative to the other. The $B = 8$ Skyrmion has a similar structure, except that the squares are replaced by hexagons with six pentagons hanging down. The top hexagon is parallel to the bottom hexagon but rotated by 30°. The $B = 7$ Skyrmion is icosahedrally symmetric [30], its symmetry group $Y_h$ being an extension of $D_{5d}$. The baryon density of the $B = 7$ Skyrmion is concentrated along the edges of a dodecahedron. The halves of the $B = 7$ Skyrmion have five pentagons hanging from a pentagon, hence the larger symmetry.

Table 6.1: Symmetry group $K$ and energy $E$ of Skyrmions for $1 \leq B \leq 8$.

| $B$ | $K$ | $E/12\pi^2 B$ |
|---|---|---|
| 1 | $O(3)$ | 1.2322 |
| 2 | $D_{\infty h}$ | 1.1791 |
| 3 | $T_d$ | 1.1462 |
| 4 | $O_h$ | 1.1201 |
| 5 | $D_{2d}$ | 1.1172 |
| 6 | $D_{4d}$ | 1.1079 |
| 7 | $Y_h$ | 1.0947 |
| 8 | $D_{6d}$ | 1.0960 |

For $B \geq 7$ it is possible to form a polyhedron from 12 pentagons and $2B - 14$ hexagons, having $2B - 2$ faces in total. Such polyhedra are (Buckminster)fullerene-like [30]. Similar fullerene structures arise in carbon chemistry, where carbon atoms sit at the vertices of such polyhedra. In particular, there is a unique icosahedrally-symmetric configuration with $B = 17$ corresponding to the famous fullerene structure of the $C_{60}$ Buckyball, and given its high symmetry it is not surprising that the minimal-energy $B = 17$ Skyrmion has this structure. This fullerene has 32 holes in its faces, the sum of the face numbers of a dodecahedron and an icosahedron.

For $B > 2$, and going up to $B = 8$ and beyond, it is found that the polyhedron associated with the Skyrmion is usually composed of almost regular polygons meeting at trivalent vertices, with the baryon density concentrated along the edges. Using the trivalent property together with Euler's formula, any one of the three parameters – the number of vertices $v$, faces $f$, or edges $e$ – determines the other two. Explicitly, $v = 4(B-2)$, $f = 2(B-1)$, $e = 6(B-2)$. This is consistent with the observation that the baryon density isosurface contains $2(B-1)$ holes. In the next section we discuss the rational map ansatz for Skyrmions. One feature of Skyrmions that it explains is why their baryon density isosurface has this number of holes.

In Table 6.1 we present, for baryon numbers $B = 1$ to $B = 8$, the symmetry group $K$ and energy $E$ of the Skyrmions with massless pions [30]. (It is convenient here and in tables below to divide the energy $E$ by $12\pi^2 B$, the Faddeev–Bogomolny lower bound on the energy, to assess the extent by which the energy exceeds the bound.)

## 6.2 The Rational Map Ansatz

The rational map ansatz [130] is motivated by the shell-like fullerene structures of Skyrmions, and the apparent similarity of Skyrmions and certain non-abelian monopoles. It is an approximate separation of variables for Skyrme fields in spherical polar coordinates. The dependence of the Skyrme field on the angular coordinates is determined by a rational map between 2-spheres, $S^2$, and the radial dependence by an independent profile function. A rational map $R(z)$ is a ratio of polynomials in the complex variable $z$ that parametrises points on $S^2$ by stereographic projection.

Recall that the stereographic coordinate $z$ can be identified with conventional spherical polar coordinates $(r, \theta, \varphi)$ by

$$z = \tan\frac{\theta}{2} e^{i\varphi}, \tag{6.1}$$

and $z = \infty$ when $\theta = \pi$. Another useful formula is

$$z = \frac{x^1 + ix^2}{1 + x^3}, \tag{6.2}$$

where $x^1, x^2, x^3$ are Cartesian coordinates on the unit 2-sphere. In reverse, the point $z$ corresponds to the unit spatial vector

$$\mathbf{n}_z = \frac{1}{1+|z|^2}\left(z + \bar{z},\ i(\bar{z} - z),\ 1 - |z|^2\right). \tag{6.3}$$

The usual area element $\sin\theta\, d\theta d\varphi$ on $S^2$ is equivalent to $2i\, dz d\bar{z}/(1+|z|^2)^2$.

Similarly, the complex value $R$ of the rational map is identified with the unit vector on the target 2-sphere

$$\mathbf{n}_R = \frac{1}{1+|R|^2}\left(R + \bar{R},\ i(\bar{R} - R),\ 1 - |R|^2\right). \tag{6.4}$$

$R$ can also be expressed in terms of polar coordinates on the target 2-sphere using a formula analogous to (6.1).

Explicitly, we write the rational map as

$$R(z) = \frac{p(z)}{q(z)}, \tag{6.5}$$

where $p$ and $q$ are polynomials in $z$. $p$ and $q$ must have no common roots, otherwise factors can be cancelled between them. If $q$ is a non-zero constant, $R$ is just a polynomial. For finite $z$, $R(z)$ can have any complex value, including infinity at points where $q$ vanishes. $R(\infty)$ is the limit as $z \to \infty$ of $p(z)/q(z)$, and can either be finite or infinite. With these conventions for dealing with infinity, rational maps become smooth maps from $S^2$ to $S^2$.

Let us now describe the ansatz in more detail. Rational maps are maps $R$ from $S^2 \mapsto S^2$, whereas Skyrmions are maps $U$ from $\mathbb{R}^3 \mapsto SU(2)$. The main idea is to identify the domain $S^2$ of the rational map with concentric 2-spheres centred at the origin in spatial $\mathbb{R}^3$, and the target $S^2$ with spheres of latitude ($\sigma$ constant) on the target 3-sphere, $S^3 = SU(2)$. These spheres of latitude degenerate at the poles $U = 1$ and $U = -1$. Let us denote a point in $\mathbb{R}^3$ by $(r, z)$, where $r$ is the radial coordinate and $z$ specifies the angular direction, i.e. a point on the unit 2-sphere. The ansatz for the Skyrme field, depending on a rational map $R(z)$ and a radial profile function $f(r)$, is

$$U(r, z) = \cos f(r) + i \sin f(r) \, \mathbf{n}_{R(z)} \cdot \boldsymbol{\tau}, \qquad (6.6)$$

where $\mathbf{n}_{R(z)}$ is the unit vector defined by (6.4) and, as usual, $\boldsymbol{\tau}$ denotes the vector of Pauli matrices. The sigma and unit pion field values are $\sigma = \cos f(r)$ which is independent of the angles, and $\hat{\pi} = \mathbf{n}_{R(z)}$ which is independent of radius. For the ansatz to be well defined at the origin, $f(0) = n\pi$ for some integer $n$. We take $n = 1$ in what follows. The boundary condition $U = 1$ at $r = \infty$ is satisfied by setting $f(\infty) = 0$. The ansatz (6.6) is a generalisation of the hedgehog ansatz (4.45), and reduces to it for $R(z) = z$.

The baryon number of this field configuration is $B = N$, where $N$ is the algebraic degree of the map $R(z)$. The algebraic degree of $R$ is the larger of the degrees of the polynomials $p$ and $q$. For example, the maps $(z-a)/(z-b)$, $1/z^2$ and $z^3+a$ have algebraic degrees 1, 2 and 3, respectively. The algebraic degree is directly related to the topological degree of the map, and in turn to the baryon number, as we now explain.

The topological degree of $R$, as a map from $S^2$ to $S^2$, is given by the expression (4.35). It is the number of preimages on the domain for a generic target point $c$, each counted with its multiplicity $\pm 1$. The preimages are found algebraically by solving the equation $R(z) = c$. By multiplying by the dominator, this becomes the polynomial equation

$$p(z) - cq(z) = 0, \qquad (6.7)$$

which is generically of degree $N$ and has $N$ simple, isolated roots. At each of these roots, the complex derivative $dR/dz$ is non-zero. By expanding in real and imaginary parts, we find that as a real map between 2-spheres, $R$ has Jacobian $|dR/dz|^2$, which is positive. More geometrically, this is because the map, being holomorphic, locally preserves orientation. Thus each preimage of $c$ occurs with positive multiplicity. Therefore, the topological degree of $R$ equals the algebraic degree $N$.

Some values of $c$ are exceptional. As $c$ varies, the roots of $p - cq$ will sometimes coalesce, but the net number of preimages doesn't change if their multiplicities are defined with care. Also $p - cq$ may sometimes have one or more leading powers of $z$ missing. But then the missing finite roots of (6.7) are regarded as being at infinity. This becomes clear if one changes $c$ a little. For example, the equation

$$\frac{1}{z^2} = c, \tag{6.8}$$

with $c$ small, has simple roots at $z = \pm 1/\sqrt{c}$, near infinity, so the equation

$$\frac{1}{z^2} = 0, \tag{6.9}$$

which degenerates if expressed in the form (6.7), is regarded as having a double root at $z = \infty$. Either way, the map $R(z) = 1/z^2$ has algebraic and topological degree 2.

The rational map ansatz for a Skyrme field extends the rational map $R$ to a map $U$ from $\mathbb{R}^3$ to $S^3$. Let us count the preimages of a given point on the target $S^3$. Most target points, except $U = 1$ and $U = -1$, are generic. Choose one of these and, as before, assume that $f'$ is negative for all $r$, although this isn't essential. Then all the preimages lie on a single 2-sphere in $\mathbb{R}^3$, at some positive radius from the origin. On this 2-sphere, the Skyrme field is effectively the same as the rational map. So the number of preimages is the degree $N$ of the rational map, and all occur with positive orientation (because this is the case for a pure rational map, and the derivative of $f$ is negative at each preimage). So the baryon number $B$ equals $N$.

A more formal argument is that the rational map ansatz is an example of a (topological) suspension; the suspension points on the (compactified) domain $\mathbb{R}^3$ are the origin and the point at infinity, and on the target $SU(2)$ they are $U = -1$ and $U = 1$. Suspension is an isomorphism between the homotopy groups $\pi_2(S^2)$ and $\pi_3(S^3)$, which implies that the 3-dimensional topological degree $B$ of the Skyrme field equals the 2-dimensional topological degree $N$ of the rational map.

An attractive feature of the rational map ansatz (6.6) is that it leads to a simple energy expression which can be separately minimised with respect to the rational map $R$ and the profile function $f$ to find approximate Skyrmions. To find the energy we exploit the elastic strain formulation of the Skyrme energy function introduced in Section 4.4.1. For the ansatz, the strain in the radial direction is orthogonal to the strains in the angular directions. Moreover, because any rational map $R(z)$ is conformal, the angular strains are isotropic. (This is one reason to work with rational maps, rather than general maps $R(z,\bar{z})$ from $S^2$ to $S^2$.) If we identify $\lambda_1^2$ with the radial strain and $\lambda_2^2$ and $\lambda_3^2$ with the angular strains, we can easily compute that

$$\lambda_1 = -f'(r), \quad \lambda_2 = \lambda_3 = \frac{\sin f}{r}\frac{1+|z|^2}{1+|R|^2}\left|\frac{dR}{dz}\right|. \qquad (6.10)$$

Each angular strain is the ratio of the radii of the target and domain 2-spheres $\frac{\sin f}{r}$, multiplied by the square root of the local ratio of the area elements of target and domain, thinking of these as unit spheres.

Substituting these strains into the baryon density (4.43) and integrating, we find that the baryon number is

$$B = -\frac{1}{2\pi^2}\int f'\left(\frac{\sin f}{r}\frac{1+|z|^2}{1+|R|^2}\left|\frac{dR}{dz}\right|\right)^2 \frac{2i\,dzd\bar{z}}{(1+|z|^2)^2}r^2\,dr. \qquad (6.11)$$

Now, the part of the integrand

$$\left(\frac{1+|z|^2}{1+|R|^2}\left|\frac{dR}{dz}\right|\right)^2 \frac{2i\,dzd\bar{z}}{(1+|z|^2)^2} \qquad (6.12)$$

is precisely the pull-back of the area form $2i\,dRd\bar{R}/(1+|R|^2)^2$ on the target sphere of the rational map; therefore its integral is $4\pi$ times the degree $N$ of $R$. So (6.11) simplifies to

$$B = -\frac{2N}{\pi}\int_0^\infty f'\sin^2 f\,dr = N, \qquad (6.13)$$

where we have used the boundary conditions $f(0) = \pi$, $f(\infty) = 0$. This verifies again that the baryon number of the Skyrme field derived from the rational map ansatz is equal to the degree of the map.

Substituting the strains (6.10) into the energy density (4.42) yields the total energy

$$E = \int\left\{f'^2 + 2\frac{\sin^2 f}{r^2}(1+f'^2)\left(\frac{1+|z|^2}{1+|R|^2}\left|\frac{dR}{dz}\right|\right)^2\right. \qquad (6.14)$$

$$\left. + \frac{\sin^4 f}{r^4}\left(\frac{1+|z|^2}{1+|R|^2}\left|\frac{dR}{dz}\right|\right)^4\right\}\frac{2i\,dzd\bar{z}}{(1+|z|^2)^2}r^2\,dr,$$

which can be simplified, using the above remarks about the baryon number integral, to

$$E = 4\pi\int_0^\infty\left(r^2 f'^2 + 2B\sin^2 f(1+f'^2) + \mathcal{I}\frac{\sin^4 f}{r^2}\right)dr, \qquad (6.15)$$

a radial integral very similar to (4.49), the energy integral for a hedgehog. Here $\mathcal{I}$ denotes the purely angular integral

$$\mathcal{I} = \frac{1}{4\pi}\int\left(\frac{1+|z|^2}{1+|R|^2}\left|\frac{dR}{dz}\right|\right)^4\frac{2i\,dzd\bar{z}}{(1+|z|^2)^2}, \qquad (6.16)$$

which only depends on the rational map $R$, and is an interesting, geometrically natural function on the space of rational maps of given degree. We will discuss its minimisation in Sections 6.4 and 6.5. With $B$ fixed and $\mathcal{I}$ minimised, we can then minimise $E$, given by (6.15). The Euler–Lagrange equation for the profile $f(r)$ is

$$(r^2 + 2B\sin^2 f)f'' + 2rf' + \sin 2f \left( Bf'^2 - B - \mathcal{I}\frac{\sin^2 f}{r^2} \right) = 0, \quad (6.17)$$

which is similar to eq.(4.50) but with modified coefficients. Its solution can be straightforwardly found numerically.

## 6.3 Symmetric Rational Maps

In the application to Skyrmions we will often be dealing with rational maps between 2-spheres that are highly symmetric, so let us explain what it means for a rational map $R(z)$ to be symmetric under a rotational symmetry group $K \subset SO(3)$ [130]. We will say something about reflection symmetries too.

Consider a spatial rotation $k \in SO(3)$ which acts on $z$ by an $SU(2)$ Möbius transformation

$$z \mapsto k(z) = \frac{\gamma z + \delta}{-\bar{\delta}z + \bar{\gamma}} \quad \text{where} \quad |\gamma|^2 + |\delta|^2 = 1. \quad (6.18)$$

Similarly, a rotation $\mu \in SO(3)$ of the target 2-sphere acts on $R$ by an $SU(2)$ Möbius transformation

$$R \mapsto \mu(R) = \frac{\Gamma R + \Delta}{-\bar{\Delta}R + \bar{\Gamma}} \quad \text{where} \quad |\Gamma|^2 + |\Delta|^2 = 1. \quad (6.19)$$

This corresponds to a rotation of the vector $\mathbf{n}_R$, and therefore extends to an isorotation of the Skyrme field. The map $R(z)$ is $K$-symmetric if, for each $k \in K$, there exists a compensating target space rotation $\mu_k$ whose effect is equivalent to the effect of the spatial rotation, that is,

$$R(k(z)) = \mu_k(R(z)). \quad (6.20)$$

The Skyrme field (6.6) constructed using such a map inherits the symmetry, i.e. a spatial rotation is equivalent to an isorotation. For consistency, $\mu_{kk'} = \mu_k \mu_{k'}$ for all $k, k' \in K$, so $k \mapsto \mu_k$ is a homomorphism.

Since the realisation of the $SO(3)$ action on the domain and target is by $SU(2)$ transformations, the group $K$ should really be replaced by its double group in $SU(2)$, which we still call $K$. This is the group with twice as many elements, obtained by including both elements of $SU(2)$ that correspond to an element of $SO(3)$. In particular, it includes both $\pm 1 \in SU(2)$. We will assume that the map $k \mapsto \mu_k$ is then a homomorphism of $K$ into $SU(2)$.

Rational maps can have reflection symmetries, or more generally, symmetries under improper rotations in $O(3)$. The simplest reflection is realised by the complex conjugation $z \mapsto \bar{z}$, and more general improper rotations are combinations of this with a Möbius transformation. Inversion is the map $z \mapsto -1/\bar{z}$; it is the operation that combines a reflection in any plane with a 180° rotation in that plane. No rational map can be invariant purely under an improper rotation of $z$ (unless it is constant), but the transformed map may be equivalent to the map obtained by a reflection or another improper rotation on the target, realised by the complex conjugation $R \mapsto \bar{R}$ or a combination of this with a target space Möbius transformation. Many of the rational maps we are interested in have symmetries that include some combined improper rotations on domain and target. The Skyrme fields they generate acquire these symmetries.

The simplest example of a symmetric map is the spherically-symmetric map $R(z) = z$ with degree 1. Use of this map gives the hedgehog ansatz. Any rotation of the domain is equivalent to the same rotation on the target. This map is also invariant under reflections; for example, $R(\bar{z}) = \overline{R(z)}$. The map $R(z) = cz$, where $c$ is any constant, is axially symmetric, as $R(e^{i\alpha}z) = e^{i\alpha}R(z)$. Also axially symmetric is the map $R(z) = z^n$ of degree $n$, as $R(e^{i\alpha}z) = e^{in\alpha}R(z)$.

The systematic way to find rotationally symmetric rational maps is to use the representation theory of $K$. This is discussed in ref.[130] and in the book [184]. Here, however, we will use a more direct approach, which is simpler but still effective for discussing most examples.

First, let us consider the derivative of a general map $R(z) = p(z)/q(z)$,

$$\frac{dR}{dz} = \frac{p'(z)q(z) - q'(z)p(z)}{q(z)^2}. \tag{6.21}$$

The numerator

$$W(z) = p'(z)q(z) - q'(z)p(z) \tag{6.22}$$

is called the *Wronskian* of the map, and its roots (zeros) on the domain are where the derivative of $R$ vanishes. The zero set of $W$ is important, as it is unchanged by Möbius transformations of the target. This is because if $R(z)$ is replaced by $(\alpha R(z) + \beta)/(\gamma R(z) + \delta)$, then $W(z)$ changes only by a non-zero multiplicative constant, as is verified by differentiation.

Generically, the Wronskian is a polynomial in $z$ of degree $2N - 2$, where $N$ is the degree of $R$. Naively, it appears from the formula (6.22) that

$W$ has degree $2N - 1$, but the leading power of $z$ always cancels. The Wronskian can have lower degree, in which case one interprets it as having zeros at infinity. In this way, $W(z)$ always has $2N - 2$ zeros, counted with multiplicity. These zeros are the locations at which the multi-valued inverse of the map $R : S^2 \mapsto S^2$ has branch points.

For a $K$-symmetric map $R$, the Wronskian $W$ is unchanged (up to a constant multiple) not only under all rotations of the target, but also under the rotations $k \in K$ of the domain (spatial rotations), because each rotation $k$ is equivalent to some target rotation. The locations of the $2N - 2$ zeros of $W$ are therefore invariant under $K$, and that often makes them easy to visualise.

The $2B - 2$ Wronskian zeros of a rational map of degree $B$ largely determine the spatial shape of the Skyrmion of baryon number $B$ derived from it. Where $W$ is zero, the derivative $dR/dz$ is zero, so the strain eigenvalues in the angular directions, $\lambda_2$ and $\lambda_3$, vanish. The baryon density, being proportional to $\lambda_1 \lambda_2 \lambda_3$, therefore vanishes along the entire radial line in the direction specified by any zero of $W$. The energy density is also low along such a radial line, since there will only be the contribution $\lambda_1^2$ from the radial strain eigenvalue. The rational map ansatz thus makes manifest why the baryon density contours of Skyrmions look like polyhedra with holes in the directions given by the zeros of $W$, and why there are $2B - 2$ such holes, precisely what is seen in all the plots in Fig. 6.1. These holes are at the polyhedral face centres.

### 6.3.1 *Platonic symmetries*

Rational maps with Platonic symmetries are particularly interesting. Recall that the Platonic solids have rotational symmetry groups $T$, $O$ and $Y$ which denote, respectively, the tetrahedral, octahedral and icosahedral groups.

There are certain important polynomials, known as Klein polynomials [153], which form 1-dimensional representations of the Platonic symmetry groups. That means that under the action of any symmetry element, such a polynomial changes by at most a constant multiple, so its zero set is invariant under the group. Klein polynomials can therefore be the Wronskians of rational maps with Platonic symmetries, and they are constructed as follows. Take the example of the tetrahedral group. Scale a regular tetrahedron so that its vertices are on the unit 2-sphere, with one vertex in the positive octant and its edge mid-points on the Cartesian axes. Using the complex coordinate $z$, the positions of the vertices are the four points

$\pm(1+i)/(\sqrt{3}+1)$, $\pm(1-i)/(\sqrt{3}-1)$. Now construct the unique monic polynomial[1] of degree 4 which has these points as roots. This yields the Klein polynomial

$$\mathcal{T}_v = z^4 + 2\sqrt{3}iz^2 + 1 \tag{6.23}$$

associated with the vertices of the tetrahedron. It is invariant under the action of any element $k$ of $T$, possibly up to a constant factor, because $k$ just permutes the roots. Applying the same procedure to the centres of the faces and to the mid-points of the edges of the tetrahedron (in the same orientation) produces the two further Klein polynomials

$$\mathcal{T}_f = z^4 - 2\sqrt{3}iz^2 + 1 \tag{6.24}$$
$$\mathcal{T}_e = z^5 - z \tag{6.25}$$

associated with the faces and edges. Note that a tetrahedron has six edges, but the polynomial $\mathcal{T}_e$ is only of degree 5. This is because in the orientation we have chosen, the mid-point of one of the edges is at $z = \infty$. The five remaining edge mid-points are at the five finite roots $z = 0, 1, i, -1, -i$. So $\mathcal{T}_e$ should really be regarded as a degree 6 polynomial with one root at infinity.

Applying the above construction to the vertices, face centres and edge mid-points of the octahedron and icosahedron produces the Klein polynomials

$$\mathcal{O}_v = z^5 - z \tag{6.26}$$
$$\mathcal{O}_f = z^8 + 14z^4 + 1 \tag{6.27}$$
$$\mathcal{O}_e = z^{12} - 33z^8 - 33z^4 + 1 \tag{6.28}$$
$$\mathcal{Y}_v = z^{11} + 11z^6 - z \tag{6.29}$$
$$\mathcal{Y}_f = z^{20} - 228z^{15} + 494z^{10} + 228z^5 + 1 \tag{6.30}$$
$$\mathcal{Y}_e = z^{30} + 522z^{25} - 10005z^{20} - 10005z^{10} - 522z^5 + 1, \tag{6.31}$$

where the notation is analogous. Recall that the cube and dodecahedron are dual to the octahedron and icosahedron, respectively. Their Klein polynomials are just as above, but vertices are exchanged with face centres. For a tetrahedron, exchanging vertices and face centres gives another tetrahedron, rotated relative to the first by 90° or equivalently by the transformation $z \to iz$. These dual tetrahedra have the same edge mid-points, and together, their vertices are those of a cube, which explains the relation $\mathcal{T}_v \mathcal{T}_f = \mathcal{O}_f$.

---

[1] A monic polynomial has leading coefficient unity.

Let us now consider a concrete example, and construct the rational map of lowest degree that is tetrahedrally symmetric. Of course, the spherically-symmetric map $R = z$ is automatically $K$-symmetric for any $K \subset SO(3)$, but we ignore this degree 1 map. We have already seen that for a degree $N$ map the $2N - 2$ zeros of the Wronskian must be strictly invariant under the symmetry group, and for $n$ points on a sphere to be invariant under the tetrahedral group $T$ requires that $n \geq 4$, with the lower limit $n = 4$ corresponding to placing the four points on the vertices of a tetrahedron. From this we see that $N \geq 3$ for a $T$-symmetric map. There is, in fact, a unique (up to orientations of the domain and target spheres) $T$-symmetric, degree 3 map $R(z)$. It can be explicitly calculated by applying the group generators directly to a general map. We begin by requiring $R(z)$ to be symmetric under two independent 180° rotations in the group, around the Cartesian 3-axis and 1-axis. These two symmetries are realised by

$$R(-z) = -R(z) \quad \text{and} \quad R\left(\frac{1}{z}\right) = \frac{1}{R(z)}. \quad (6.32)$$

The first condition implies that the numerator of $R$ is even in $z$ and the denominator is odd, or vice versa. These two possibilities are related by a Möbius transformation, so we choose the former and ignore the latter. Imposing the second condition then gives us maps of the form

$$R(z) = \frac{\sqrt{3}az^2 - 1}{z(z^2 - \sqrt{3}a)} \quad (6.33)$$

with $a$ complex. The inclusion of the $\sqrt{3}$ factor here is a convenience. Tetrahedral symmetry is obtained by imposing the further condition

$$R\left(\frac{iz+1}{-iz+1}\right) = \frac{iR(z)+1}{-iR(z)+1} \quad (6.34)$$

which is satisfied by (6.33) if $a = \pm i$, the two choices being related by the 90° spatial rotation $z \mapsto iz$ followed by a rotation of the target sphere. Note that $z \mapsto (iz+1)/(-iz+1)$ sends $0 \mapsto 1 \mapsto i \mapsto 0$ and hence generates the 120° rotation cyclically permuting the three Cartesian axes.

The $T$-symmetric map

$$R(z) = \frac{\sqrt{3}iz^2 - 1}{z^3 - \sqrt{3}iz} \quad (6.35)$$

also has a reflection symmetry, represented by the relation $R(i\bar{z}) = \overline{iR(z)}$. This reflection extends the symmetry group $T$ to $T_d$, where the subscript $d$ denotes that the plane of the reflection symmetry contains a $C_2$ axis, which is the case for a tetrahedron. Alternatively, the group $T$ could be extended by inversion $z \mapsto -1/\bar{z}$, which produces the group $T_h$, but there are no $T_h$-symmetric maps of degree 3. The rotation groups $O$ and $Y$ can also be extended by inversion to produce the groups $O_h$ and $Y_h$, which are the full symmetry groups of an octahedron (or cube) and an icosahedron (or dodecahedron).

It is interesting to look at the Wronskian of maps of the form (6.33),

$$W(z) = -\sqrt{3}a(z^4 + \sqrt{3}(a - a^{-1})z^2 + 1). \tag{6.36}$$

For $a = i$, $W$ is proportional to the tetrahedral Klein polynomial $\mathcal{T}_v$, and for $a = -i$ it is proportional to $\mathcal{T}_f$. In both cases the zeros of the Wronskian are tetrahedrally symmetric, as anticipated. Thus, calculating the Wronskian is an alternative way of fixing the coefficient $a$ in the family of maps (6.33), easier than imposing the 120° rotation symmetry (6.34) directly.

Let us now turn to $O$-symmetric maps. The $2N - 2$ zeros of the Wronskian must be located on the sphere with octahedral symmetry, which requires at least six points. Thus the lowest possible degree for the map is $N = 4$. A computation along the lines above produces the map

$$R(z) = \frac{z^4 + 2\sqrt{3}iz^2 + 1}{z^4 - 2\sqrt{3}iz^2 + 1} \tag{6.37}$$

which is in fact $O_h$-symmetric, due to the additional inversion symmetry $R(-1/\bar{z}) = 1/\overline{R(z)}$. The Wronskian of this map is proportional to the Klein polynomial $\mathcal{O}_v$, so it is zero on the six face centres of a cube (a more useful characterisation than the six vertices of an octahedron).

There is a 1-parameter family of degree 4 maps with tetrahedral symmetry, of which the map with octahedral symmetry is a special case. This is the family

$$R(z) = c\,\frac{z^4 + 2\sqrt{3}iz^2 + 1}{z^4 - 2\sqrt{3}iz^2 + 1} \tag{6.38}$$

where $c$ is a complex constant. By imposing a reflection symmetry, $c$ can be restricted to be real, and the maps are $T_d$-symmetric. The transformation $c \mapsto 1/c$ is equivalent to the 90° rotation $z \mapsto iz$, and when $c = 1$ the map (6.38) becomes the octahedrally-symmetric map (6.37).

Finally, for the icosahedral symmetry group the $2N - 2$ zeros of the Wronskian can be placed on the twelve face centres of a dodecahedron if $N = 7$. There is indeed a unique $Y_h$-symmetric degree 7 map,

$$R(z) = \frac{z^7 - 7z^5 - 7z^2 - 1}{z^7 + 7z^5 - 7z^2 + 1}, \tag{6.39}$$

Table 6.2: Rational map Skyrmions: symmetry $K$, angular integral $\mathcal{I}$, and energy $E$.

| $B$ | $K$ | $\mathcal{I}$ | $E/12\pi^2 B$ |
|---|---|---|---|
| 1 | $O(3)$ | 1.0 | 1.232 |
| 2 | $D_{\infty h}$ | 5.8 | 1.208 |
| 3 | $T_d$ | 13.6 | 1.184 |
| 4 | $O_h$ | 20.7 | 1.137 |
| 5 | $D_{2d}$ | 35.8 | 1.147 |
| 6 | $D_{4d}$ | 50.8 | 1.137 |
| 7 | $Y_h$ | 60.9 | 1.107 |
| 8 | $D_{6d}$ | 85.6 | 1.118 |

and its Wronskian is proportional to the Klein polynomial $\mathcal{Y}_v$. In the orientation we have chosen, the $C_5$ symmetry is (fairly) manifest, as the rotation $z \mapsto e^{2\pi i/5}z$ is equivalent to a Möbius transformation of $R$. The map also has the inversion symmetry $R(-1/\bar{z}) = -1/\overline{R(z)}$ that combines inversion in space with inversion in isospace.

## 6.4 Skyrmions from Rational Maps

Optimising the rational map ansatz, through the choice of both the rational map itself and the profile function, leads to good approximations to Skyrmions for a range of baryon numbers. The Skyrmions inherit all the symmetries of the rational map. To minimise the energy $E$ one should first minimise $\mathcal{I}$, the angular integral (6.16), over all maps of degree $B$. The profile function $f(r)$ minimising the energy (6.15) is then found by solving eq.(6.17) with $B$ and $\mathcal{I}$ as fixed parameters. In Section 6.5 we discuss the results of a numerical search for $\mathcal{I}$-minimising maps among all rational maps of degree $B$, but in this section we consider the simpler problem where the map is restricted to a given symmetric form, with symmetries corresponding to a known Skyrmion solution. If the map still contains a few free parameters, $\mathcal{I}$ can be minimised with respect to these. This procedure is appropriate for all baryon numbers up to $B = 8$. Table 6.2 lists the energies $E$ (divided again by $12\pi^2 B$) of the approximate Skyrmion solutions obtained this way, together with the values of the optimised integral $\mathcal{I}$. The optimised profile functions are shown in Fig. 6.2.

For $B = 2, 3, 4, 7$ the symmetries of the Skyrmions are $D_{\infty h}, T_d, O_h, Y_h$ respectively. In each of these cases, we have seen that there is a unique

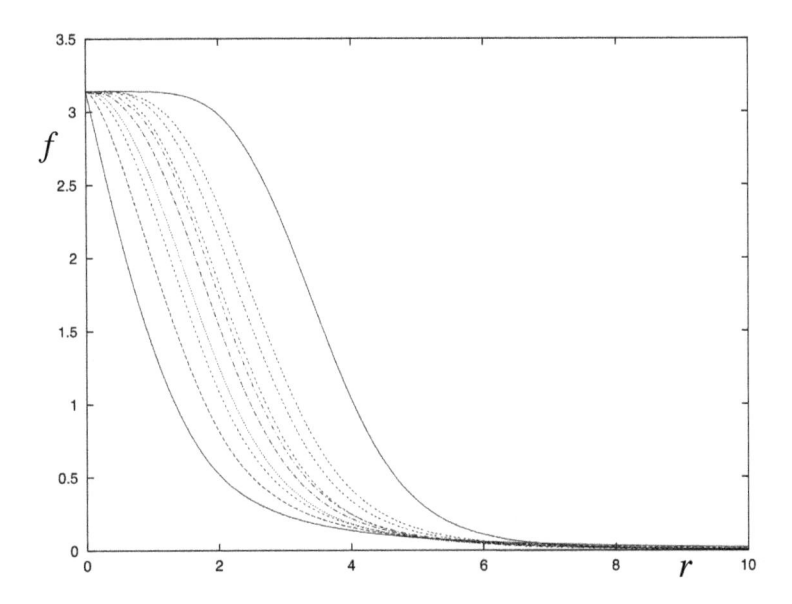

Fig. 6.2: Optimal profile functions for the rational map ansatz. Left to right: $B$ increasing from 1 to 9, and additionally $B = 17$.

rational map of degree $B$ with the given symmetry. They are

$$R(z) = z^2, \quad R(z) = \frac{z^3 - \sqrt{3}iz}{\sqrt{3}iz^2 - 1},$$

$$R(z) = \frac{z^4 + 2\sqrt{3}iz^2 + 1}{z^4 - 2\sqrt{3}iz^2 + 1}, \quad R(z) = \frac{z^7 - 7z^5 - 7z^2 - 1}{z^7 + 7z^5 - 7z^2 + 1}. \quad (6.40)$$

Using these maps with the optimal profile functions, we obtain Skyrme fields whose baryon density isosurfaces are indistinguishable from those shown in Fig. 6.1.

The $B = 3$ Skyrmion is asymptotically like a $B = -1$ antiSkyrmion. This can be understood from the symmetry of the rational map (6.35), but more naively by a simple addition of the pion dipole moments of its constitutent $B = 1$ Skyrmions. First consider two single Skyrmions brought together along the $x^1$-axis. They are in the attractive channel if the first is in standard orientation and the second is rotated by 180° about the $x^3$-direction. The first has a triplet of dipole moments $\mathbf{p} = 4\pi C(\mathbf{e}_1, \mathbf{e}_2, \mathbf{e}_3)$ and the second $\mathbf{q} = 4\pi C(-\mathbf{e}_1, -\mathbf{e}_2, \mathbf{e}_3)$. Their sum is $4\pi C(\mathbf{0}, \mathbf{0}, 2\mathbf{e}_3)$, implying that the toroidal $B = 2$ Skyrmion has only a single dipole, with roughly double strength. Now bring in a third Skyrmion along the $x^3$-axis, rotated by

180° about the $x^1$-direction, with dipole moments $\mathbf{r} = 4\pi C(\mathbf{e}_1, -\mathbf{e}_2, -\mathbf{e}_3)$. The total of the dipoles is $\mathbf{p} + \mathbf{q} + \mathbf{r} = 4\pi C(\mathbf{e}_1, -\mathbf{e}_2, \mathbf{e}_3)$, precisely those of an antiSkyrmion.

Similarly, the $B = 4$ cubic Skyrmion has no dipoles, since it can be constructed from two $B = 2$ tori whose single dipoles can be oriented so as to cancel. The symmetry of the degree 4 map (6.37) is consistent with this cancellation. The absence of dipoles explains why the $B = 4$ Skyrmion is so tightly bound and why it interacts only weakly with other Skyrmions, through its quadrupole and octupole asymptotic fields. The dodecahedral $B = 7$ Skyrmion has neither asymptotic dipole nor asymptotic quadrupole strength.

For the remaining low baryon numbers, $B = 5, 6, 8$, the Skyrmions have extended dihedral symmetries $D_{nd}$. Constructing $D_n$-symmetric maps is straightforward [184]. One can explicitly apply the two generators of $D_n$ to a map. These generators may be taken to be $z \mapsto e^{2\pi i/n} z$ and $z \mapsto 1/z$. The reflection $z \mapsto 1/\bar{z}$ extends the symmetry to $D_{nh}$, whereas $z \mapsto e^{\pi i/n} \bar{z}$ extends the symmetry to $D_{nd}$.

For example, there is a family of degree 5 maps with $D_{2d}$ symmetry,

$$R(z) = \frac{z(z^4 + ibz^2 + a)}{az^4 + ibz^2 + 1} \tag{6.41}$$

with $a$ and $b$ real. Enhanced symmetry occurs if $b = 0$; $R(z)$ then has $D_{4h}$ symmetry, the symmetry of a square. There is octahedral symmetry $O_h$ if, in addition, $a = -5$. This value ensures the 120° rotational symmetry

$$R\left(\frac{iz+1}{-iz+1}\right) = \frac{iR(z)+1}{-iR(z)+1}. \tag{6.42}$$

The $O_h$-symmetric map $R(z) = z(z^4 - 5)/(-5z^4 + 1)$ has Wronskian

$$W(z) = -5(z^8 + 14z^4 + 1) \tag{6.43}$$

which is proportional to $\mathcal{O}_f$, the face polynomial of an octahedron.

Using (6.41) in the rational map ansatz gives a field structure that is a polyhedron with eight faces. In the special case $a = -5, b = 0$, this polyhedron is a regular octahedron, and the angular integral is $\mathcal{I} = 52.1$; however, a numerical search over the parameters $a$ and $b$ finds that $\mathcal{I}$ is minimised when $a = -3.07, b = 3.94$, taking the value $\mathcal{I} = 35.8$. The approximate Skyrmion generated from the map with these parameter values has a baryon density isosurface virtually identical to that of the numerically computed $B = 5$ Skyrmion shown in Fig. 6.1. There is an exact $O_h$-symmetric $B = 5$ solution of the Skyrme field equation, but it is a saddle point with higher energy. Although many minimal-energy Skyrmions are highly symmetric, symmetry is not always the most important factor in determining the structure, and less symmetric configurations sometimes have lower energy.

A further example of a symmetric saddle point is the $B = 7$ solution of the field equation with octahedral symmetry. The relevant rational map is $R(z) = (7z^4 + 1)/(z^7 + 7z^3)$, having Wronskian $W(z) = -21z^2(z^4 - 1)^2$ proportional to the square of $\mathcal{O}_v$. Each root of this Wronskian is a double root (including the one at infinity), and they lie at the face centres of a cube. A baryon density isosurface for this solution therefore resembles that of the $B = 4$ Skyrmion, but the overall scale is larger.

## 6.5 Skyrmions up to $B = 22$

In the preceding section, for each $B \leq 8$, the map $R$ was selected so that the symmetry of the resulting Skyrme field matched that of the numerically computed Skyrmion. There is an alternative approach based purely on energy minimisation; this has been applied for all $B \leq 22$ [30]. In this approach, no assumption is made as to the possible symmetry of the Skyrmion. The main task is to search for the rational map of degree $B$ that minimises the angular integral $\mathcal{I}$. This task has been performed using a simulated annealing algorithm, a Monte Carlo based method whose major advantage over conventional minimisation techniques in that changes that increase the energy are allowed, enabling the algorithm to escape from local minima towards the global minimum. For $B \leq 8$ the simulated annealing algorithm reproduces the rational maps and approximate Skyrmions discussed above.

The results of the simulated annealing algorithm applied to general rational maps of degrees $9 \leq B \leq 22$ are shown in Table 6.3. In each case, we tabulate the identified symmetry group $K$, the optimised value of $\mathcal{I}$, and the energy $E$ for the profile function that solves eq.(6.17). Some of these rational maps, for example the $D_{3h}$-symmetric map of degree 11 and the $T_d$-symmetric map of degree 12, will be useful later in the context of the multi-layer rational map ansatz.

By minimising within certain symmetric families of rational maps whose symmetries are not shared by the optimal map, one finds other critical points of $\mathcal{I}$. Table 6.4 presents the results, denoted by a star; these support the conclusion that the maps presented in Table 6.3 are the global minima for $\mathcal{I}$. However, in a few cases the $\mathcal{I}$-minimising map does not represent

Table 6.3: Annealed Skyrmions: symmetry $K$, angular integral $\mathcal{I}$ and energy $E$.

| $B$ | $K$ | $\mathcal{I}$ | $E/12\pi^2 B$ |
|---|---|---|---|
| 9 | $D_{4d}$ | 109.3 | 1.116 |
| 10 | $D_{4d}$ | 132.6 | 1.110 |
| 11 | $D_{3h}$ | 161.1 | 1.109 |
| 12 | $T_d$ | 186.6 | 1.102 |
| 13 | $O$ | 216.7 | 1.098 |
| 14 | $D_2$ | 258.5 | 1.103 |
| 15 | $T$ | 296.3 | 1.103 |
| 16 | $D_3$ | 332.9 | 1.098 |
| 17 | $Y_h$ | 363.4 | 1.092 |
| 18 | $D_2$ | 418.7 | 1.095 |
| 19 | $D_3$ | 467.9 | 1.095 |
| 20 | $D_{6d}$ | 519.7 | 1.095 |
| 21 | $T$ | 569.9 | 1.094 |
| 22 | $D_{5d}$ | 621.6 | 1.092 |

Table 6.4: Alternative critical points of the angular integral $\mathcal{I}$.

| $B$ | $K$ | $\mathcal{I}$ | $E/12\pi^2 B$ |
|---|---|---|---|
| 9* | $T_d$ | 112.8 | 1.123 |
| 10* | $D_3$ | 132.8 | 1.110 |
| 10* | $D_{3d}$ | 133.5 | 1.111 |
| 10* | $D_{3h}$ | 143.2 | 1.126 |
| 13* | $D_{4d}$ | 216.8 | 1.098 |
| 13* | $O_h$ | 265.1 | 1.140 |
| 15* | $T_d$ | 313.7 | 1.113 |
| 16* | $D_2$ | 333.4 | 1.098 |
| 17* | $O_h$ | 367.2 | 1.093 |
| 19* | $T_h$ | 469.8 | 1.096 |
| 22* | $D_3$ | 623.4 | 1.092 |

the true Skyrmion, since some of the $\mathcal{I}$ values in Tables 6.4 and 6.3 are very close. In particular, the $\mathcal{I}$ values for the $B = 10$ configurations with $D_3$ and $D_{3d}$ symmetry, $B = 13$ with $D_{4d}$ symmetry, $B = 16$ and $B = 22$ are extremely close to the corresponding values in Table 6.3, suggesting the possibility of alternative minima or low-lying saddle points of the Skyrme energy.

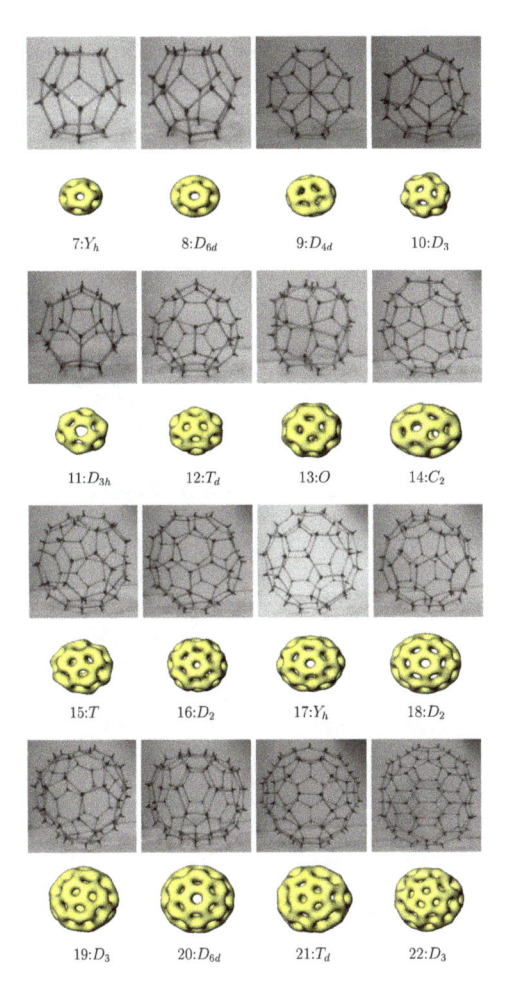

Fig. 6.3: Baryon density isosurfaces for Skyrmions with $7 \leq B \leq 22$ (to scale), their symmetry groups, and polyhedral models (not to scale)

For most baryon numbers there is a sufficient gap between the minimal value of $\mathcal{I}$ and that of any other critical point to be confident that the optimal map corresponds to the Skyrmion. However, for baryon numbers $B = 10, 16$ and $22$ a glance at Tables 6.3 and 6.4 reveals that there are different maps (with different symmetries) whose associated Skyrme fields have energies that differ by less than 0.1%. Full field simulations suggest that for these three baryon numbers the maps presented in Table 6.4, rather

Table 6.5: Symmetries and energies of exact Skyrmions up to $B = 22$.

| $B$ | $K$ | $E/12\pi^2 B$ | $E/12\pi^2$ |
|---|---|---|---|
| 1  | $O(3)$      | 1.2322 | 1.2322 |
| 2  | $D_{\infty h}$ | 1.1791 | 2.3582 |
| 3  | $T_d$       | 1.1462 | 3.4386 |
| 4  | $O_h$       | 1.1201 | 4.4804 |
| 5  | $D_{2d}$    | 1.1172 | 5.5860 |
| 6  | $D_{4d}$    | 1.1079 | 6.6474 |
| 7  | $Y_h$       | 1.0947 | 7.6629 |
| 8  | $D_{6d}$    | 1.0960 | 8.7680 |
| 9  | $D_{4d}$    | 1.0936 | 9.8424 |
| 10 | $D_3$       | 1.0904 | 10.9040 |
| 11 | $D_{3h}$    | 1.0889 | 11.9779 |
| 12 | $T_d$       | 1.0856 | 13.0272 |
| 13 | $O$         | 1.0834 | 14.0842 |
| 14 | $C_2$       | 1.0842 | 15.1788 |
| 15 | $T$         | 1.0825 | 16.2375 |
| 16 | $D_2$       | 1.0809 | 17.2944 |
| 17 | $Y_h$       | 1.0774 | 18.3158 |
| 18 | $D_2$       | 1.0788 | 19.4184 |
| 19 | $D_3$       | 1.0786 | 20.4934 |
| 20 | $D_{6d}$    | 1.0779 | 21.5580 |
| 21 | $T_d$       | 1.0780 | 22.6380 |
| 22 | $D_3$       | 1.0766 | 23.6852 |

than in Table 6.3, represent the minimal-energy Skyrmions.

Taking into account the above comments, Table 6.5 lists the symmetry $K$ and energy per baryon for all true, minimal-energy Skyrmions with massless pions, for $B \leq 22$; also listed is the total energy. These values were computed by relaxation of the Skyrme energy function with initial conditions created from the corresponding rational map ansatz. Fig. 6.3 shows baryon density isosurfaces for these Skyrmions with $7 \leq B \leq 22$; also shown are models of the associated polyhedra. For most baryon numbers except $B = 9$ and $B = 13$ the polyhedra have only trivalent vertices, as these seem to be the most stable.

A particularly interesting example is the $B = 17$ Skyrmion, which has the icosahedrally-symmetric structure of the famous $C_{60}$ Buckyball. It is formed from 12 pentagons and 20 hexagons and is the structure with isolated pentagons having the least number of vertices. It can be constructed

using the unique $Y_h$-symmetric, degree 17 rational map [26],
$$R(z) = \frac{z^{17} + 17z^{15} + 119z^{12} - 187z^{10} + 187z^7 + 119z^5 + 17z^2 - 1}{z^{17} - 17z^{15} + 119z^{12} + 187z^{10} + 187z^7 - 119z^5 + 17z^2 + 1}. \tag{6.44}$$

## 6.6 Rigorous Investigation of Skyrmions

In this section we bring together a number of results and studies that have relied on a rigorous analysis of the Skyrme field equation and energy function.

Recall that an improper spatial rotation (an element of $O(3)$ that is not in $SO(3)$) reverses the baryon number of a Skyrme field, but the baryon number is restored by an improper isorotation. In particular, any pure spatial reflection symmetry of a Skyrmion must be realised by combining the reflection with some compensating improper isorotation. Generally, an improper isorotation combines a reflection in isospace with a proper isorotation in the plane fixed by the reflection (i.e. is a rotary reflection in isospace), but rather remarkably, it has been proved that the proper isorotation part has to be trivial, so for a Skyrmion a pure spatial reflection is always compensated by a pure isospatial reflection [100]. The proof uses the Skyrme field equation, and in particular, properties of the conserved energy-momentum tensor on the spatial plane that is fixed by the reflection, so it applies to any static solution, whether stable or not.

For example, the hedgehog Skyrmion in its standard orientation has a reflection symmetry in any plane through the origin. Here, the compensating improper isorotation is the same pure reflection in isospace. The finitely-many reflection symmetries of the $B = 3$ tetrahedral Skyrmion, whose symmetry group is $T_d$, are similar, because the asymptotic field approaches that of a $B = -1$ hedgehog. We have seen further examples of reflection symmetries earlier in this chapter, and these will be important when we discuss the rigid-body quantization of Skyrmions, and calculate the parities of the quantum states.

After the baryon number, symmetry and energy, the most significant characteristic of a static Skyrmion is its asymptotic, long-range tail field, which satisfies the linearised form of the field equation. To leading order, each of the three components of a massless pion field $\boldsymbol{\pi}$ obeys Laplace's equation, and $\sigma$ can be taken to be unity. As the triplet of pion tail fields $\boldsymbol{\pi}$ asymptotically approach zero, the leading-order term is a multipole, an inverse power of $r = |\mathbf{x}|$, say $r^{-(l+1)}$, times a triplet of angular functions.

To satisfy Laplace's equation, the angular functions must be linear combinations of the spherical harmonics $Y_{l,m}(\theta, \varphi)$, with $m$ taking integer values in the range $-l \leq m \leq l$. These spherical harmonics can also be expressed in Cartesian coordinates, which often leads to more convenient and elegant formulae for the tail fields. Sometimes, one component of the pion field (or possibly two) has a vanishing multipole to leading order, and its asymptotic form is determined by a subleading multipole. We have previously discussed the example of the asymptotic dipole fields ($l = 1$) of a hedgehog Skyrmion, and have mentioned the asymptotic dipoles, asymptotic quadrupoles ($l = 2$) and asymptotic octupoles ($l = 3$) of Skyrmions with higher $B$.

A rigorous result concerning solutions of the Skyrme field equation is that the leading pion multipole cannot be a monopole – a term with $l = 0$. The leading-order term must be a dipole or higher multipole. The proof is as follows [177]. For a static field, the field equation (4.18) implies that the spatial current

$$\tilde{R}_i = R_i + \frac{1}{4}[R_j, [R_j, R_i]] \tag{6.45}$$

has zero divergence and no singularity. Therefore the flux of $\tilde{R}_i$ through a large 2-sphere $S_r^2$ of radius $r$ centred at the origin vanishes, i.e.

$$\int_{S_r^2} \tilde{R}_i n^i \, dS = 0 \tag{6.46}$$

where $\mathbf{n} = \hat{\mathbf{x}}$ is the unit outward normal. Now, in the asymptotic region, $R_i$ is small so $\tilde{R}_i$ can be replaced by $R_i$, which in turn simplifies to $i(\partial_i \boldsymbol{\pi}) \cdot \boldsymbol{\tau}$. For a monopole asymptotic field,

$$\boldsymbol{\pi} = \frac{\mathbf{c}}{r} \tag{6.47}$$

where $\mathbf{c}$ is a constant vector, so $\tilde{R}_i$ has the leading asymptotic behaviour $-i\mathbf{c} \cdot \boldsymbol{\tau} x^i / r^3$. Then $\tilde{R}_i n^i = -i\mathbf{c} \cdot \boldsymbol{\tau}/r^2$ and the flux through the sphere is $-4\pi i \mathbf{c} \cdot \boldsymbol{\tau}$. This vanishes only if $\mathbf{c} = \mathbf{0}$.

It has also been rigorously proved [183] that for any non-vacuum solution of the Skyrme field equation, with finite $B$ and $E$, the multipole expansion is non-trivial. In other words, the pion field $\boldsymbol{\pi}$ does not vanish to all orders in $l$, and the leading term is a finite-$l$ multipole.

Knowing about the asymptotic multipoles of Skyrmions is potentially helpful for proving rigorously that Skyrmions exist for all baryon numbers. Unfortunately such a proof remains elusive but some progress has been made, and is outlined here.

The existence of a Skyrmion (an energy minimiser) with baryon number $B$ has been established by Esteban [81] under the assumption that

$$E_B < E_{B'} + E_{B-B'} \qquad (6.48)$$

for all $B' \in \mathbb{Z} - \{0, B\}$, where $E_B$ denotes the infimum of the energy (4.36) within $\mathcal{C}_B$, the space of Skyrme field configurations with baryon number $B$. Esteban was able to prove the weaker inequality

$$E_B \leq E_{B'} + E_{B-B'} \qquad (6.49)$$

but the strict inequality (6.48) is not yet proved in general. The interpretation of the strict inequality is that it would prevent the breakup of a sufficiently low-energy field configuration with baryon number $B$ into infinitely-separated clusters with baryon numbers $B'$ and $B - B'$. The energy infimum $E_B$ would then be attained by a true Skyrmion solution with baryon number $B$, and not just approached by Skyrmion clusters with this total baryon number moving off to infinity (bubbling off).

Castillejo and Kugler [63] gave a physical perspective on these inequalities in terms of the asymptotic forces between well-separated Skyrmions, and proposed an argument that almost proves Estaban's strict inequality. They noted that if the asymptotic interaction energy of two well-separated Skyrmion clusters of any baryon number is positive, then it can be made negative by performing an appropriate isorotation on one of the clusters. We have already explicitly seen that this is true in the case of two $B = 1$ Skyrmions, as illustrated by eq.(4.57). It may appear that this result constitutes a proof of (6.48), since it suggests it is always possible to arrange for two clusters to have a negative interaction energy, and hence a total energy that is lower than the sum of their individual energies. However, the flaw in this argument is that the asymptotic interaction energy may vanish. In this case, it cannot be made negative by an isorotation. So it wouldn't be possible to conclude that the interaction energy is negative, only that it is non-positive. This is another manifestation of the fact that the weaker energy inequality (6.49) has been proved but the strict inequality (6.48) has not.

However, further progress on this problem has been made. Recall that it has been established that any Skyrmion has a finite, non-zero leading multipole. It has also been shown that a pair of well-separated Skyrmions can be oriented and positioned so that through their leading multipoles they have negative interaction energy and do attract [183]. Unfortunately, there is still a loophole. The required energy inequality does not follow if the leading multipole of one of the Skyrmions is of high order, or more precisely, if the orders of the leading multipoles of the pair of Skyrmions differ by more than two. This is because of the uncontrolled effect of nonlinear terms, which can compete with the very small interaction energy obtained from the leading multipoles in this situation. Nevertheless, as Schroers has shown [211], some rigorous conclusions about the existence of Skyrmions with higher baryon numbers are possible.

Of course, Skyrmions believed to be of minimal energy have been discovered numerically for many values of $B$. Their energies are listed in Table 6.5, and these energies do satisfy the strict inequality (6.48) for all $B \leq 22$. This can be checked by simple addition for all $B'$ in the natural range $0 < B' < B$. But note that it is also necessary to consider $B'$ outside this range, describing the possibility of a Skyrmion of baryon number $B$ breaking up into a Skyrmion of baryon number $B + n$ (with $n$ a positive integer) and an antiSkyrmion of baryon number $-n$. Fortunately, one can use the Faddeev–Bogomolny inequality $E_B > 12\pi^2|B|$ to rule out this implausible scenario for the tabulated energies. The energy of such a Skyrmion-antiSkyrmion pair is at least $12\pi^2(B+2n)$, according to this inequality. From the table, one sees that $E_B$ is always less than $12\pi^2(B+2)$ for $B \leq 22$, so all the unnatural breakups are ruled out.

These considerations do not rigorously prove that Skyrmions exist for all baryon numbers up to $B = 22$ because the numerical search for the energy minimum may have missed the true minimum (or infimum if clusters separate to infinity). However, if we claim that the numerical search has been successful in all these cases, then the results are at least consistent because they satisfy the strict Esteban inequality. It is then at least plausible that the solutions that have been found are the true Skyrmions, the global energy minima.

## 6.7 The Skyrmion Crystal

So far we have only discussed Skyrmions with a finite baryon number, but the lowest known value of the energy per baryon actually occurs for an infinite crystal of Skyrmions. At some baryon number $B$ greater than 22, it is favoured for the minimal-energy Skyrmion with massless pions to be more compact than a hollow polyhedral structure, and to acquire a crystalline form as $B$ increases further.

To study the Skyrmion crystal, one imposes periodic boundary conditions on the Skyrme field and works within a unit cell (equivalently, a 3-

torus). The first attempted construction of a crystal was by Klebanov [152] using a simple cubic lattice of $B = 1$ Skyrmions. After relaxation, Klebanov's crystal had an energy per baryon $E/12\pi^2 B = 1.08$. Other symmetries were proposed which led to slightly lower-energy crystals. Finally, Castillejo et al. [62] and Kugler and Shtrikman [158] discovered that to minimise the energy, the Skyrmions should initially be arranged as a face-centred cubic (FCC) lattice, with their orientations chosen to give maximal attraction between all nearest neighbours. Explicitly, the Skyrme field is strictly periodic after translation by $2L$ in the $x^1$-, $x^2$- or $x^3$-directions, where $L$ can be freely chosen at first, and the unit cell is a cube of side length $2L$. There are hedgehog-like, $B = 1$ Skyrmions in standard orientation at the vertices of this unit cell, and the Skyrmions at each face centre are rotated by 180° about the direction normal to the face. Every Skyrmion then has twelve nearest neighbours, all in the attractive channel. Inside a cubic unit cell the total baryon number is $B = 4$. If we fix the origin at the centre of one of the unrotated Skyrmions, this configuration has the combined spatial and isospatial symmetries generated by

$$(x^1, x^2, x^3) \mapsto (-x^1, x^2, x^3), \ (\sigma, \pi_1, \pi_2, \pi_3) \mapsto (\sigma, -\pi_1, \pi_2, \pi_3); \quad (6.50)$$

$$(x^1, x^2, x^3) \mapsto (x^2, x^3, x^1), \ (\sigma, \pi_1, \pi_2, \pi_3) \mapsto (\sigma, \pi_2, \pi_3, \pi_1); \quad (6.51)$$

$$(x^1, x^2, x^3) \mapsto (x^1, x^3, -x^2), \ (\sigma, \pi_1, \pi_2, \pi_3) \mapsto (\sigma, \pi_1, \pi_3, -\pi_2); \quad (6.52)$$

$$(x^1, x^2, x^3) \mapsto (x^1 + L, x^2 + L, x^3), \ (\sigma, \pi_1, \pi_2, \pi_3) \mapsto (\sigma, -\pi_1, -\pi_2, \pi_3). \quad (6.53)$$

Symmetry (6.50) is a reflection in a face of the cube, (6.51) is a 120° rotation around a cube diagonal, (6.52) is a 90° rotation around a cube edge, and (6.53) is a translation from a vertex of the cube to a face centre.

At modest densities, this FCC arrangement of Skyrmions minimises the crystal energy. The individual Skyrmions retain an approximately spherically-symmetric shape, so the surfaces where $\sigma = 0$ are approximately spheres. The field in most of the crystal is close to the vacuum value, $\sigma = 1$.

As the density increases, i.e. as $L$ is reduced, the energy decreases and there is a remarkable (zero-temperature) phase transition to a crystal of half-Skyrmions at $2L \approx 7.6$. At this point the crystal symmetry is spontaneously enhanced by the addition of the generator

$$(x^1, x^2, x^3) \mapsto (x^1 + L, x^2, x^3), \quad (\sigma, \pi_1, \pi_2, \pi_3) \mapsto (-\sigma, -\pi_1, \pi_2, \pi_3), \quad (6.54)$$

a translation half-way along the unit cell edge. Note that this symmetry involves a chiral $SO(4)$ rotation, rather than just an $SO(3)$ isorotation as before. This is allowed because a crystal does not have to satisfy the usual boundary conditions. The previous translational symmetry (6.53) is obtained by combining the new generator with this generator rotated by 90°. The solution becomes a primitive cubic crystal of half-Skyrmions with a cubic unit cell of side length $L$. Each of these smaller unit cells has similar pion field distributions and baryon number $\frac{1}{2}$. The $\sigma < 0$ and $\sigma > 0$ regions alternate between neighbouring unit cells, with $\sigma = 0$ on all the faces. The field still has strict periodicity $2L$, but the energy and baryon densities have periodicity $L$. Because $\langle \sigma \rangle = 0$ after averaging over the crystal, there is a partial restoration of chiral symmetry.

This half-Skyrmion phase is the phase realised by the Skyrmion crystal when $L$ is allowed to vary freely. The minimum of the energy occurs at $2L \approx 4.7$. Here, a variational method based on a truncated Fourier series expansion of the fields [158] estimates the energy per baryon to be $E/12\pi^2 B = 1.038$, remarkably close to the Faddeev–Bogomolny lower bound. A purely numerical calculation [31] gives $E/12\pi^2 B = 1.036$. Fig. 6.4 shows a baryon density isosurface for the Skyrmion crystal. Each lump represents a half-Skyrmion and the total baryon number shown is 4.

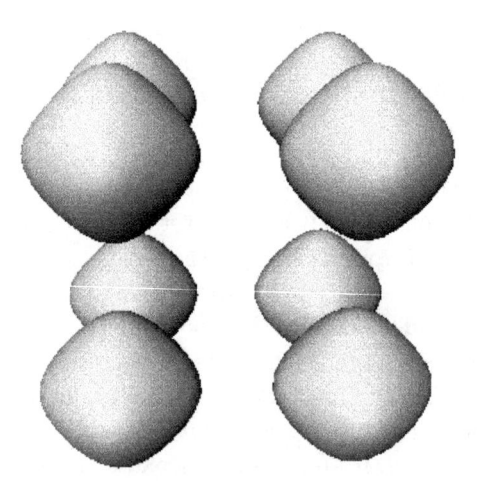

Fig. 6.4: A baryon density isosurface for a portion of the Skyrmion crystal

The fields obtained either numerically or by optimising the Fourier series are very well approximated by the formulae [62]

$$\sigma = -c_1 c_2 c_3,$$

$$\pi_1 = s_1 \sqrt{1 - \frac{s_2^2}{2} - \frac{s_3^2}{2} + \frac{s_2^2 s_3^2}{3}} \quad \text{and cyclic}, \quad (6.55)$$

where $c_i = \cos(\pi x^i/L)$ and $s_i = \sin(\pi x^i/L)$. This approximation to the half-Skyrmion crystal has the right symmetries and is motivated by an exact solution for a crystal in the 2-dimensional $O(3)$ sigma model.

Chapter 7

# Rigid-Body Skyrmion Quantization

Here, we apply rigid-body quantization to Skyrmions, starting with its general formulation, and then turning to examples with baryon numbers up to $B = 7$. We will see later that for this limited range of baryon numbers the Skyrmions and their quantum states are qualitatively the same whether the pions are massless or have their physical mass.

In rigid-body quantization, the dynamical coordinates of the Skyrmions – their collective coordinates – all arise from the action of the symmetry group of Skyrme theory. States are classified by their spin and isospin, and as before we ignore the centre of mass motion and spatial momentum. The spin and isospin are constrained by the intrinsic symmetry of the particular Skyrmion being quantized, and much of this chapter is concerned with solving these constraints – for further details see refs.[29, 254]. Rigid-body quantization has its limitations, particularly for nuclei with small binding energies like the deuteron, and we will need to consider both here, and more so in later chapters, the role of collective, vibrational excitations of Skyrmions.

## 7.1 Collective Coordinate Quantization

Consider a static Skyrmion $U_0(\mathbf{x})$ with baryon number $B$ centred at the origin. By acting with the product of the rotation group $SO(3)$ and the isorotation group $SO(3)$, we generate a set of related Skyrmions, all with the same energy. It is convenient to think of rotations and isorotations as given by $SU(2)$ matrices $A$ and $A^{\text{iso}}$. To make the rotations explicit we use the $SO(3)$ matrix $M(A)_{ij} = \frac{1}{2}\text{Tr}(\tau_i A \tau_j A^{-1})$, which is the same for $A$ and $-A$. The action of $A$ on $U_0(\mathbf{x})$ is then to rotate $x^i$ to $M(A)_{ij}x^j$, or in vector notation $\mathbf{x}$ to $M(A)\mathbf{x}$. The action of $A^{\text{iso}}$ is to replace $U_0(\mathbf{x})$ by

$A^{\text{iso}} U_0(\mathbf{x}) (A^{\text{iso}})^{-1}$. As $A^{\text{iso}}$ and $-A^{\text{iso}}$ have the same effect, this is also an $SO(3)$ action.

In classical rigid-body dynamics, the parameters $A(t)$ and $A^{\text{iso}}(t)$ are the dynamical variables. The time-dependent field is

$$U(t,\mathbf{x}) = A^{\text{iso}}(t) \, U_0(M(A(t))\mathbf{x}) \, (A^{\text{iso}}(t))^{-1} \quad (7.1)$$

and the Skyrmion acquires 3-vector angular velocities $\boldsymbol{\omega}$ in physical space and $\boldsymbol{\omega}^{\text{iso}}$ in isospace whose components are, respectively,

$$\omega_j = -i\,\text{Tr}(\tau_j \dot{A} A^{-1}), \quad \omega_j^{\text{iso}} = i\,\text{Tr}(\tau_j (A^{\text{iso}})^{-1} \dot{A}^{\text{iso}}). \quad (7.2)$$

From the Skyrme theory Lagrangian, we can derive a reduced, rigid-body Lagrangian quadratic in these angular velocities, of the general form

$$L = \frac{1}{2}\omega_i^{\text{iso}} U_{ij} \omega_j^{\text{iso}} - \omega_i^{\text{iso}} W_{ij} \omega_j + \frac{1}{2}\omega_i V_{ij} \omega_j. \quad (7.3)$$

$U_{ij}$ is the inertia tensor for isorotations, $V_{ij}$ is the more familiar inertia tensor for spatial rotations, and $W_{ij}$ is the tensor that couples them. These are spatial integrals of certain densities that depend on the detailed shape of the Skyrmion field $U_0(\mathbf{x})$. Explicitly [51],

$$U_{ij} = -\int \text{Tr}\left(T_i T_j + \frac{1}{4}[R_k, T_i][R_k, T_j]\right) d^3x,$$

$$V_{ij} = -\int \epsilon_{ilm}\,\epsilon_{jnp}\,x_l x_n \, \text{Tr}\left(R_m R_p + \frac{1}{4}[R_k, R_m][R_k, R_p]\right) d^3x,$$

$$W_{ij} = \int \epsilon_{jlm}\,x_l\,\text{Tr}\left(T_i R_m + \frac{1}{4}[R_k, T_i][R_k, R_m]\right) d^3x, \quad (7.4)$$

where $R_i = \partial_i U_0\, U_0^{-1}$ is the $su(2)$ current that appeared before in the integrals for the energy and baryon number, and

$$T_i = \frac{i}{2}[\tau_i, U_0]\, U_0^{-1} \quad (7.5)$$

is a further $su(2)$ current. The rigid-body Lagrangian $L$ is purely kinetic, because the potential energy is a constant, unaffected by the action of rotations and isorotations. There is no explicit dependence on the pion mass parameter $m$, but it implicitly affects the detailed form of $U_0(\mathbf{x})$ and the values of the inertia tensors. When calculating these tensors numerically, it is convenient to use the expressions for the integrands in terms of the sigma and pion fields, given in ref.[163].

The tensors greatly simplify if the Skyrmion has a substantial amount of intrinsic symmetry. In particular, for the $B = 1$ hedgehog Skyrmion with profile $f(r)$, the $O(3)$ symmetry implies that $U_{ij} = V_{ij} = W_{ij} = \lambda \delta_{ij}$, where

$$\lambda = \frac{16\pi}{3} \int_0^\infty \sin^2 f \left(1 + f'^2 + \frac{\sin^2 f}{r^2}\right) r^2 \, dr. \tag{7.6}$$

Adkins, Nappi and Witten found that $\lambda = 107$ for massless pions [7], a value we used in Chapter 5.

The quantization of the Lagrangian (7.3) is similar to the quantization of rigid-body rotational motion of a molecule, as discussed by Landau and Lifshitz [160], for example. The quantized momenta corresponding to $\boldsymbol{\omega}$ and $\boldsymbol{\omega}^{\text{iso}}$ are the body-fixed spin and isospin angular momentum operators $\mathbf{L}$ and $\mathbf{K}$. These triplets of operators satisfy $su(2)$ commutation relations, and they mutually commute [51]. The rigid-body Hamiltonian is quadratic in these operators, having the general form

$$H = \frac{1}{2}(\mathbf{K}, \mathbf{L}) \Lambda^{-1} (\mathbf{K}, \mathbf{L})^{\mathrm{T}}, \tag{7.7}$$

where $\Lambda^{-1}$ is the inverse of the complete $6 \times 6$ inertia tensor

$$\Lambda = \begin{pmatrix} U & -W \\ -W & V \end{pmatrix}. \tag{7.8}$$

Generally, this is a complicated expression. Mathematically, one is quantizing a "molecule" acted on by two coupled $SO(3)$ rotation groups.

By a choice of orientation of the Skyrmion $U_0(\mathbf{x})$ in both space and isospace, one can align the body-fixed axes with the principal axes of inertia, ensuring that the inertia tensors $U_{ij}$ and $V_{ij}$ are diagonal. The Hamiltonian can still be complicated if the tensor $W_{ij}$ is non-zero. However, simplifications occur if the Skyrmion has, for example, dihedral symmetry $D_{3h}$ or a larger symmetry. Some of the diagonal elements of $U_{ij}$ and $V_{ij}$ are then equal, and $W_{ij}$ simplifies or is zero. We will sometimes neglect the tensor $W_{ij}$, because for $B \geq 4$ it is numerically small. A more accurate treatment allowing for non-zero $W_{ij}$ is also possible [29]. When $W_{ij}$ is neglected, the spin and isospin terms in the Hamiltonian separate, and just involve the inverses of the inertia tensors $V_{ij}$ and $U_{ij}$, respectively. However, a link between spin and isospin remains, through the symmetry constraints on quantum states.

A basis for the Hilbert space of states is given by

$$|J, L_3\rangle \otimes |I, K_3\rangle \tag{7.9}$$

with $-J \leq L_3 \leq J$ and $-I \leq K_3 \leq I$. $J$ is the total spin quantum number and $I$ the isospin. $L_3$ and $K_3$ are body-fixed projections of spin and isospin.

As we explained in Section 5.3, the space-fixed projections can take any of their usual values, and are not shown. Concretely, these basis states are products of Wigner D-functions of the Euler angles parametrising $A$ and $A^{\text{iso}}$. The energy eigenstates we seek have definite $J$ and $I$ values.

After diagonalisation, the rotational part of the Hamiltonian is

$$H = \frac{1}{2}\left(\frac{L_1^2}{V_{11}} + \frac{L_2^2}{V_{22}} + \frac{L_3^2}{V_{33}}\right). \tag{7.10}$$

The simplest case is the symmetric top, where the inertia tensor is isotropic, so $V_{11} = V_{22} = V_{33}$. Then the Hamiltonian simplifies to

$$H = \frac{1}{2V_{33}}\mathbf{L}\cdot\mathbf{L}, \tag{7.11}$$

and a state $|J, L_3\rangle$ with spin $J$ and any value of $L_3$ in the allowed range has energy

$$E_J = \frac{1}{2V_{33}}J(J+1). \tag{7.12}$$

The next-simplest case is the asymmetric top, where $V_{11} = V_{22}$, and $V_{33}$ has a distinct value. Then (7.10) simplifies to

$$H = \frac{1}{2V_{11}}\mathbf{L}\cdot\mathbf{L} + \left(\frac{1}{2V_{33}} - \frac{1}{2V_{11}}\right)L_3^2. \tag{7.13}$$

As $L_3$ commutes with $\mathbf{L}\cdot\mathbf{L}$, the eigenstates of $H$ have definite values of $J$ and $L_3$, and have energy

$$E_{J,L_3} = \frac{1}{2V_{11}}J(J+1) + \left(\frac{1}{2V_{33}} - \frac{1}{2V_{11}}\right)L_3^2, \tag{7.14}$$

where the $L_3$ operator has been replaced by its eigenvalue. This dependence of the energy on $L_3$ will be significant in some of the examples we consider later.

The intrinsic symmetry of a Skyrmion constrains its allowed states. Acting with an element of the symmetry group, the classical Skyrmion does not change at all, so the transformed quantum state can differ at most by a phase. For each symmetry element there is what is known as a *Finkelstein–Rubinstein (FR) constraint* [51, 156]. In practice it is sufficient to find and solve these for a set of generators of the symmetry group. The constraints imply that the allowed states are generally linear superpositions of the basis states (7.9), with restricted values of $L_3$ and $K_3$, and certain combinations of $J$ and $I$ may be forbidden.

Finkelstein–Rubinstein constraints generally involve rotations and isorotations simultaneously. Before considering such combinations, let us recall the simplest examples of how pure rotations act on basis states $|J, L_3\rangle$. The effect of a rotation by $\theta$ about the body-fixed $x^3$-axis is just a change of phase, resulting in $e^{i\theta L_3}|J, L_3\rangle$. The effect of a rotation by $\pi$ about the body-fixed $x^1$-axis is

$$e^{i\pi L_1}|J, L_3\rangle = (-1)^J |J, -L_3\rangle, \qquad (7.15)$$

and an important case of this is

$$e^{i\pi L_1}|J, 0\rangle = (-1)^J |J, 0\rangle. \qquad (7.16)$$

Pure isorotations act similarly on the basis states $|I, K_3\rangle$.

Next, recall that we need to carry out a fermionic quantization, because individual nucleons are spin $\frac{1}{2}$ fermions. That means that under a $2\pi$ rotation or isorotation of a $B = 1$ Skyrmion, the wavefunction changes sign. In rigid-body quantization of a higher-$B$ Skyrmion, one cannot easily isolate a $B = 1$ Skyrmion, but there is still the constraint that under a $2\pi$ rotation or isorotation the wavefunction changes sign if $B$ is odd, but not if $B$ is even. So spin and isospin are half-integer if $B$ is odd and integer if $B$ is even. More profoundly, because $\pi_1(\mathcal{C}_B) = \mathbb{Z}_2$ for each sector $\mathcal{C}_B$ of the Skyrme field configuration space, i.e. because the simply-connected covering space is a double cover, the phase in an FR constraint equation can only be a sign factor $\chi_{\text{FR}} = \pm 1$. This sign then depends on whether the loop associated with the symmetry element is contractible or not, and for its determination, it is very helpful to have available the approximation for the Skyrmion using the rational map ansatz, as we clarify next.

## 7.2 Rational Maps and Finkelstein–Rubinstein Signs

We assume here that a Skyrmion with baryon number $B$ is approximated by the rational map ansatz with rational map $R(z)$, and that the Skyrmion and rational map have the same intrinsic symmetry group $K$. Suppose one of the symmetries of $R(z)$ is

$$R(k(z)) = \mu_k(R(z)) \qquad (7.17)$$

for some particular $SU(2)$ Möbius transformations $k$ and $\mu_k$, where $k$ represents a spatial rotation by $\theta$ about the body-fixed axis $\mathbf{n}$ and $\mu_k$ represents an isospace rotation by $\theta^{\text{iso}}$ about the body-fixed axis $\mathbf{n}^{\text{iso}}$. For $\theta$ and $\theta^{\text{iso}}$ unequal to $2\pi$ times an integer, $k$ leaves fixed only the antipodal points $z_{\pm \mathbf{n}}$ corresponding to $\mathbf{n}$,

$$z_{\mathbf{n}} = \frac{\mathbf{n}^1 + i\mathbf{n}^2}{1 + \mathbf{n}^3} \quad \text{and} \quad z_{-\mathbf{n}} = -\frac{\mathbf{n}^1 + i\mathbf{n}^2}{1 - \mathbf{n}^3}, \qquad (7.18)$$

and similarly, $\mu_k$ leaves fixed only $R_{\pm \mathbf{n}^{\mathrm{iso}}}$, where

$$R_{\mathbf{n}^{\mathrm{iso}}} = \frac{(\mathbf{n}^{\mathrm{iso}})^1 + i(\mathbf{n}^{\mathrm{iso}})^2}{1 + (\mathbf{n}^{\mathrm{iso}})^3} \quad \text{and} \quad R_{-\mathbf{n}^{\mathrm{iso}}} = -\frac{(\mathbf{n}^{\mathrm{iso}})^1 + i(\mathbf{n}^{\mathrm{iso}})^2}{1 - (\mathbf{n}^{\mathrm{iso}})^3}. \quad (7.19)$$

Choosing $z = z_\mathbf{n}$, for which $k(z) = z$, we see that eq.(7.17) requires

$$R(z_\mathbf{n}) = \mu_k(R(z_\mathbf{n})), \quad (7.20)$$

so $R(z_\mathbf{n}) = R_{\mathbf{n}^{\mathrm{iso}}}$ or $R_{-\mathbf{n}^{\mathrm{iso}}}$, one of the fixed points of $\mu_k$. Then, by reversing the signs of both $\mathbf{n}^{\mathrm{iso}}$ and $\theta^{\mathrm{iso}}$ if necessary, which doesn't change $\mu_k$, we may arrange that

$$R(z_\mathbf{n}) = R_{\mathbf{n}^{\mathrm{iso}}}. \quad (7.21)$$

This condition clarifies the relative signs of $\theta$ and $\theta^{\mathrm{iso}}$, and is exploited below.

Acting on the Skyrmion, the pair $(k, \mu_k)$ generates a loop in configuration space $\mathcal{C}_B$ if we let the rotation angle increase continuously from 0 to $\theta$, and simultaneously the isorotation angle increase continuously from 0 to $\theta^{\mathrm{iso}}$. The configurations at the beginning and end of this process are the same, hence it is a closed loop. This leads to the following FR constraint on the wavefunction:

$$e^{i\theta \mathbf{n} \cdot \mathbf{L}} e^{i\theta^{\mathrm{iso}} \mathbf{n}^{\mathrm{iso}} \cdot \mathbf{K}} |\Psi\rangle = \chi_{\mathrm{FR}} |\Psi\rangle, \quad (7.22)$$

where $\mathbf{L}$ and $\mathbf{K}$ are the body-fixed spin and isospin operators, and the Finkelstein–Rubinstein (FR) sign factor,

$$\chi_{\mathrm{FR}} = \begin{cases} +1 \text{ if the symmetry-generated loop is contractible,} \\ -1 \text{ if the symmetry-generated loop is non-contractible,} \end{cases} \quad (7.23)$$

enforces the fermionic quantization condition.

It can be verified that the assignment of FR signs to elements of the intrinsic symmetry group $K$ is a homomorphism, i.e. a $\mathbb{Z}_2$-valued representation. Sometimes, this information is enough to imply that all FR signs are $+1$. Even if not, it implies that it is sufficient to impose the FR constraints for a set of generators of $K$. Then automatically the constraints will be satisfied for the whole group. This observation saves much work when one investigates particular examples.

The topology of the space of rational maps of degree $B$ is known [213]. Zeros and poles of rational maps cannot coincide (otherwise the degree decreases) and roughly speaking, the first homotopy group is generated by a simple zero encircling a simple pole once. Using this, the following general result was proved by Krusch [156]: The value of the FR sign $\chi_{\mathrm{FR}}$ for a given symmetry element of a rational map only depends on the rotation angle $\theta$ and the isorotation angle $\theta^{\text{iso}}$, where the angles are defined such that (7.21) holds, and is given by

$$\chi_{\text{FR}} = (-1)^{\mathcal{N}} \quad \text{where} \quad \mathcal{N} = \frac{B}{2\pi}(B\theta - \theta^{\text{iso}}). \tag{7.24}$$

The proof relies on tracking the motions of the zeros and poles of the rational map as the angles increase from zero.

Evaluating $\mathcal{N}$ using this formula is generally straightforward, and one doesn't have to be too careful about the side condition (7.21) determining the relative sign of the rotation angles. This is because $\mathcal{N}$ is often an integer for one choice but not the other. However, in some cases, and in particular where a rotation angle could be $\pi$ or $-\pi$, care is needed.

Rotations and isorotations by $2\pi$ are separately symmetries for any axis $\mathbf{n}$, and using formula (7.24) we recover the conditions

$$e^{2\pi i \mathbf{n} \cdot \mathbf{L}} |\Psi\rangle = e^{2\pi i \mathbf{n} \cdot \mathbf{K}} |\Psi\rangle = (-1)^B |\Psi\rangle. \tag{7.25}$$

Therefore for even $B$ the spin $J$ and isospin $I$ are integers, and for odd $B$ they are half-integers, as we stated before.

For general $B > 1$, the moments of inertia are larger for rotations than for isorotations. Since these appear in the denominator of the quantum Hamiltonian, spin excitations of Skyrmions require less energy than isospin excitations. The quantum states of lowest energy are therefore those with minimal isospin, and a range of spins. In particular, for even $B$ the lowest-energy state has zero isospin but (because of the FR constraints) not necessarily zero spin. For odd $B$ the lowest-energy state has isospin $\frac{1}{2}$. These observations match what is seen experimentally for numerous nuclei up to baryon number $B$ near 40.

## 7.3 Parity of States

The Skyrme Lagrangian is invariant under spatial inversion $\mathbf{x} \mapsto -\mathbf{x}$. It is also invariant under an inversion in isospace $\boldsymbol{\pi} \mapsto -\boldsymbol{\pi}$, which sends $U$ to $U^{-1}$. These operations preserve the energy but each separately reverses the sign of the baryon number. So the useful version of inversion in Skyrme theory is the combination

$$\mathscr{I} : \mathbf{x} \mapsto -\mathbf{x} \quad \text{and} \quad \boldsymbol{\pi} \mapsto -\boldsymbol{\pi}, \tag{7.26}$$

which takes a classical Skyrme field configuration to a configuration that is generically different, but has the same baryon number. This operation also preserves the energy, so it commutes with the Hamiltonian in the quantum theory. Note that the square of this operation is the identity.

Suppose a classical Skyrmion has an intrinsic reflection symmetry or some other improper rotational symmetry, meaning that it is invariant under a combination of improper rotations in space and isospace. (We saw earlier that a pure reflection in space must occur combined with a pure reflection in isospace.) Let us denote this combined symmetry operation by $\mathcal{O}$. The inversion operation $\mathcal{I}$ is then the product of $\mathcal{O}$ with a combination $\mathcal{R}$ of proper $SO(3)$ rotations in space and isospace whose form can be found from the classical Skyrmion or from its rational map approximation.

We now lift these geometrical operations to operators acting on quantum states. The lift of $\mathcal{I}$ is the parity operator $\mathcal{P}$. Because $\mathcal{P}$ commutes with the Hamiltonian, and its square is the identity, it acts as $\pm 1$ on energy eigenstates. The eigenvalue $\pm 1$, usually shortened to simply $\pm$ or positive/negative, is called the parity and denoted $P$. As $\mathcal{O}$ is a symmetry leaving the Skyrmion unchanged, we can assume the Skyrmion's quantum state is unchanged by $\mathcal{O}$. The effect of $\mathcal{P}$ on the quantum state is therefore the same as the effect of the operator $\mathcal{R}$, which only involves rotations. For a Skyrmion state expressed in terms of the basis states (7.9), the parity is then straightforwardly evaluated as the eigenvalue of $\mathcal{R}$. It generally differs from state to state, depending on the spin and isospin, and on their body-fixed projections. The spin/parity is denoted $J^P$. The isospin is usually clear from the context.

Occasionally a Skyrmion is invariant under $\mathcal{I}$. For example, the $B = 1$ hedgehog has this symmetry. The operation $\mathcal{R}$ is then trivial, with eigenvalue $+1$. Hedgehog quantization therefore correctly determines that the proton, neutron and delta resonances all have positive parity.

Sometimes there is an overall sign ambiguity in the parity. If so, this applies simultaneously to all rigid-body excitations based on a particular Skyrmion. The ambiguity can arise from the need to choose a particular operation $\mathcal{O}$, with two different choices leading to opposite parities because of a negative Finkelstein–Rubinstein sign associated with the combined rotations connecting these choices. In this case, the parity of one state (usually the ground state) needs to be fixed by comparison with published data. There is never an ambiguity in the relative parities.

If a Skyrmion has no symmetry at all under combined improper rotations, then $\mathcal{I}$ creates a new Skyrmion, unrelated by any rotations to the original one. The orbits of these two Skyrmions under space and isospace rotations are distinct, and there is parity doubling of quantum states. Positive and negative parity states with the same spin then occur in pairs. If the orbits are far apart in field configuration space, these parity doubles will have a very small energy splitting, but if they are close together, then the energy splitting will be larger. The splitting depends on how the orbits are connected in field configuration space via deformations of the Skyrmion, but this goes beyond rigid-body quantization.

## 7.4 The Quantized Toroidal $B = 2$ Skyrmion

In this and the following sections, we investigate a number of examples of rigid-body quantization of Skyrmions and determine the allowed quantum numbers of the ground and excited states. We make a few remarks about the excitation energies of the states, but detailed discussion of energies is deferred until we have considered the inertia tensors of Skyrmions with massive pions, in Chapter 9.

The $B = 2$ Skyrmion is axially symmetric and has a toroidal shape. It is shown in Fig. 7.1, coloured using the Runge colour sphere. In [51], Braaten and Carson performed the collective coordinate quantization of this Skyrmion (see also [154]). The symmetry group is $D_{\infty h}$. There is a continuous rotation symmetry about one axis, which we take to be the body-fixed $x^3$-axis. There is also a discrete 180° rotation symmetry that turns the Skyrmion over, and finally an inversion symmetry.

Fig. 7.1: $B = 2$ Skyrmion

The rational map ansatz for this Skyrmion uses $R(z) = z^2$. This has corresponding symmetries

$$R(e^{i\alpha}z) = e^{2i\alpha}R(z), \quad R\left(\frac{1}{z}\right) = \frac{1}{R(z)}, \qquad (7.27)$$

showing that a rotation by $\alpha$ about the $x^3$-axis is equivalent to an isorotation by $2\alpha$ about the 3-axis in isospace (so each colour on the equator occurs twice), and a 180° rotation about the $x^1$-axis is equivalent to a 180° rotation about the 1-axis in isospace. Because of the continuous axial symmetry, the $B = 2$ Skyrmion has only five collective coordinates for rotations and isorotations. The quantum states can still be expressed in terms of the standard basis states (7.9) depending on six Euler angles, but one combination of the angles doesn't appear. When translations are included, the space of $B = 2$ Skyrmion solutions is an 8-dimensional manifold.

The axial symmetry requires a rigid-body state to satisfy the constraint

$$e^{i\alpha L_3}e^{2i\alpha K_3}|\Psi\rangle = |\Psi\rangle. \qquad (7.28)$$

As $\alpha$ can be reduced continuously to zero, the FR sign is $+1$ here. By considering $\alpha$ infinitesimal, we see that this constraint simplifies to

$$(L_3 + 2K_3)|\Psi\rangle = 0. \qquad (7.29)$$

The constraint associated with the discrete 180° rotation is

$$e^{i\pi L_1}e^{i\pi K_1}|\Psi\rangle = -|\Psi\rangle. \qquad (7.30)$$

Here $\chi_{\text{FR}} = -1$, as follows from Krusch's formula (7.24), setting both angles to $\pi$ and $B = 2$.

One can also understand the last FR sign in another way. First, consider stretching the $B = 2$ torus along the $x^1$-axis so that it becomes a pair of separated $B = 1$ Skyrmions, with one rotated about the $x^3$-direction by 180° relative to the other. Next, recall that acting on a $B = 1$ Skyrmion, an isorotation is always equivalent to a spatial rotation by the same angle about some axis that depends on the Skyrmion's orientation. For one of the above pair of $B = 1$ Skyrmions, the simultaneous rotations (7.30) by 180° about the $x^1$-axis in space and the 1-axis in isospace cancel; for the other these rotations add, producing a net rotation by 360°. This explains the negative FR sign.

For quantum states with isospin 0, $K_3$ and $K_1$ act trivially so the constraint (7.29) requires that $L_3 = 0$. Then, because of the identity $e^{i\pi L_1}|J, 0\rangle = (-1)^J|J, 0\rangle$, the constraint (7.30) implies that $J$ must be odd. The ground state of the $B = 2$ Skyrmion is therefore the spin 1, isospin 0 state $|1,0\rangle \otimes |0,0\rangle$, which has the quantum numbers of the deuteron. For isospin 1, a similar argument shows that the state $|0,0\rangle \otimes |1,0\rangle$ with spin 0 is allowed. This has higher energy and may be identified with the isovector $^1S_0$ resonance state of the two-nucleon system. It is a success of Skyrme theory that a state with spin 0 and isospin 0 is forbidden. Such a state is not seen experimentally.

The rational map $R(z) = z^2$ has the improper symmetry

$$R\left(-\frac{1}{\bar{z}}\right) = \frac{1}{\overline{R(z)}} \tag{7.31}$$

so the $B = 2$ Skyrmion is invariant under spatial inversion combined with a reflection in the 3-axis in isospace. The combined inversion operation $\mathscr{I}$ is therefore equivalent to a rotation by 180° about the 3-axis in isospace, and the parity operator is $\mathcal{P} = e^{i\pi K_3}$. States with even $K_3$ have positive parity and those with odd $K_3$ have negative parity. The physical states mentioned above both have positive parity.

These rigid-body quantum states based on the toroidal $B = 2$ Skyrmion are spatially compact, and in the Adkins–Nappi calibration they have energy more than 100 MeV below the energy of two well-separated, quantized $B = 1$ Skyrmions, representing a proton and a neutron. This very strong binding energy should be thought of as comparable to the depth of the classical potential well used in simple models of the deuteron [43]. Physically, the deuteron is not compact, but instead a loosely bound state of a proton and neutron, with a large spatial extent and a binding energy of just 2.2 MeV relative to the proton-neutron breakup threshold. The final binding energy is much less than 100 MeV because of the quantum mechanical energy associated with the dynamics of the proton-neutron separation.

Because of this problem with rigid-body Skyrmion quantization, it was suggested in [166] that the deuteron is better described as a quantum state on a 10-dimensional manifold of $B = 2$ Skyrme field configurations. This manifold includes the toroidal $B = 2$ Skyrmions of minimal energy (i.e. the 8-dimensional manifold just considered), and also deformations of these into spatially larger configurations that are approximately the product of two separated $B = 1$ Skyrmions in the most attractive relative orientation. This goes beyond rigid-body quantization. The ten coordinates of this manifold are the separation parameter, and the nine coordinates of the full symmetry group of Skyrme theory, because the separated configurations no longer have an axial symmetry. A unique bound quantum state is again found with the quantum numbers of the deuteron. After calibrating this model, the calculated deuteron binding energy is about 6 MeV, closer to the experimentally determined binding energy.

In this model there is also a state with isospin 1 and spin 0, corresponding to the isovector resonance in the two-nucleon system. Its calculated energy is about 10 MeV above the deuteron energy – the right order of magnitude. The physical dineutron resonance, with $I_3 = -1$, is estimated to have an energy about 0.6 MeV above the energy of two free neutrons, and therefore about 3 MeV above the deuteron energy [66]. This is inferred from nucleon-nucleon scattering data.

A more refined model for these very weakly bound states would need to allow for free motion of two $B = 1$ Skyrmions, and not just motion in the attractive channel. This requires a 12-dimensional manifold of Skyrme field configurations. It has not yet been possible to construct such a manifold in any detail, despite some attempts [16, 237].

## 7.5 The Quantized $B = 3$ Skyrmion

The $B = 3$ Skyrmion has tetrahedral symmetry $T_d$, and was quantized as a rigid body by Carson [61]. The symmetry is realised in the same way as for the tetrahedral subgroup of the full $O(3)$ symmetry group of the $B = 1$ hedgehog. This is consistent with the asymptotic fields being those of a $B = -1$ hedgehog. The Skyrmion is shown in Fig. 7.2.

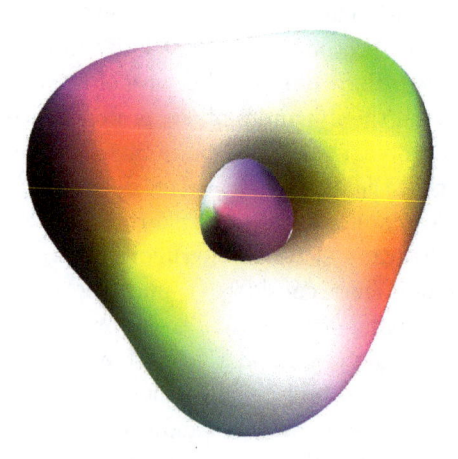

Fig. 7.2: $B = 3$ Skyrmion

The rotational symmetries are generated by a 180° and a 120° rotation, and the corresponding FR constraints are

$$e^{i\pi L_3}e^{i\pi K_3}|\Psi\rangle = |\Psi\rangle,$$
$$e^{\frac{2\pi i}{3\sqrt{3}}(L_1+L_2+L_3)}e^{\frac{2\pi i}{3\sqrt{3}}(K_1+K_2+K_3)}|\Psi\rangle = |\Psi\rangle. \tag{7.32}$$

Krusch's formula shows that both FR signs are positive.

These constraints can be simplified by using the grand spin operators $\mathbf{M} = \mathbf{L} + \mathbf{K}$. States with grand spin 0 are all allowed, and there are unique states with $J = I = \frac{1}{2}$ and with $J = I = \frac{3}{2}$ having grand spin 0. The FR constraints require, more precisely, that $|\Psi\rangle$ is a tetrahedrally-symmetric state of grand spin, so the next allowed grand spin state is $|3, 2\rangle - |3, -2\rangle$, the unique possibility with grand spin $M = 3$. This can be constructed from basis states with $J = \frac{3}{2}$ and $I = \frac{3}{2}$. Rigid-body quantization of the $B = 3$ Skyrmion therefore gives in total a single state with spin and isospin $\frac{1}{2}$, and a pair of states with spin and isospin $\frac{3}{2}$. The appearance of the combination $|3, 2\rangle - |3, -2\rangle$ is not surprising; it also occurs if one takes the tetrahedrally-symmetric function of Cartesian coordinates $xyz$ and expresses it in terms of spherical harmonics.

In the quantum Hamiltonian, the $B = 3$ Skyrmion behaves as a pair of spherical rotors in space and isospace coupled together, so the total energy depends on spin and isospin through positive multiples of $J(J + 1)$ and $I(I + 1)$, and additionally depends on the grand spin through a negative multiple of $M(M + 1)$ [61].

Inversion is not a symmetry of the $B = 3$ Skyrmion, but because of the Skyrmion's reflection symmetry, inversion is equivalent to the 90° rotation in space and isospace that takes the tetrahedron to its dual. The parity operator is therefore $\mathcal{P} = e^{i\frac{\pi}{2}L_3}e^{i\frac{\pi}{2}K_3}$, a 90° rotation in grand spin space. The parity is positive for both states with grand spin $M = 0$, and negative for the state with grand spin $M = 3$ (because a 90° rotation about a Cartesian axis sends $xyz$ to $-xyz$).

The $J = I = \frac{1}{2}$ state has the lowest energy. Physically, this is identified with the isospin doublet of the nuclei Helium-3/Hydrogen-3 in their ground states. The prediction of spin $\frac{1}{2}$ is correct. (The Hydrogen-3 nucleus, with one proton and two neutrons, is called a triton, and is the nucleus of a tritium atom. A triton beta decays to Helium-3 with a lifetime of just over 12 years. This lifetime makes tritium usable in engineering applications like fusion reactors, but dangerous and difficult to handle. Helium-3, like a deuteron, is stable.) Of the states with $J = I = \frac{3}{2}$, the $M = 3$ state with negative parity is the lower of the two. This would most likely be seen in the three-neutron system, rather than as an excited state of Helium-3 or Hydrogen-3, where the excitation energy would be well above the 8 MeV breakup threshold for either nucleus. Various experiments have searched for bound states or resonances in the three-neutron system, but there is no confirmed observed state [77]. Theoretically, no bound state or resonance is expected in the three-neutron system, but the extra three-body binding needed for a $J^P = \frac{3}{2}^-$ bound state is significantly less than for a $J^P = \frac{3}{2}^+$ state [164], matching the energy ordering of the Skyrmion states.

## 7.6 The $B = 4$ Skyrmion and the $\alpha$-Particle

The $B = 4$ Skyrmion is shown in Fig. 7.3. The Runge colouring is particularly well adapted to its octahedral symmetry. This Skyrmion was first quantized by Walhout [239, 240], using the numerically obtained solution and its $O_h$ symmetries.

Fig. 7.3: $B = 4$ Skyrmion

The rational map ansatz for this Skyrmion uses the unique map of degree 4 with octahedral symmetry,

$$R(z) = \frac{z^4 + 2\sqrt{3}iz^2 + 1}{z^4 - 2\sqrt{3}iz^2 + 1}. \qquad (7.33)$$

The FR constraints corresponding to generators of the symmetry group are
$$e^{i\frac{\pi}{2}L_3}e^{i\pi K_1}|\Psi\rangle = |\Psi\rangle,$$
$$e^{\frac{2\pi i}{3\sqrt{3}}(L_1+L_2+L_3)}e^{i\frac{2\pi}{3}K_3}|\Psi\rangle = |\Psi\rangle. \quad (7.34)$$

The first generator is a rotation by 90° about a face centre of the cube and the second is a rotation by 120° about a body diagonal, both accompanied by isorotations. The square of the first generator is a pure 180° rotational symmetry, unaccompanied by an isorotation, and this extends to a $D_2$ of pure rotational symmetries. The FR signs are all $+1$, something almost unavoidable whenever the baryon number is a multiple of 4. The ground state, in the basis introduced previously, is $|0,0\rangle \otimes |0,0\rangle$ which trivially satisfies the constraints. This state with spin 0 and isospin 0 may be identified with the $\alpha$-particle, the ground state of Helium-4.

The next allowed state with isospin 0 has spin 4, a large spin for an excited $\alpha$-particle. The state is [135, 239]
$$|\Psi\rangle = \left(|4,4\rangle + \sqrt{\frac{14}{5}}|4,0\rangle + |4,-4\rangle\right) \otimes |0,0\rangle. \quad (7.35)$$

The moment of inertia of the $B = 4$ Skyrmion is relatively small, and one finds that the excitation energy of this state is 40 – 50 MeV, with the precise value depending on the pion mass parameter. This is well above the energy for the complete breakup of a physical $\alpha$-particle into two protons and two neutrons, and such a high-energy, high-spin state of the $\alpha$-particle has not yet been observed. It is a success of Skyrme theory that the $B = 4$ Skyrmion's octahedral symmetry forbids rigid-body, isospin 0 states with spins 1, 2 or 3, as their relatively-low energies would not be compatible with the observed Helium-4 spectrum. A spin 2 state, for example, would have an excitation energy of about 15 MeV and would have been seen long ago, if it existed. There are a number of known excited states of an $\alpha$-particle with excitation energies in the range 20 – 30 MeV. These states can be modelled as rotational/vibrational excitations of the $B = 4$ Skyrmion. We will discuss them in Chapter 10.

The rational map (7.33) again has the improper symmetry (7.31), so $\mathcal{P} = e^{i\pi K_3}$. The $\alpha$-particle and its rigid-body, isospin 0 excitations therefore have positive parity. The lowest-energy allowed state with isospin 1 is the negative-parity, $J^P = 2^-$ state [135, 174]
$$|\Psi\rangle = \left(|2,2\rangle + \sqrt{2}i|2,0\rangle + |2,-2\rangle\right) \otimes |1,1\rangle$$
$$- \left(|2,2\rangle - \sqrt{2}i|2,0\rangle + |2,-2\rangle\right) \otimes |1,-1\rangle. \quad (7.36)$$

This matches the isotriplet combining the lowest-energy states in Lithium-4 and Hydrogen-4 with the lowest-energy isospin 1 state in Helium-4 (all these observed states are resonances). No rigid-body excitations with isospin 1 and spins 0 or 1 are allowed.

## 7.7  $B = 5$ and $B = 7$

The nuclei Lithium-5 and Helium-5 are regarded as paradigms for the shell model (independent-particle model) [43, 221]. A single nucleon orbits an inert, spin 0 Helium-4 core. As the S-wave orbitals are occupied, the single nucleon is in a P-wave state with orbital angular momentum 1 and negative parity, and because the nucleon has spin $\frac{1}{2}$ the total spin/parity of the nucleus is $J^P = \frac{3}{2}^-$ or $\frac{1}{2}^-$. A fundamental ingredient of the shell model is a strong *spin-orbit coupling* that favours the orbital angular momentum and spin of the nucleon being aligned, so the $\frac{3}{2}^-$ state has lowest energy. Therefore the ground states of Lithium-5/Helium-5 form an isospin doublet with $J^P = \frac{3}{2}^-$. In practice, these states are seen as resonances with energy about 0.6 MeV above the sum of the Helium-4 and nucleon masses. So the nucleon is very far from strongly bound. The expected higher-energy states with $J^P = \frac{1}{2}^-$ are seen as broader resonances [226].

The Skyrmion with $B = 5$ has $D_{2d}$ symmetry. When quantized as a rigid body, the lowest-energy states have spin $\frac{1}{2}$ and isospin $\frac{1}{2}$. This is an example where the parity has some ambiguity, but if the lowest-energy states are assigned negative parity then all states have negative parity. The rigid-body approach does not work here, because it does not give a $J^P = \frac{3}{2}^-$ ground state. What is needed is a model with a dynamical $B = 1$ Skyrmion orbiting a $B = 4$ core [107], although for the moment there is still no good understanding of $B = 5$ nuclei using Skyrme theory.

Quantization of the $B = 7$ Skyrmion is more interesting, and physically promising. Recall that the $B = 7$ Skyrmion has $Y_h$ symmetry and a dodecahedral shape, as shown in Fig. 7.4 (left). When quantized as a rigid body, the lowest-energy states have spin $\frac{7}{2}$ and isospin $\frac{1}{2}$. This result is quite tricky to establish, because of the complicated form of the FR constraints for the icosahedral group $Y_h$ [135], but it is directly related to the existence of the rational map $R(z)$ with $Y_h$ symmetry and degree 7. The reason is that a rational map can be regarded as a doublet of polynomials (numerator and denominator) that transforms as an isospin $\frac{1}{2}$ doublet under $SU(2)$ Möbius transformations of the target 2-sphere. Also, degree 7 polynomials of $z$ have eight coefficients and therefore transform under spatial $SU(2)$ Möbius transformations as an 8-dimensional representation of $SU(2)$, i.e. spin $\frac{7}{2}$. So the existence of the degree 7 map with $Y_h$ symmetry, where the spatial transformations of $z$ are compensated by mixing of the numerator and denominator, ensures the existence of a $Y_h$-symmetric quantum state with spin $\frac{7}{2}$ and isospin $\frac{1}{2}$, where rotations are compensated by isorotations [173].

Fig. 7.4: $B = 7$ Skyrmion and its deformation into clusters

The rational map for the $Y_h$-symmetric $B = 7$ Skyrmion is invariant under combined inversions in space and isospace. The parity operator therefore appears to be the identity operator, in which case all rigid-body quantum states would have positive parity. But here again there is an ambiguity, because we can extend the inversion by combining it with a spatial $2\pi$ rotation about any axis. Acting on quantum states, the $2\pi$ rotation has eigenvalue $-1$ because the baryon number is odd, so now all states have negative parity. This is the choice we make, and it is consistent with experimental observations, because all established states of nuclei having baryon number 7, with isospin both $\frac{1}{2}$ and $\frac{3}{2}$, have negative parity. In particular, rigid-body quantization leads to a $J^P = \frac{7}{2}^-$ ground state with isospin $\frac{1}{2}$.

Superficially, this disagrees with the observed $J^P = \frac{3}{2}^-$ for the isodoublet of Beryllium-7/Lithium-7 ground states. However, this problem has a resolution when one allows for a deformation of the Skyrmion into $B = 4$ and $B = 3$ clusters [113], as shown in Fig. 7.4 (right). The deformation requires exciting a vibrational mode, and because of the reduced symmetry, the ground state spin/parity can now be $\frac{3}{2}^-$. The energy gain from reducing the spin from $\frac{7}{2}$ to $\frac{3}{2}$ is substantial (because the moment of inertia is relatively small, and $\frac{7}{2}$ is a large spin for a small nucleus) and outweighs the vibrational energy cost. A prediction from this scenario is that there should be an excited $J^P = \frac{7}{2}^-$ state based on the undeformed Skyrmion, not far above the $\frac{3}{2}^-$ ground state. Such a state is indeed observed at a relatively low mean energy of 4.6 MeV in Beryllium-7/Lithium-7. We will describe the details of this model for $B = 7$ nuclei in Chapter 10, where deformations and vibrations of Skyrmions will be considered more extensively.

## 7.8 Quantization of the $B = 6$ Skyrmion

Nuclei with $B = 6$ are interesting as these are the smallest nuclei that have several observed excited states with energies just a few MeV above the ground state. Lithium-6, with isospin 0, is common in nature, though less common than Lithium-7. It may be used in the blanket of a fusion reactor, as it can become a source of tritium when bombarded with the high-energy products of the fusion, thus replenishing the fuel. The lowest triplet of isospin 1 states includes the Helium-6 ground state. Helium-6 is stable to nuclear breakup, but it beta decays to Lithium-6 with a lifetime of about 1 s. Experiments can be performed with radioactive beams of Helium-6.

The $B = 6$ Skyrmion, shown in Fig. 7.5, has $D_{4d}$ symmetry and is well approximated by the rational map ansatz. The optimised rational map, in a convenient orientation, is

$$R(z) = \frac{z^2(iaz^4 + 1)}{z^4 + ia}, \quad a = 0.16. \tag{7.37}$$

The $D_4$ symmetry group is generated by a 90° rotation about the $x^3$-axis combined with a 180° rotation about the 3-axis in isospace, and a 180° rotation about the $x^1$-axis combined with a 180° rotation about the 1-axis in isospace. The corresponding symmetries of the rational map are

$$R(iz) = -R(z), \quad R\left(\frac{1}{z}\right) = \frac{1}{R(z)}, \tag{7.38}$$

which lead to the FR constraints

$$e^{i\frac{\pi}{2}L_3}e^{i\pi K_3}|\Psi\rangle = |\Psi\rangle,$$
$$e^{i\pi L_1}e^{i\pi K_1}|\Psi\rangle = -|\Psi\rangle. \tag{7.39}$$

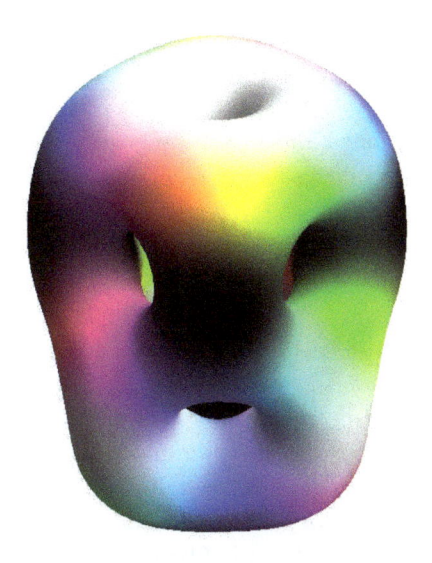

Fig. 7.5: $B = 6$ Skyrmion

These are rather similar to the FR constraints for the $B = 2$ toroidal Skyrmion, but the continuous axial symmetry has become discrete. These constraints can also be derived by deforming the $B = 6$ Skyrmion into a stack of three well-separated $B = 2$ tori [135].

The first constraint requires that $L_3 + 2K_3 = 0$ mod 4. The effect of the second can be determined by using the identity $e^{i\pi L_1}|J, L_3\rangle = (-1)^J|J, -L_3\rangle$, and the analogous identity for the operator $e^{i\pi K_1}$. A list of allowed states is given in Table 9.8 in Chapter 9. The list includes all the allowed states for the $B = 2$ torus, with $L_3 + 2K_3 = 0$, and further states with $L_3 + 2K_3 = \pm 4$.

In particular, the ground state is $|1, 0\rangle \otimes |0, 0\rangle$, a state with spin 1 and isospin 0. These are the quantum numbers of the Lithium-6 nucleus in its ground state. The next allowed state, also with isospin 0, is the spin 3 state $|3, 0\rangle \otimes |0, 0\rangle$, matching the first-excited state of Lithium-6, which is marginally unbound. The lowest state with isospin 1 is $|0, 0\rangle \otimes |1, 0\rangle$, the isotriplet consisting of the spin 0 ground states of Beryllium-6 and Helium-6, together with an excited spin 0 state of Lithium-6.

The Skyrmion and its rational map have reflection symmetries. One reflection plane contains the $x^3$-axis and is at an angle $\frac{\pi}{8}$ to the $x^1$-axis. This is slightly awkward to deal with, but one can show that the parity operator is the combined rotation operator

$$\mathcal{P} = e^{i\frac{\pi}{4}L_3} e^{-i\frac{\pi}{2}K_3} \, . \tag{7.40}$$

The three states mentioned above all have positive parity.

A state that is not allowed for the $B = 2$ torus but is theoretically allowed here is an isospin 0 state with spin 4. It is necessary that $|L_3| = 4$, so the parity is negative. The $B = 6$ Skyrmion is prolate, as it is a stack of three $B = 2$ tori, so states with a non-zero spin projection $L_3$ along the body-fixed 3-axis tend to have high energy. The ground state and first-excited state of Lithium-6 with $J^P = 1^+$ and $3^+$ are clearly seen, but a $4^-$ state with isospin 0 has not been observed.

This completes our discussion of quantized Skyrmions for now. For baryon numbers 8 and above we need an understanding of the novel shapes that classical Skyrmions acquire when the pion mass is at or close to its physical value, and not zero. This is the topic of the next chapter, and in Chapter 9 we will discuss the rigid-body quantization of a selection of these Skyrmions. Later, we will also allow for Skyrmion deformations and vibrations.

# Chapter 8

# Skyrmions with Higher $B$ – Massive Pions

## 8.1 Massive Pions

In Chapter 4 we explained how to include a pion mass term in Skyrme theory. For static fields, the energy including this mass term is

$$E = \int \left\{ -\frac{1}{2}\text{Tr}(R_i R_i) - \frac{1}{16}\text{Tr}([R_i, R_j][R_i, R_j]) + m^2\text{Tr}(1-U) \right\} d^3x, \quad (8.1)$$

and the vacuum is $U = 1$. The parameter $m$ is the pion mass in Skyrme units. Clearly, the energy of a Skyrmion with $m > 0$ will be slightly higher than with $m = 0$, because the pion mass term is positive for all fields. Motivated by the results in refs. [27, 33], we shall sometimes assume that $m = 1$. Alternatively, $m$ can be allowed to depend on the baryon number, to get better quantitative results for a range of nuclei. One finds that the optimal $m$ values remain fairly close to $m = 1$. The physical pion mass is kept fixed, so varying $m$ means varying the length scale. Skyrme theory should be calibrated to fit, if possible, the masses and sizes of a few nuclei like Carbon-12. (This approach means giving up the fit to the mass of the delta resonance, which is no great loss as the delta is broad, radiates pions rapidly, and is highly excited on the usual energy scale of nuclear states.)

One of the most important effects of the pion mass is to make the $B = 1$ Skyrmion, and also Skyrmions with higher $B$, exponentially localised. At large distances from the Skyrmion centre, the field $U$ approaches the vacuum exponentially fast, in contrast to the algebraic decay in the massless pion case. In particular, the equation for the hedgehog profile function,

$$(r^2 + 2\sin^2 f)f'' + 2rf' + \sin 2f \left( f'^2 - 1 - \frac{\sin^2 f}{r^2} \right) - m^2 r^2 \sin f = 0, \quad (8.2)$$

has the Yukawa-type dipole solution for large $r$,

$$f(r) \sim C \left( \frac{1}{r^2} + \frac{m}{r} \right) e^{-mr}. \quad (8.3)$$

More generally, Skyrmions have tail fields that can be interpreted as pion multipoles, and these tail fields have exponential decay.

Skyrmions for $B$ less than 8 are fairly insensitive to an increase of $m$ from 0 to 1, but it was a crucial discovery that for larger baryon numbers, starting with $B = 8$, increasing $m$ to 1 has an important qualitative effect [32,33,129]. The resulting Skyrmions do not have the hollow core structures derived from the rational map ansatz, as the hollow Skyrmions are no longer stable. This is not really surprising, as the field in the core is very close to the antivacuum $U = -1$ and here the pion mass term gives the field a maximal potential energy. Consequently, the solutions become unstable to squashing modes. The region where $U \approx -1$ tends to pinch off and separate into smaller subregions.

In this chapter we present many Skyrmion solutions with massive pions, concentrating on baryon numbers $B$ which are multiples of four. As the $B = 4$ Skyrmion is cubic and relatively tightly bound, it makes sense to glue copies of this Skyrmion together. These field configurations can then be relaxed to find true solutions. One finds a number of low-energy Skyrmions this way that are analogous to "molecules" of $\alpha$-particles in the $\alpha$-particle model of nuclei [43,55]. Recall that $\alpha$-particles are Helium-4 nuclei, having zero spin and isospin, and baryon number 4. The $\alpha$-particle model treats these as point particles subject to a phenomenological attractive potential, forming molecules in their quantized states. It has considerable success describing the ground and excited states of nuclei with baryon number a multiple of 4 and isospin 0 (i.e. equal, even numbers of protons and neutrons). Skyrme theory gives a new perspective on this old model.

The landscape of solutions is actually quite complicated as $B$ increases. For example, several stable or metastable Skyrmions with $10 \leq B \leq 16$ can be interpreted as clusters of $B = 4$ cubes and $B = 3$ tetrahedra in a planar layer [33]. However, for larger baryon numbers the global minima tend to be more compact and closer to spherical, while still partly composed of $B = 4$ subunits. They can also have $B = 3$ tetrahedra as constituents, as we will see.

## 8.2 The Double Rational Map Ansatz

One can intuitively understand the intrinsic symmetry of a Skyrme field configuration obtained by gluing $B = 4$ cubes together in various orientations. However, for a more precise control of the symmetry, it is very helpful to exploit a multi-layer generalisation of the rational map ansatz.

This allows one to fill in the hollow core of the configuration generated using a single rational map of high degree, and avoid the large region of antivacuum. For $m \approx 1$, many Skyrmions with higher baryon numbers can be obtained by relaxing fields constructed using the multi-layer ansatz. The construction has some use for $m = 0$ too. The simplest version is the double rational map ansatz with two layers [181]; this is useful for baryon numbers from $B = 12$ up to about $B = 40$. The three-layer extension is useful for larger baryon numbers including those around $B = 100$.

The double rational map ansatz exploits two rational maps, $R_1(z)$ in the inner layer and $R_2(z)$ in the outer layer, and a radial profile function $f(r)$ with boundary conditions $f(0) = 2\pi$ and $f(\infty) = 0$. It is assumed that $f$ decreases monotonically as $r$ increases, passing through $\pi$ at some radius $r_0$. The initial ansatz for the Skyrme field is eq.(6.6), as before, with the understanding that for $r < r_0$, $R(z) = R_1(z)$, and for $r > r_0$, $R(z) = R_2(z)$. $U = 1$ both at the origin and at spatial infinity, and $U = -1$ on the entire sphere $r = r_0$. The field is continuous at $r = r_0$. It is clear by adding the contributions of the inner and outer layers that the total baryon number is $B = B_1 + B_2$, the sum of the degrees of the maps $R_1$ and $R_2$. To find Skyrmion solutions, it helps if $R_1$ and $R_2$ share a substantial intrinsic symmetry. This symmetry is usually preserved, and sometimes enhanced, during numerical relaxation to a true Skyrmion.

Along a radial half-line from the origin through $r_0$ to infinity in the angular direction defined by $z$, $R_1(z)$ and $R_2(z)$ are generically different. Therefore, in the initial ansatz, the field value evolves radially from $U = 1$ along a meridian (half a great circle) in the target $S^3$ to the antivacuum $U = -1$, and then returns to $U = 1$ along a different meridian making an angle with the first. Only if $R_1(z) = R_2(z)$ do these meridians combine smoothly into a complete great circle. The picture for the relaxed Skyrmion solution is rather different. Along the radial half-lines where $R_1(z) = R_2(z)$, $U$ still passes through or close to $-1$, but along generic radial half-lines the field can smoothly take a short-cut, and avoid going near $U = -1$. Consequently, after relaxation, $U$ takes the antivacuum value $-1$ at discrete points only. These points are all at approximately the same distance from the origin and angularly located exactly where $R_1(z) = R_2(z)$ if there is a symmetry reason for this; otherwise they will be approximately at these angular locations. The radial derivative of the Skyrme field along the special half-lines tends to be higher than where $U$ takes a short-cut, so the field has higher energy density. There is a distorted version of a $B = 1$ Skyrmion here, centred where $U = -1$.

There is a nice consistency check on this geometrical picture. If $R_1 = p_1/q_1$ and $R_2 = p_2/q_2$, then the condition $R_1(z) = R_2(z)$ reduces to the polynomial equation

$$p_1(z)q_2(z) - p_2(z)q_1(z) = 0, \qquad (8.4)$$

which is an equation of degree $B$, the total baryon number. There are $B$ solutions, precisely the number expected on topological grounds, because a Skyrme field of baryon number $B$ must have at least $B$ locations where $U = -1$, by preimage counting.

A curious example combines the unique $Y_h$-symmetric map (6.39) of degree 7 as inner map $R_1(z)$ with the same map rotated through 90° as outer map $R_2(z)$. $R_2(z)$ is obtained by replacing $z$ by $iz$ in $R_1(z)$. The combination retains $T_d$ symmetry and after relaxing the field, the symmetry is enhanced to $O_h$. This way, one obtains a highly symmetric solution of the Skyrme field equation with baryon number 14, shown in Fig. 8.1. Here, the equation $R_1 = R_2$ can be reexpressed as $\mathcal{O}_v \mathcal{O}_f = 0$, where $\mathcal{O}_v(z)$ and $\mathcal{O}_f(z)$ are the octahedral Klein polynomials (6.26) and (6.27), so the 14 constituent $B = 1$ Skyrmions are on the six vertices and eight face centres of an octahedron. An alternative way to construct this solution combines an outer $O_h$-symmetric degree 13 map with the standard inner degree 1 map. Although this solution is interesting, it has rather high energy and is unfortunately not stable, at least when $m = 0$.

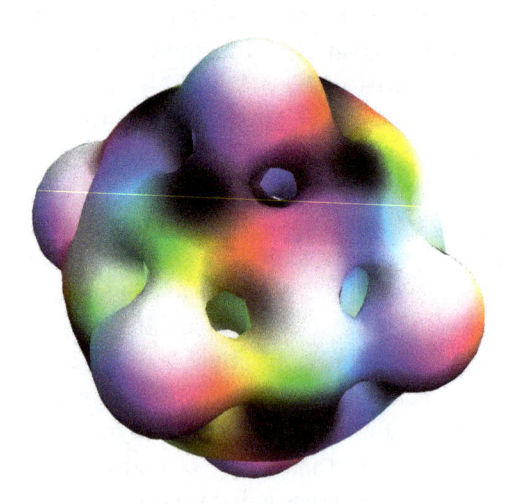

Fig. 8.1: $B = 14$ solution

More generally, the double rational map ansatz is not very helpful as a starting point for finding stable Skyrmions when $m = 0$, but is much more useful when $m \approx 1$, the physical case [85]. To increase the likelihood of the solution being stable it helps if the outer map has considerably higher degree than the inner map.

## 8.3 Skyrmions from $B = 8$ to $B = 32$

### 8.3.1 $B = 8$

For $m = 0$, the $B = 8$ Skyrmion is a hollow polyhedron with $D_{6h}$ symmetry. However, motivated by the $\alpha$-particle model, we expect that when $m \approx 1$, the lowest-energy solution will resemble a molecule of two cubic $B = 4$ Skyrmions.

The $B = 4$ Skyrmion has no asymptotic pion dipoles. Two of the pion field components have leading asymptotic quadrupoles and the remaining component has a leading octupole [177], so $B = 4$ cubes interact relatively weakly until they almost touch. Two $B = 4$ cubes placed in the same orientation and next to each other do attract, because of the quadrupole-quadrupole interaction, and there is an associated static solution of the Skyrme equation with $D_{4h}$ symmetry, but it is only a saddle point, not an energy minimum. Because of the significant octupole-octupole interaction, it is better to twist one cube by $90^0$ relative to the other around the line joining their centres [83]. The resulting stable Skyrmion also has $D_{4h}$ symmetry, and can be regarded as a diagonally stretched version of the $D_{6d}$-symmetric Skyrmion. Both the twisted and untwisted 2-cube solutions are shown in Fig. 8.2. The untwisted solution has energy $E/12\pi^2 B = 1.284$, but this decreases to $E/12\pi^2 B = 1.280$ for the twisted Skyrmion. For comparison, the $B = 4$ Skyrmion has energy $E/12\pi^2 B = 1.295$ when $m = 1$.

There is a simpler explanation of the attraction of the $B = 4$ cubes in the same or twisted orientation. The pion field component that dominates along the line joining them is the same for each cube, and has the same value (i.e. the same colour) on the face centres that are closest to touching. The gradient of the pion field along this line is therefore small, and the total energy decreases as the faces approach each other, until the repulsion due to other field components sets in. The faces can be interpreted as having a pion charge, and like charges attract. The slight preference for the 90° twist is because the most stable solution is where the field values on the closest corners of the nearly touching cubes match, rather than those on

Fig. 8.2: Left: $B = 8$ Skyrmion with twisted cubes; Right: $B = 8$ solution with untwisted cubes

the closest edges.

We will see later that the lowest-energy rotational quantum state for the twisted 2-cube Skyrmion has spin 0 and isospin 0, consistent with the quantum numbers of Beryllium-8. This fits the physical picture that Beryllium-8, although not a stable nucleus, is an almost bound state of two $\alpha$-particles. The spectrum of Beryllium-8 resonances is well enough known that this nucleus can be treated as a molecule of two $\alpha$-particles in an attractive potential, with one $\alpha - \alpha$ bond that is not quite strong enough to produce a bound state. The $B = 8$ Skyrmion should be thought of as the classical solution corresponding to the minimum of this potential. The classical energy required to break it into two well-separated $B = 4$ clusters is small. When quantum mechanical kinetic effects are included, the molecule marginally unbinds.

The $D_{6d}$-symmetric $B = 8$ Skyrmion also survives for $m \approx 1$, becoming a little squashed in the direction of the main symmetry axis. For $m = 1$, its energy is $E/12\pi^2 B = 1.290$, slightly exceeding the energies of both $D_{4h}$-symmetric solutions. Note that all the $B = 8$ Skyrmions we have discussed have 14 outer holes in their energy and baryon densities, as expected from the zeros of the Wronskian of a rational map with degree 8. These holes persist even when a Skyrmion is squashed or stretched.

### 8.3.2 $B = 12$

In the $\alpha$-particle model, the classical minima of the potential energy for three and four $\alpha$-particles occur, respectively, for an equilateral triangle with three $\alpha - \alpha$ bonds and a regular tetrahedron with six $\alpha - \alpha$ bonds [43, 55, 255]. These clusters model Carbon-12 and Oxygen-16. Skyrmion analogues of these configurations also exist.

There is a $B = 12$ Skyrmion with the $D_{3h}$ symmetry of an equilateral triangle, composed of three $B = 4$ cubes. It is shown in Fig. 8.3. Each cube is related to its neighbour through a spatial rotation by 120° combined with an isorotation by 120°. The isorotation cyclically permutes the values of the pion fields on the faces of the cube, so that these values match on the (nearly) touching faces. Each pair of cubes therefore attracts and forms a bent version of the 90°-twisted $B = 8$ solution described above.

Fig. 8.3: $B = 12$ Skyrmion with $D_{3h}$ symmetry

This arrangement of three cubes can also be viewed as a $B = 11$ Skyrmion (see Fig. 6.3) with a $B = 1$ Skyrmion inside. Such a field configuration can be constructed with exact $D_{3h}$ symmetry using the double rational map ansatz. This involves a $D_{3h}$-symmetric outer map $R_2$ of degree 11, and the standard spherically-symmetric degree 1 inner map $R_1$. Explicitly the maps are (as in ref.[28], but differently oriented)

$$R_2(z) = \frac{z^9 + az^6 + bz^3 + c}{z^2(cz^9 + bz^6 + az^3 + 1)}, \tag{8.5}$$

$$R_1(z) = z, \tag{8.6}$$

where the optimised coefficients are $a = -2.47$, $b = -0.84$ and $c = -0.13$. Note that the orientation of the outer map has to be chosen so that its $D_{3h}$ symmetry is realised compatibly with the $D_{3h}$ subgroup of the full $O(3)$ symmetry group of the inner map. Numerical relaxation yields the solution modelling three $B = 4$ cubes shown in Fig. 8.3, retaining exact $D_{3h}$ symmetry.

This $D_{3h}$-symmetric solution is actually slightly unstable. The central $B = 1$ Skyrmion prefers to merge into either the upper or lower triangular layer, breaking the up-down reflection symmetry and leaving only $C_{3v}$ symmetry. The energy is $E/12\pi^2 B = 1.277$. However, the central Skyrmion cannot escape further, so in a low-energy quantum state it will not be strongly localised in either layer and its mean position will be central. The intrinsic symmetry relevant for quantum states is therefore still $D_{3h}$. In models of Carbon-12 and its isobars (nuclei with the same baryon number, like Nitrogen-12) it is a convenient simplification to ignore this small instability, and to focus only on the degrees of freedom where the triangle of $B = 4$ Skyrmions rotates and isorotates rigidly, or possibly changes its overall shape.

Some of these shape changes are analogous to the rearrangements of $\alpha$-particles that model excited states of Carbon-12. An example is the Skyrmion analogue of the configuration of three $\alpha$-particles in a straight chain [28]. This Skyrmion is important for understanding the Hoyle state, the first-excited $J^P = 0^+$ state of Carbon-12 [131, 189]. It is a generalisation of the $B = 8$ Skyrmion and consists of a straight chain of three $B = 4$ cubes, with the middle cube twisted relative to the other two by 90° around the chain axis. This $B = 12$ Skyrmion is shown in Fig. 8.4. It has energy $E/12\pi^2 B = 1.275$, slightly less than that of the triangular Skyrmion. It is a member of a family of chain solutions made from any number of $B = 4$ cubes [123], with a 90° twist between each neighbouring pair. The limiting, infinitely-long chain has an enhanced 45° twist symmetry between neighbouring $B = 2$ subunits that is visible as an approximate symmetry in the $B = 12$ chain solution.

Note that the Hoyle state excitation energy is 7.65 MeV, less than 0.1% of the total energy of a Carbon-12 nucleus, and therefore smaller than the uncertainties in numerical energy computations for Skyrmions. Unlike in other nuclear models, in Skyrme theory it is hard to separate off the classical binding energies from the overall rest energies of Skyrmions. It is the total energy that is calculated numerically.

Between the triangular and chain Skyrmions there is a further $B = 12$

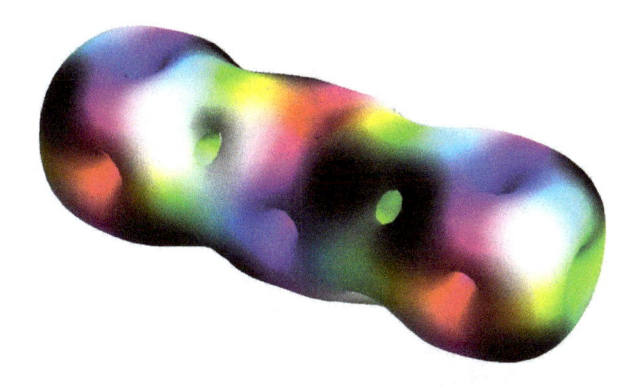

Fig. 8.4: $B = 12$ Skyrmion with $D_{4h}$ symmetry

solution – a slightly obtuse L-shaped configuration of three $B = 4$ cubes, where each touching pair has a relative 90° twist [85]. For this L-shaped solution, the neighbouring empty space is bounded by two faces of the same colour, so a further $B = 4$ cube cannot be inserted. In Chapters 9 and 10 we will discuss models of the quantum states of Carbon-12 that incorporate both the triangular and straight chain Skyrmions, and configurations like the L-shaped solution that interpolate between them.

### 8.3.3 $B = 16$

As one would anticipate from the $\alpha$-particle model, there is a $B = 16$ Skyrmion with tetrahedral symmetry that is an arrangement of four $B = 4$ cubic Skyrmions. It can be approximated within the double rational map ansatz by combining an outer map $R_2(z)$ of degree 12 and an inner map $R_1(z)$ of degree 4, with compatible symmetries. This construction fills the hollow $T_d$-symmetric $B = 12$ Skyrmion (stable for $m = 0$) with the $O_h$-symmetric $B = 4$ Skyrmion, giving $T_d$ symmetry overall. To write down these rational maps, and several others that will appear later, it is helpful to introduce the notation

$$p_{\pm}(z) = z^4 \pm 2\sqrt{3}\, iz^2 + 1. \tag{8.7}$$

These polynomials are the same as the tetrahedral Klein polynomials $\mathcal{T}_v(z)$ and $\mathcal{T}_f(z)$. Where convenient, we suppress the argument $z$. Then the optimised outer and inner maps are

$$R_2 = \frac{ap_+^3 + bp_-^3}{p_+^2 p_-}, \tag{8.8}$$

$$R_1 = \frac{p_+}{p_-}, \tag{8.9}$$

with the real coefficients $a = -0.53$ and $b = 0.78$. Numerically, the field relaxes to the $B = 16$ Skyrmion displayed in Fig. 8.5 (left), for which $U = -1$ at 16 points clustered into groups of four close to the centre of each $B = 4$ cube. The energy of this solution is $E/12\pi^2 B = 1.276$, and the four cubes are clearly visible and surprisingly distinct, in comparison to the earlier $B = 8$ and $B = 12$ solutions in which the cubes merge to a greater extent.

Fig. 8.5: $B = 16$ Skyrmions composed of four $B = 4$ cubes. Left: Tetrahedral arrangement; Middle: Bent square; Right: Flat square.

Similarly to the $B = 12$ case, this tetrahedral $B = 16$ solution is part of a rather flat potential-energy landscape of configurations of four $B = 4$ cubes. It is marginally favourable for the two cubes on a pair of opposite edges of the tetrahedron to open out, leading to the $D_{2d}$-symmetric Skyrmion shown in Fig. 8.5 (middle), with the lower energy $E/12\pi^2 B = 1.271$. This Skyrmion resembles a bent square (non-planar rhombus) having cubes on its vertices. A configuration similar to this bent square has been found as a low-energy state in an $\alpha$-cluster model, where it is termed a bent rhomb [35]. A stable tetrahedral solution would be preferable, because it has been shown in a version of the shell model that the closed shell structure of Oxygen-16, just slightly perturbed, is compatible with clustering into a tetrahedral arrangement of four $\alpha$-particles [195].

There is a further Skyrmion with slightly higher energy in which four $B = 4$ cubes, all with the same orientation, merge together to form the $D_{4h}$-symmetric, flat square solution shown in Fig. 8.5 (right). This has been obtained by using an initial configuration derived from the product ansatz for four cubes, and also by using a rational map ansatz with $D_{4h}$ symmetry. The solution has energy $E/12\pi^2 B = 1.272$, so one might expect that it is only a saddle point, having an unstable mode towards the bent square. However, perturbations of this solution have failed to excite such a mode, because the landscape is so flat.

In Chapter 11 we will discuss a model for the ground and excited states of Oxygen-16 that takes into account the degrees of freedom by which a tetrahedral $B = 16$ Skyrmion can deform through a bent square into a flat square (in three independent ways). The $J^P = 0^+$ ground state of Oxygen-16 and its excited states at 6.1 MeV and 10.4 MeV, with $J^P = 3^-$ and $4^+$, look convincingly like the rotational band of a tetrahedral intrinsic structure [70, 206], but deformations of the tetrahedral shape are needed to account for further states.

It is a general observation that a rearrangement of clusters often has just a small effect on the total energy of a Skyrmion, so as $B$ increases there is an ever-enlarging landscape of shapes with several local energy minima separated by relatively-low saddle points. The overall geometry of the landscape appears to be more important than the precise energies of the minima and saddle points. Selected directions in this landscape can be interpreted as linear or nonlinear vibrational modes of Skyrmions and their subclusters, and it is necessary to quantize some of these degrees of freedom to obtain a comprehensive understanding of nuclear spectra.

### 8.3.4  $B = 24$ and $B = 32$

In the Skyrmion crystal with massive pions, the half-Skyrmion symmetry is slightly broken. There is no longer a symmetry connecting locations where $U = -1$ and $U = 1$. The true unit cell remains a cube with four units of baryon number, and the field configuration in this unit cell becomes very similar to the isolated $B = 4$ Skyrmion. Many Skyrmions with large baryon numbers are crystal chunks, but at the same time they look like clusters of mostly $B = 4$ Skyrmions glued together, analogous to what is expected in the $\alpha$-particle model. $U = -1$ at the centre of each $B = 4$ cube.

Even for relatively small values of the pion mass parameter $m$, the minimal-energy $B = 32$ Skyrmion has the compact cubic structure obtained by placing eight $B = 4$ Skyrmions at the vertices of a larger cube. The orientations in space and isospace all need to be the same, to ensure that the pion fields on touching faces match. This Skyrmion can also be created by cutting out a cubic $B = 32$ chunk from the Skyrmion crystal [22].

Fig. 8.6: $B = 32$ Skyrmion

Another way to obtain the $B = 32$ Skyrmion is to start with an $O_h$-symmetric double rational map ansatz. A $B = 4$ cube can be placed inside a $B = 28$ hollow cube by using the maps

$$R_2 = \frac{p_+(ap_+^6 + bp_+^3 p_-^3 - p_-^6)}{p_-(p_+^6 - bp_+^3 p_-^3 - ap_-^6)}, \qquad (8.10)$$

$$R_1 = \frac{p_+}{p_-}, \qquad (8.11)$$

with the optimised coefficients $a = 0.33$ and $b = 1.64$, and $p_\pm$ as before. Numerical relaxation yields the solution shown in Fig. 8.6, with energy $E/12\pi^2 B = 1.266$. If the initial rational maps are perturbed to break the octahedral symmetry, the final result is unchanged, showing the local stability of this solution. Slicing this solution in half produces the $B = 16$ flat square shown in Fig. 8.5. Recall that a Skyrme field created using the single rational map ansatz with a degree $B$ map has a polyhedral energy and baryon density with $2B - 2$ holes. As is easily seen from Fig. 8.6, and even more clearly in the solution for $m = 0$ [22], the $B = 32$ crystal chunk has 54 exterior holes, corresponding to the outer degree 28 map. This motivated the above rational map construction using maps of degrees 28 and 4.

One can obtain a $B = 24$ solution, as a crystal chunk, by removing two $B = 4$ cubes from diagonally opposite corners of the $B = 32$ Skyrmion. This solution is a non-planar ring of six $B = 4$ cubes, all with the same orientation. However, because of the two missing corner cubes, the energy can be reduced by reorienting the six cubes so that each touching pair has a 90° relative twist around their separation axis. The new solution can be interpreted (in more than one way) as two L-shaped $B = 12$ Skyrmions glued together. The untwisted and twisted $B = 24$ Skyrmions have $E/12\pi^2 B = 1.273$ and $E/12\pi^2 B = 1.269$, respectively, and both are shown in Fig. 8.7. Another $B = 24$ Skyrmion with similar energy is obtained by stacking a pair of triangular $B = 12$ Skyrmions one on top of the other with colours matching, as shown in Fig. 8.8. The binding energy of these various configurations can be estimated by counting the number of short bonds between the $B = 4$ Skyrmions – the analogue of $\alpha$-particle bonds. The non-planar rings have six strong bonds, but the stack of $B = 12$ Skyrmions has nine bonds, some strong and some less so.

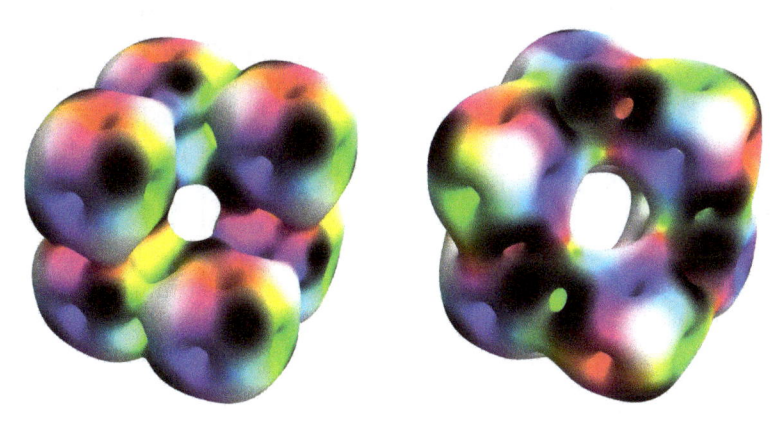

Fig. 8.7: Left: $B = 24$ crystal chunk Skyrmion; Right: $B = 24$ Skyrmion with twisted cubes

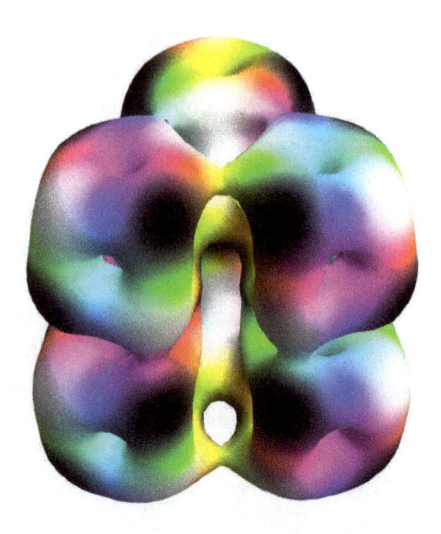

Fig. 8.8: Triangular $B = 24$ Skyrmion

## 8.4 Geometrical Construction of Rational Maps

Symmetric rational maps of low degree can be found using symmetry generators to fix the coefficients of the numerator and denominator. Symmetry may leave one or two coefficients undetermined, but these can be found numerically by minimising $\mathcal{I}$, the angular integral (6.16) that contributes to the Skyrme energy [130]. For degrees above about 20, however, this minimisation become intractible and a new approach is needed.

In particular, for Skyrmions with baryon numbers just below $B = 32$, having little symmetry, it is not practical to precisely optimise the many coefficients of an outer rational map with degree just below 28. Instead, to fix these, a geometrical construction of rational maps has been developed that distributes the exterior holes in the baryon density rather uniformly, as required for a Skyrmion crystal chunk [85]. Recall that the Skyrmion crystal, for massless pions, is a primitive cubic lattice of half-Skyrmions. The field values at lattice points of the crystal are $\sigma = \pm 1$ (i.e. $U = \pm 1$) in the orientation presented previously. However, to construct a crystal chunk with the correct boundary conditions, and no half-Skyrmion at the centre, we need $U = 1$ at infinity and $U = \pm 1$ at the centre. This requires a chiral reorientation of the crystal, and it is convenient to arrange that $\pi_3 = \pm 1$ at the half-Skyrmion centres.

In this orientation, a finite chunk of the crystal can be well approximated by the (multi-layer) rational map ansatz. The values $\pi_3 = \pm 1$ occur at angular locations corresponding to the zeros and poles of the rational map $R = p/q$, i.e. to the roots of the polynomials $p(z)$ and $q(z)$, together with radial locations where the profile function satisfies $\cos f(r) = 0$. This observation is the basis for a construction of crystal chunks with baryon numbers $B = 4n^3$, for any integer $n$. A key advantage of this approach is that the maps and their degrees can be manipulated to construct further Skyrmions with less symmetry and other baryon numbers.

Here is how the $B = 32$ Skyrmion is rederived. Fig. 8.9 shows the $4 \times 4 \times 4$ grid of half-Skyrmion locations in a cubic chunk of the crystal. The (black) circles and (blue) squares are used as the zeros and poles of the rational maps. The grid has two layers; the outer layer has a total of 56 points and the inner (hidden) layer has $2 \times 2 \times 2 = 8$ points. The corresponding rational maps have degrees 28 and 4.

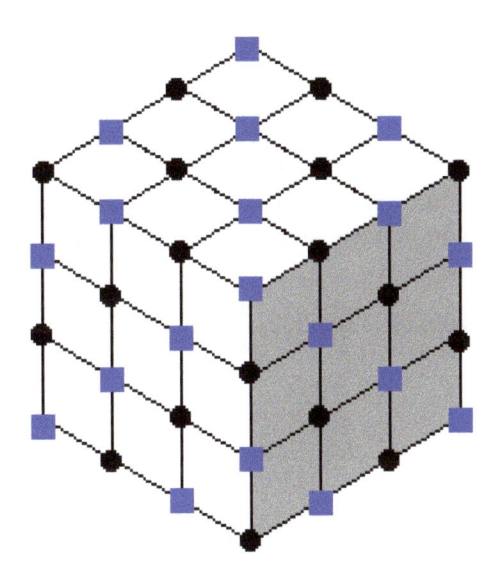

Fig. 8.9: $4 \times 4 \times 4$ cubic grid. The outer layer defines the zeros (black circles) and poles (blue squares) of a degree 28 rational map.

To find these maps explicitly, we introduce scaled Cartesian coordinates $(y^1, y^2, y^3)$ and radial coordinate $\rho = \sqrt{(y^1)^2 + (y^2)^2 + (y^3)^2}$, with the origin at the centre of the grid and the grid points having half-integer coordinates $\pm\frac{1}{2}, \pm\frac{3}{2}, \pm\frac{5}{2}, \ldots$. The inner layer has its eight points at $(\pm\frac{1}{2}, \pm\frac{1}{2}, \pm\frac{1}{2})$, with $\rho = \frac{1}{2}\sqrt{3}$. The outer corner points are at $(\pm\frac{3}{2}, \pm\frac{3}{2}, \pm\frac{3}{2})$, with $\rho = \frac{3}{2}\sqrt{3}$. Other points in the outer layer are at distances $\rho = \frac{1}{2}\sqrt{19}$ (points on edges, e.g. $(\frac{3}{2}, \frac{3}{2}, \frac{1}{2})$) and $\rho = \frac{1}{2}\sqrt{11}$ (interior face points, e.g. $(\frac{3}{2}, \frac{1}{2}, \frac{1}{2})$). The complex (stereographic) coordinate for any of these points is

$$z = \frac{y^1 + iy^2}{\rho + y^3}, \qquad (8.12)$$

a variant of the formula (6.2), and this specifies the angular location of a zero or pole of a rational map.

From the inner layer of points, one constructs the degree 4 map whose numerator $p(z)$ has roots at $z = \pm(1+i)/(\sqrt{3}+1), \pm(1-i)/(\sqrt{3}-1)$ and whose denominator $q(z)$ has roots at $z = \pm(1-i)/(\sqrt{3}+1), \pm(1+i)/(\sqrt{3}-1)$. This gives

$$R_1(z) = \frac{(z + \frac{1+i}{\sqrt{3}+1})(z - \frac{1+i}{\sqrt{3}+1})(z + \frac{1-i}{\sqrt{3}-1})(z - \frac{1-i}{\sqrt{3}-1})}{(z + \frac{1-i}{\sqrt{3}+1})(z - \frac{1-i}{\sqrt{3}+1})(z + \frac{1+i}{\sqrt{3}-1})(z - \frac{1+i}{\sqrt{3}-1})}$$

$$= \frac{z^4 + 2\sqrt{3}iz^2 + 1}{z^4 - 2\sqrt{3}iz^2 + 1}, \qquad (8.13)$$

the familiar map with octahedral symmetry used to generate the $B = 4$ Skyrmion.

A similar procedure gives the degree 28 rational map of the outer layer, whose numerator and denominator are expressed as products of their linear factors. The map is

$$R_2(z) = \frac{(z - \frac{3-i}{\sqrt{19}+3})(z - \frac{1-3i}{\sqrt{19}+3})(z - \frac{-3+i}{\sqrt{19}+3})(z - \frac{-1+3i}{\sqrt{19}+3})(z - \frac{3+3i}{\sqrt{19}-1})(z - \frac{-3+3i}{\sqrt{19}+1})}{(z - \frac{3+i}{\sqrt{19}+3})(z - \frac{1+3i}{\sqrt{19}+3})(z - \frac{-3-i}{\sqrt{19}+3})(z - \frac{-1-3i}{\sqrt{19}+3})(z - \frac{3-3i}{\sqrt{19}+1})(z - \frac{-3+3i}{\sqrt{19}-1})}$$

$$\times \frac{(z - \frac{-3-3i}{\sqrt{19}+1})(z - \frac{3-3i}{\sqrt{19}-1})(z - \frac{3+i}{\sqrt{19}-3})(z - \frac{-1-3i}{\sqrt{19}-3})(z - \frac{-3-i}{\sqrt{19}-3})(z - \frac{1+3i}{\sqrt{19}-3})}{(z - \frac{-3-3i}{\sqrt{19}+1})(z - \frac{3-3i}{\sqrt{19}-1})(z - \frac{3-i}{\sqrt{19}-3})(z - \frac{-1+3i}{\sqrt{19}-3})(z - \frac{-3+i}{\sqrt{19}-3})(z - \frac{1-3i}{\sqrt{19}-3})}$$

$$\times \frac{(z - \frac{1+i}{\sqrt{11}+3})(z - \frac{-1-i}{\sqrt{11}+3})(z - \frac{3+i}{\sqrt{11}+1})(z - \frac{3-i}{\sqrt{11}-1})(z - \frac{1+3i}{\sqrt{11}+1})(z - \frac{-1+3i}{\sqrt{11}-1})}{(z - \frac{1-i}{\sqrt{11}+3})(z - \frac{-1+i}{\sqrt{11}+3})(z - \frac{3+i}{\sqrt{11}-1})(z - \frac{3-i}{\sqrt{11}+1})(z - \frac{1+3i}{\sqrt{11}-1})(z - \frac{-1+3i}{\sqrt{11}+1})}$$

$$\times \frac{(z - \frac{-3-i}{\sqrt{11}+1})(z - \frac{-3+i}{\sqrt{11}-1})(z - \frac{-1-3i}{\sqrt{11}+1})(z - \frac{1-3i}{\sqrt{11}-1})(z - \frac{-1+i}{\sqrt{11}-3})(z - \frac{1-i}{\sqrt{11}-3})}{(z - \frac{-3-i}{\sqrt{11}-1})(z - \frac{-3+i}{\sqrt{11}+1})(z - \frac{-1-3i}{\sqrt{11}-1})(z - \frac{1-3i}{\sqrt{11}+1})(z - \frac{1+i}{\sqrt{11}-3})(z - \frac{-1-i}{\sqrt{11}-3})}$$

$$\times \frac{(z - \frac{-1-i}{\sqrt{3}+1})(z - \frac{1+i}{\sqrt{3}+1})(z - \frac{-1+i}{\sqrt{3}-1})(z - \frac{1-i}{\sqrt{3}-1})}{(z - \frac{1-i}{\sqrt{3}+1})(z - \frac{-1+i}{\sqrt{3}+1})(z - \frac{1+i}{\sqrt{3}-1})(z - \frac{-1-i}{\sqrt{3}-1})}. \qquad (8.14)$$

Requiring octahedral symmetry and matching to the inner map fixes the overall coefficient to be 1. The linear factors could be multiplied out; however, there is little reason to do this. The linear factor representation makes it easier to check the symmetry and to recall the coordinates of the half-Skyrmions. It also avoids problems with overflowing numerics.

The map $R_2$ is not optimal energetically in the sense of minimising $\mathcal{I}$, but it is close to optimal because the zeros and poles are evenly spread and rather well separated from each other. Using these maps of degrees 4 and 28 in the double rational map ansatz as initial data, one can successfully recover the cubic $B = 32$ Skyrmion by numerical relaxation.

### 8.4.1 $B = 24$ to $B = 31$ *solutions by corner cutting*

By cutting single Skyrmions from the corners of the $B = 32$ Skyrmion, it is fairly easy to generate new solutions with baryon numbers from $B = 31$ down to $B = 24$ [85, 178]. The corner cutting is performed on the outer degree 28 rational map (8.14); it is best understood in terms of the zeros and poles. The inner degree 4 map is unchanged.

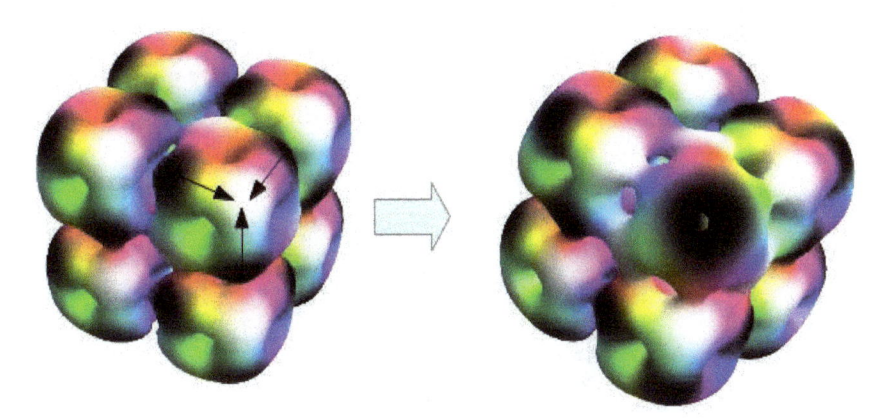

Fig. 8.10: $B = 31$ Skyrmion obtained by corner cutting

The simplest way to remove one unit of baryon number is to decrease the degree of the rational map by one by merging a zero and a pole. Linear factors cancel in the numerator $p(z)$ and denominator $q(z)$. Applied to the map (8.14), this method destroys all the symmetry, which is not desirable. Rather, to preserve as much symmetry as possible, three neighbouring zeros are moved simultaneously towards a corner pole, as shown in Fig. 8.10. The pole cancels against one of the zeros, leaving a double zero at the corner, and after numerical relaxation a new stable Skyrmion with $B = 31$ is obtained. Notice that what had been a white corner becomes a black one with a hole. The $O_h$ symmetry is broken down to $C_{3v}$. This corner-cutting procedure can be repeated up to eight times. At each corner, three poles are merged with one zero, or three zeros with one pole, and the result is either a double pole or a double zero. These persist in the Skyrmion solutions as holes in the baryon density at the cut-off corners.

Solutions from $B = 31$ to $B = 24$ have been obtained using this method. In the $B = 31$ Skyrmion, one of the $B = 4$ cubes becomes a slightly deformed $B = 3$ tetrahedral Skyrmion. The remaining seven cubes hardly change, because the interaction between Skyrmions near distinct corners is weak. When this corner cutting is repeated, further $B = 4$ cubes are replaced by $B = 3$ tetrahedra. In order to preserve as much symmetry as possible, for $B = 30$ a pair of diagonally opposite corners are cut and $D_{3d}$ symmetry remains. For $B = 29$, three corners, with each pair face-diagonally opposite, are cut. For $B = 28$, four corners forming a tetrahedron are removed and $T_d$ symmetry remains. One further corner is removed to generate the $B = 27$ Skyrmion.

Fig. 8.11: Unstable $B = 24$ solution

For $B = 26$, just two diagonally opposite corners are left uncut, but now the anticipated final $D_{3d}$-symmetric structure is not stable, and it relaxes to a $D_{2h}$-symmetric configuration having two similar $B = 13$ subclusters. Similarly, the relaxed $B = 25$ Skyrmion does not have the expected shape. The case $B = 24$ is interesting. Cutting all eight corners from the $B = 32$ Skyrmion produces the $O_h$-symmetric solution shown in Fig. 8.11 that is best thought of as six $B = 4$ cubes at the vertices of the octahedron dual to the original $B = 32$ cube, rather than eight $B = 3$ tetrahedra at the corners. However, this solution is again unstable. After further relaxation it approaches one of the stable $B = 24$ Skyrmions with less symmetry mentioned in Section 8.3.4.

Table 8.1 (at the end of this chapter) includes the energies of all the stable Skyrmions found using this corner-cutting technique. The energy per baryon increases as $B$ decreases from 32 to 27, but the instability and change of structure for $B < 27$ results in a discontinuity in this trend.

## 8.5 Rational Maps with $O_h$ and $T_d$ Symmetry

Skyrmions in the form of crystal chunks having either the full octahedral point symmetry $O_h$ of the crystal, or the slightly smaller tetrahedral symmetry $T_d$, are particularly interesting. Recall the tetrahedrally symmetric polynomials

$$p_\pm(z) = z^4 \pm 2\sqrt{3}\, iz^2 + 1. \tag{8.15}$$

Their ratio $R(z) = p_+(z)/p_-(z)$ is the degree 4, $O_h$-symmetric rational map of the $B = 4$ Skyrmion. The symmetry under a 90° rotation sends $z$ to $iz$, and hence $R(z)$ to $1/R(z)$. $p_+/p_-$ is the key ingredient in the geometrical construction of $O_h$- and $T_d$-symmetric rational maps. For example, the 56 points in the outer layer of the $4 \times 4 \times 4$ cubic grid shown in Fig. 8.9 can be separated into three subsets, those on the face interiors, those on the edges and those on the vertices. Each subset has $O_h$ symmetry.

The degree 4 rational map constructed from the vertex subset is $R_v = p_+/p_-$, and the degree 12 maps constructed from the face interior and edge subsets are

$$R_f = \frac{c_1 p_+^3 + p_-^3}{p_+^3 + c_1 p_-^3}, \quad R_e = \frac{c_2 p_+^3 + p_-^3}{p_+^3 + c_2 p_-^3}, \tag{8.16}$$

with $c_1 = -2.873$ and $c_2 = 0.178$. (Both $c_1$ and $c_2$ have analytic expressions, but these are complicated.) The degree 28 rational map (8.10) has the same form as the product

$$R_v \times R_f \times R_e = \frac{p_+}{p_-} \left( \frac{c_1 c_2\, p_+^6 + (c_1 + c_2) p_+^3 p_-^3 + p_-^6}{p_+^6 + (c_1 + c_2) p_+^3 p_-^3 + c_1 c_2\, p_-^6} \right). \tag{8.17}$$

Note that $O_h$ symmetry does not uniquely fix 12 zeros and 12 poles; there is a family of degree 12 rational maps with one real parameter $c$ having this symmetry,

$$R_{(c)} = \frac{c p_+^3 + p_-^3}{p_+^3 + c p_-^3}. \tag{8.18}$$

The rational maps $R_v$, $R_f$ and $R_e$ have also been used to seek a stable $B = 20$ Skyrmion. The $\alpha$-particle model of nuclei suggests that the $B = 20$ Skyrmion is a $D_{3h}$-symmetric bipyramid formed from five $B = 4$ Skyrmions, but only a slightly unstable, saddle point solution with this structure has been found [111]. Instead, two useful rational maps are $R = R_v \times R_f$ and $R = R_v \times R_e$. Each has degree 16 and can be used as an outer map. The inner map is the familiar map $R_v$ of degree 4. Using these pairs of maps in the double rational map ansatz, and relaxing, gives two candidate $B = 20$ Skyrmions shown in Fig. 8.12, but neither has a bipyramid shape. The $O_h$ symmetry is not rigorously enforced by the numerics, and ends up broken. The first Skyrmion has $T_d$ symmetry and the second only $D_{2h}$ symmetry. They both appear to be stable, and separated by another saddle point solution.

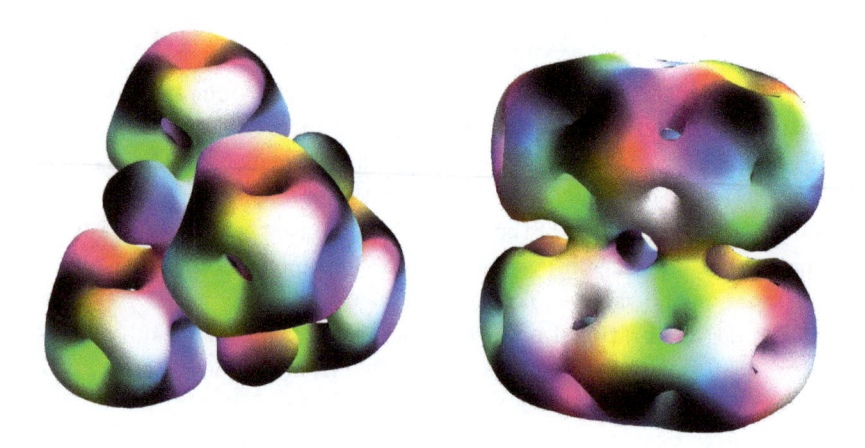

Fig. 8.12: Left: Tetrahedral, $T_d$-symmetric $B = 20$ Skyrmion; Right: $D_{2h}$-symmetric $B = 20$ Skyrmion

The $T_d$-symmetric $B = 20$ Skyrmion has energy $E/12\pi^2 B = 1.277$. It can be interpreted as four slightly distorted $B = 4$ cubes at the vertices of a tetrahedron and four $B = 1$ Skyrmions at the face centres, resembling the $4\alpha + 4n$ cluster structure suggested for Oxygen-20 [99]. The orientations are such that the $B = 1$ and $B = 4$ Skyrmions are all attracting. This solution and its quantization have been discussed in ref.[163]. The $D_{2h}$-symmetric solution has slightly lower energy, $E/12\pi^2 B = 1.274$, and appears to be the minimal-energy solution for $B = 20$. This Skyrmion consists of two parallel clusters loosely bound together, each in the form of the $B = 10$

Skyrmion [33], which resembles two $B = 4$ cubes bound together by a pair of $B = 1$ Skyrmions (see Fig. 9.3). The tetrahedral solution can also be interpreted as two $B = 10$ Skyrmions loosely bound together, obtained from the $D_{2h}$-symmetric solution by a relative 90° twist.

Unsurprisingly, to obtain the $T_d$-symmetric solution, it is preferable to start with an outer rational map with $T_d$ symmetry. A suitable degree 16 map is

$$R = \left(\frac{1+c_2}{1+c_1}\right) \frac{p_+}{p_-} \left(\frac{c_1 p_+^3 + p_-^3}{p_+^3 + c_2 p_-^3}\right), \qquad (8.19)$$

where $c_1$ and $c_2$ are as in (8.16).

## 8.6 Skyrmions up to Baryon Number 256

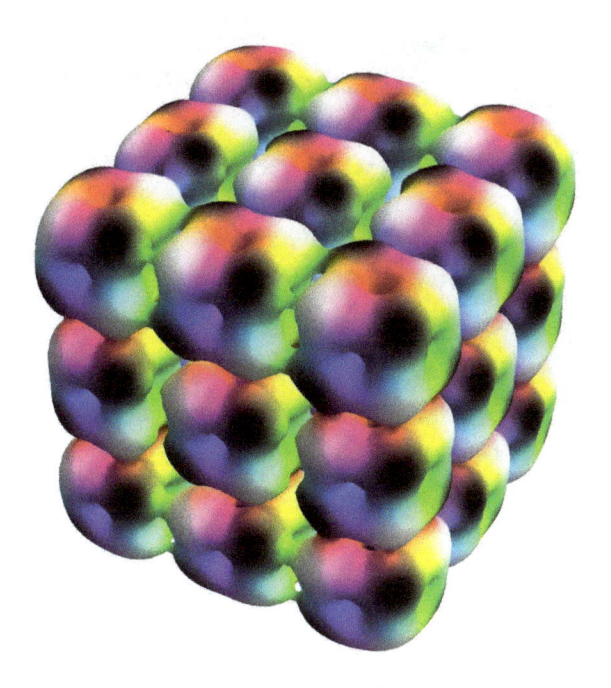

Fig. 8.13: $B = 108$ Skyrmion

Beyond $B = 32$, the next cubic crystal chunk contains 27 $B = 4$ cubes and therefore has baryon number $B = 108$. This can be created using a triple rational map ansatz, wrapping a third layer around the degree 28 and degree 4 layers. The outer map needs to have degree 76, which produces

150 holes, precisely the number of exterior holes in the $B = 108$ crystal chunk. Maps of degree 76 with $O_h$ symmetry have six coefficients, and optimising these numerically is demanding. The geometrical construction described above is more practical.

Using a $6 \times 6 \times 6$ half-Skyrmion grid, one can construct a degree 76 outer rational map (analogous to (8.14) but too long to write out). This is combined with the degree 28 and degree 4 maps from the $B = 32$ Skyrmion as middle and inner maps. A profile function $f(r)$ running from $3\pi$ to 0 is used. After relaxation, the stable Skyrmion shown in Fig. 8.13 is obtained [85]. It has the familiar structure of touching $B = 4$ subcubes all with the same orientation.

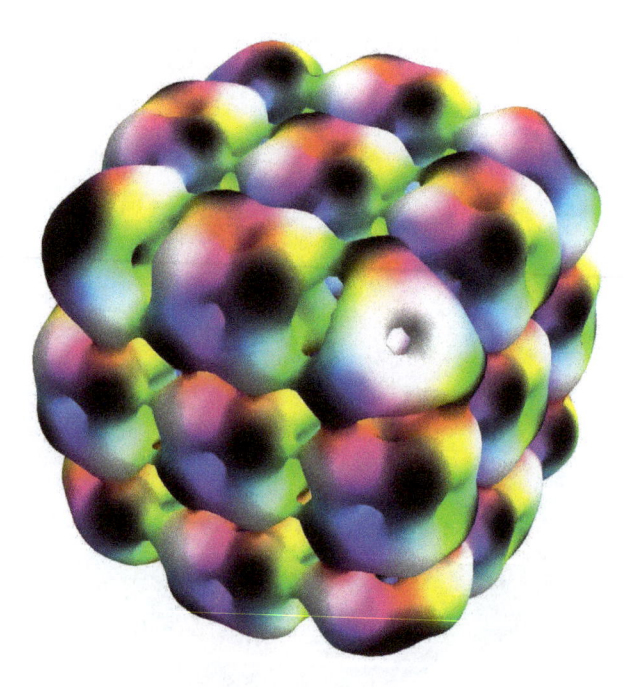

Fig. 8.14: $B = 100$ Skyrmion

The next step is to remove single Skyrmions from the corners, as in the $B = 32$ case, by merging three zeros (poles) of the outer map with a pole (zero) at each corner. This procedure generates Skyrmions with all baryon numbers from $B = 107$ down to $B = 100$. Energies are again listed in Table 8.1. The crystal chunk structure is locally retained, except that the $B = 4$ cubes at the corners are replaced by $B = 3$ tetrahedra. Recall that for the $B = 32$ Skyrmion there was some structural change when six or more corner Skyrmions were removed. The $B = 108$ Skyrmion is more stable to corner cutting as the corners are further from each other. Removing all eight corners gives the cubic $B = 100$ Skyrmion shown in Fig. 8.14.

The $B = 256$ cubic crystal chunk containing 64 $B = 4$ cubes has also been constructed [84]. This is the largest, explicitly known Skyrmion.

There has been some limited exploration of multi-layer Skyrmions with icosahedral symmetry. Battye, Houghton and Sutcliffe have constructed a sequence of rational maps with $Y_h$ symmetry, having degrees 7, 17, 37, 47, 67, 97, and compatible orientations [26]. The degree 7 and 17 maps were given in Chapter 6. For a multi-layer Skyrmion to have low energy, the degrees need to increase substantially from the inside to the outside. The maps of degrees 7 and 37 combine well to form a $B = 44$ Skyrmion. A larger $Y_h$-symmetric Skyrmion with $B = 208$ has been constructed by Halcrow, by combining the maps of degrees 7, 37, 67 and 97 [114]. This is shown (in only a partially relaxed state) in Fig. 8.15. The right-hand figure shows the layering. The energies of the $B = 44$ and $B = 208$ Skyrmions are not known accurately, and are omitted from Table 8.1.

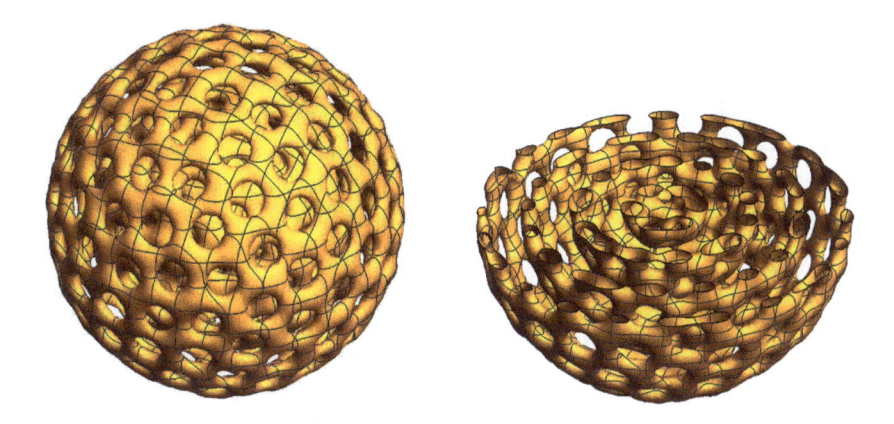

Fig. 8.15: $B = 208$ Skyrmion (complete, and sliced in half to show the layering)

It has been proposed that a quantized, icosahedral rotational band of states, with spin/parities $0^+, 6^+, 10^+, 12^+, \ldots$, matches a selection of excited states of the doubly-magic nucleus Lead-208 that cannot be understood as particle-hole excitations [128]. This nucleus is very far from having isospin zero, because it has 82 protons and 126 neutrons. Curiously, these numbers arise if the Skyrmion is quantized so that the inner two layers consist purely of neutrons, and the outer layers have isospin zero and an equal number of protons and neutrons. Having protons near the outside helps to minimise the electrostatic energy. However, the numerology here may be accidental.

## 8.7 Summary

We have seen that Skyrmions with pion mass $m \approx 1$, for baryon numbers $B \geq 8$, are not hollow polyhedra but more compact, multi-layer structures that often have a clear clustering into $B = 4$ subcubes, the Skyrmion analogue of $\alpha$-particles. This gives confidence that Skyrme theory is a true competitor to other successful models of nuclei. We will investigate this further in the following chapters by looking at the energy spectra of quantum states of $\alpha$-particle molecules in the context of Skyrme theory. The molecules can rotate and vibrate but the individual $B = 4$ Skyrmions, like $\alpha$-particles, cannot easily be excited. The Skyrmion energies for $m = 1$ and for all the baryon numbers we have discussed are presented in Table 8.1.

The numerical methods used to find almost all these Skyrmions and their energies were developed by Feist [84]. The Skyrme energy is discretised on a finite lattice in a numerical box, and to find an energy minimum the nonlinear conjugate gradient method is used. This can be seen as a geometrically enhanced version of gradient descent. To get accurate Skyrmion energies, the number of lattice points is gradually increased and the lattice spacing reduced. It is advantageous that for massive pions the Skyrme field approaches the vacuum $U = 1$ exponentially fast at large distances from the core, as the box size can be smaller than in the case of massless pions.

Another way to find a minimum of the energy is to let an initially static configuration develop in time using a numerical version of the dynamical Skyrme field equation. First, the field accelerates towards the energy minimum. To avoid an overshoot, kinetic energy could be taken out by introducing friction, but this eliminates the real advantage over gradient descent. Instead, a more advanced method is to suddenly remove all kinetic energy when the potential energy starts to increase [30]. This method is called arrested Newtonian dynamics and when iterated, gives rapid convergence to a Skyrmion.

Table 8.1: Energies and symmetries of Skyrmions for $m = 1$. The scaled energy $E/12\pi^2$ is accurate to $\pm 0.01$. Most of the data is from Feist's thesis [84], updating [85]. (The entries for the $D_{6d}$-symmetric $B = 8$ solution, the $B = 16$ solutions, and the $D_{3h}$-symmetric $B = 24$ solution are from refs.[30, 32, 111], respectively, and are less accurate.)

| $B$ | $E/12\pi^2$ | $E/12\pi^2 B$ | $K$ | Comment |
|---|---|---|---|---|
| 1 | 1.465 | 1.465 | $O(3)$ | Hedgehog |
| 2 | 2.77 | 1.385 | $D_{\infty h}$ | Toroidal |
| 3 | 4.02 | 1.340 | $T_d$ | Tetrahedral |
| 4 | 5.18 | 1.295 | $O_h$ | Cubic |
| 8 | 10.24 | 1.280 | $D_{4h}$ | Two $B = 4$ cubes with 90° twist |
|  | 10.27 | 1.284 | $D_{4h}$ | Two $B = 4$ cubes without twist |
|  | 10.32 | 1.290 | $D_{6d}$ |  |
| 10 | 12.80 | 1.280 | $D_{2h}$ |  |
| 12 | 15.30 | 1.275 | $D_{4h}$ | Chain |
|  | 15.32 | 1.277 | $C_{3v}$ | Equilateral triangle |
|  | 15.33 | 1.278 | $C_{2v}$ | L-shape |
| 16 | 20.34 | 1.271 | $D_{2d}$ | Bent square |
|  | 20.35 | 1.272 | $D_{4h}$ | Flat square |
|  | 20.42 | 1.276 | $T_d$ | Tetrahedral |
| 20 | 25.47 | 1.274 | $D_{2h}$ | Two $B = 10$ clusters |
|  | 25.53 | 1.277 | $T_d$ |  |
| 24 | 30.47 | 1.269 | $D_{3d}$ | Six $B = 4$ cubes with twists |
|  | 30.57 | 1.273 | $D_{3d}$ | Six $B = 4$ cubes without twists |
|  | 30.57 | 1.273 | $D_{3h}$ | Stack of $B = 12$ triangles |
|  | 30.80 | 1.283 | $O_h$ | Octahedral cluster of $B = 4$ cubes |
| 25 | 31.79 | 1.272 | $C_{1h}$ |  |
| 26 | 33.05 | 1.271 | $C_{2h}$ | Two $B = 13$ clusters |
| 27 | 34.39 | 1.274 | $C_{3v}$ |  |
| 28 | 35.57 | 1.270 | $T_d$ |  |
| 29 | 36.78 | 1.268 | $C_{3v}$ |  |
| 30 | 38.00 | 1.267 | $D_{3d}$ |  |
| 31 | 39.25 | 1.266 | $C_{3v}$ |  |
| 32 | 40.51 | 1.266 | $O_h$ | Cubic cluster of 8 $B = 4$ cubes |
| 100 | 125.68 | 1.257 | $O_h$ |  |
| 101 | 126.86 | 1.256 | $C_{3v}$ |  |
| 102 | 128.05 | 1.255 | $D_{3d}$ |  |
| 103 | 129.26 | 1.255 | $C_{3v}$ |  |
| 104 | 130.47 | 1.255 | $T_d$ |  |
| 105 | 131.71 | 1.254 | $C_{3v}$ |  |
| 106 | 132.95 | 1.254 | $D_{3d}$ |  |
| 107 | 134.21 | 1.254 | $C_{3v}$ |  |
| 108 | 135.47 | 1.254 | $O_h$ | Cubic cluster of 27 $B = 4$ cubes |
| 256 | 319.71 | 1.249 | $O_h$ | Cubic cluster of 64 $B = 4$ cubes |
| $\infty$ | $\infty$ | 1.238 | $O_h$ | Skyrmion crystal [83] |

## Chapter 9

# Quantized Skyrmions with Even $B \leq 12$

In this chapter we consider the rigid-body quantization of Skyrmions with massive pions for baryon numbers $B = 4, 6, 8, 10$ and $12$ [29], and compare the results with known nuclear energy spectra. The spin/parity states compatible with the Skyrmion symmetries are determined first, for spin and isospin values up to and just beyond what is experimentally accessible. The quantization of the $B = 4$ and $6$ Skyrmions is essentially the same as in the massless pion case, because the Skyrmion symmetries are unchanged, but we now also present the energy spectra. For $B = 8$, 10 and 12, the rigid-body quantization has not been discussed before, and we give more details.

For the smaller baryon numbers, the energy predictions are not very good. This is because the moments of inertia are small, so the excitation energies are large, sufficient to distort or breakup the Skyrmions and the corresponding nuclei. The predictions are better for the larger baryon numbers, and especially for $B = 12$. Rigid-body quantization of a Skyrmion can only give one $0^+$ state with isospin 0, the ground state, so some deformation of the Skyrmion is needed to explain higher-energy $0^+$ states. Two distinct $B = 12$ Skyrmions will be used here to model the Carbon-12 ground state and the excited Hoyle state, both of which have $J^P = 0^+$.

## 9.1 Masses, Charge Radii and Calibration

The Skyrme Lagrangian is defined in terms of the three parameters $F_\pi$, $e$ and $m$. The factor $F_\pi/4e$ converts the Skyrme energy unit to MeV; the factor $2/eF_\pi$ converts the Skyrme length unit to $(\text{MeV})^{-1}$, and $\hbar = 197.3$ MeV fm converts this to fm. The physical pion mass is $m_\pi = meF_\pi/2$, depending just on $m$ and the length scale factor.

For each even baryon number $B$, these three parameters can be calibrated to the physical pion mass and the ground state mass and charge radius of the nucleus with isospin 0. This means that the parameters vary to some extent with baryon number. In particular, $m$ is different for each $B$.

The Skyrmion masses (energies) $\mathcal{M}_B$, charge radii $\langle r^2 \rangle^{\frac{1}{2}}$, and inertia tensor elements in Skyrme units are tabulated in the sections below for pion mass parameters $m = 0.5$, 1 and 1.5, and values for intermediate $m$ can be obtained by quadratic interpolation [29]. For $m = 1$, the masses $\mathcal{M}_B$ match the energies in Table 8.1 obtained in more recent calculations (except that $\mathcal{M}_4$ is 2% larger).

In Skyrme units, the root-mean-square charge radius of a nucleus with isospin 0 is

$$\langle r^2 \rangle^{\frac{1}{2}} = \left( \frac{\int r^2 \mathcal{B}(x)\, d^3x}{\int \mathcal{B}(x)\, d^3x} \right)^{\frac{1}{2}}, \qquad (9.1)$$

since the electric charge density is half the baryon density $\mathcal{B}$ [54]. The charge radius is converted to physical units by multiplying by the length scale factor $2/eF_\pi$. The physical pion mass and the charge radius of the isospin 0 nucleus for each $B$ fix both $m$, which varies between about 0.7 and 1.1, and $2/eF_\pi$. These optimised $m$ values are shown in the final columns of the tables below together with the corresponding inertia tensor elements that determine the energies of excited nuclear states.

The energy scale factor $F_\pi/4e$ is then fitted to convert the Skyrmion mass $\mathcal{M}_B$ to the physical nuclear mass. Here we ignore the small mass contribution arising from spin in the Skyrme picture. Lithium-6 and Boron-10 are examples of isospin 0 nuclei having non-zero spin in their ground states, but their spin energy is estimated to be less than 0.1% of the nuclear mass.

The experimental charge radii and nuclear masses, the fitted value of $m$, and the length and energy scale factors are summarised in Table 9.1. The length scale $2/eF_\pi$ is always fairly close to 1 fm, and the energy scale $F_\pi/4e$ is almost constant at close to 6 MeV. More meaningful in Skyrme theory is $12\pi^2$ times the energy scale, which is close to 700 MeV. This would be the energy per baryon of a nucleus if the classical Skyrmions satisfied the Faddeev–Bogomolny bound, and the nucleus had no rotational or isorotational kinetic energy. $F_\pi$ and $e$ separately vary quite a lot, so one must regard $F_\pi$ as a renormalised pion decay constant, but this is similar to how this quantity was treated in refs.[5, 7].

Table 9.1: Experimental nuclear data and calibration for each $B$.

| $B$ | Nucleus | Charge radius (fm) | Mass (MeV) | $m$ | Length scale $2/eF_\pi$ (fm) | Energy scale $F_\pi/4e$ (MeV) | Quantum energy scale $e^3 F_\pi$ (MeV) |
|---|---|---|---|---|---|---|---|
| 4 | $^4$He | 1.71 | 3727 | 0.820 | 1.173 | 6.169 | 4588 |
| 6 | $^6$Li | 2.55 | 5601 | 1.153 | 1.648 | 5.752 | 2492 |
| 8 | $^8$Be | 2.51 | 7455 | 0.832 | 1.190 | 6.336 | 4339 |
| 10 | $^{10}$B | 2.58 | 9327 | 0.830 | 1.187 | 6.348 | 4354 |
| 12 | $^{12}$C | 2.46 | 11178 | 0.685 | 0.980 | 6.525 | 6216 |

Experimentally, the charge radii of nuclei with $6 \leq B \leq 12$ are approximately constant[1], whereas the Skyrmion charge radius in Skyrme units increases with $B$. With varying parameters, the nuclear charge radii can be kept approximately constant. For example, the small length scale for $B = 12$ reflects the relatively compact physical size of the Carbon-12 nucleus. The larger length scale for $B = 6$ takes into account the looser vibrational motion of the $B = 6$ Skyrmion, in particular the modes that lead to breakup into a $B = 4$ and $B = 2$ Skyrmion, or three $B = 2$ Skyrmions. This gives larger physical moments of inertia, making the splitting between spin excitations of Lithium-6 smaller and improving the fit to experimental data.

Recall that the quantum Hamiltonian for a rigidly rotating body is schematically the squared angular momentum operator divided by twice the moment of inertia of the body. The moment of inertia has units of the energy (mass) scale multiplied by the square of the length scale: $(F_\pi/4e) \times (2/eF_\pi)^2 = 1/e^3 F_\pi$. The quantum energy scale is its reciprocal, $e^3 F_\pi$, and this is used to convert rotational and isorotational energies to physical units. Its value is shown in the final column of Table 9.1.

Having separate parameter sets for each $B$ also gives a better fit for the ground state rotational band of Beryllium-8 relative to Carbon-12. These nuclei have $0^+$ ground states and $2^+$ and $4^+$ excited states, at 3.0 MeV and 11.4 MeV for Beryllium-8, and 4.4 MeV and 14.4 MeV for Carbon-12. One might expect Carbon-12 to have a larger moment of inertia than Beryllium-8, since in Skyrme units the $B = 12$ Skyrmion is spatially larger and has a larger mass than the $B = 8$ Skyrmion. This would lead to Carbon-12

---
[1]For updated nuclear charge radii, see ref.[13].

Table 9.2: $B = 4$ Skyrmion properties.

| $m$ | 0.5 | 1 | 1.5 | 0.820 |
|---|---|---|---|---|
| $U_{11}$ | 201 | 151 | 124 | 167 |
| $U_{33}$ | 241 | 180 | 146 | 198 |
| $V_{33}$ | 928 | 701 | 576 | 771 |
| $\langle r^2 \rangle^{\frac{1}{2}}$ | 1.679 | 1.360 | 1.185 | 1.458 |
| $\mathcal{M}_4$ | 569 | 624 | 681 | 604 |

having smaller rotational band splittings than Beryllium-8, the opposite of what is observed. This problem is dealt with by using different length scales, which allows for the loose nature of the slightly unbound Beryllium-8 ground state, compared with the more tightly bound Carbon-12 ground state.

## 9.2 $B = 4$

The $B = 4$ Skyrmion has $O_h$ symmetry and a cubic shape. Its low-lying quantum states were discussed in Chapter 7, and here we consider their energies. The $O_h$ symmetry implies that the inertia tensors are diagonal, with $U_{11} = U_{22}$ and $U_{33}$ different, $V_{ij}$ proportional to the identity matrix, and $W_{ij} = 0$. The rigid-body Hamiltonian is therefore the uncoupled sum of a spherical top in space and a symmetric top in isospace,

$$H = \frac{1}{2V_{33}}\mathbf{J}^2 + \frac{1}{2U_{11}}\mathbf{I}^2 + \left(\frac{1}{2U_{33}} - \frac{1}{2U_{11}}\right)K_3^2. \quad (9.2)$$

(Recall that $\mathbf{L}$ and $\mathbf{K}$ are the body-fixed spin and isospin operators, and $\mathbf{J}^2 = \mathbf{L}^2$ and $\mathbf{I}^2 = \mathbf{K}^2$.) Energy eigenstates have definite quantum numbers $J$, $I$ and $|K_3|$, and energies (in Skyrme units)

$$E = \frac{1}{2V_{33}}J(J+1) + \frac{1}{2U_{11}}I(I+1) + \left(\frac{1}{2U_{33}} - \frac{1}{2U_{11}}\right)|K_3|^2. \quad (9.3)$$

The relevant inertia tensor elements for the optimised pion mass parameter $m$ are in the final column of Table 9.2.

The lowest state is a $0^+$ state with isospin 0, the quantum numbers of the $\alpha$-particle – the Helium-4 ground state. As noted earlier, the first-excited state with isospin 0 is a $4^+$ state, which has not been experimentally observed because of its high energy. The lowest state with isospin 1 is the $J^P = 2^-$ state (7.36), matching the quantum numbers of the observed

Table 9.3: Energy levels of the quantized $B = 4$ Skyrmion.

| $I$ | $J^P$ | $E \times 10^4$ | $E$ (MeV) |
|---|---|---|---|
| 0 | $0^+$ | 0.0 | 0.0 |
|   | $4^+$ | 129.7 | 59.5 |
| 1 | $2^-$ | 94.1 | 43.2 |

Table 9.4: $B = 6$ Skyrmion properties.

| $m$ | 0.5 | 1 | 1.5 | 1.153 |
|---|---|---|---|---|
| $U_{11}$ | 305 | 228 | 186 | 211 |
| $U_{33}$ | 329 | 245 | 199 | 227 |
| $V_{11}$ | 2195 | 1658 | 1362 | 1542 |
| $V_{33}$ | 1927 | 1451 | 1190 | 1349 |
| $W_{33}$ | −105 | −84 | −71 | −79 |
| $\langle r^2 \rangle^{\frac{1}{2}}$ | 1.948 | 1.620 | 1.430 | 1.547 |
| $\mathcal{M}_6$ | 858 | 946 | 1036 | 973 |

isotriplet comprising the Lithium-4 and Hydrogen-4 ground states and an excited Helium-4 state. The excitation energies $E$ in Skyrme units are given in Table 9.3. The final column shows the excitation energies in physical units, using the $B = 4$ conversion factor $e^3 F_\pi = 4588$ MeV from Table 9.1. The excitation energy of the $2^-$ isotriplet is overestimated as 43.2 MeV compared with the experimental value (averaged over the three nuclei) of 23.7 MeV [229]. This is not too surprising, as these states are unbound resonances with an effective spatial moment of inertia larger than the rigid-body estimate. If we drop the first term from (9.3), the excitation energy reduces to 25.3 MeV, much closer to the experimental value.

## 9.3 $B = 6$

The $B = 6$ Skyrmion has $D_{4d}$ symmetry, and its quantization was also discussed in Chapter 7. The Hamiltonian is that of a system of rather

Table 9.5: Wavefunctions and energy levels of the quantized $B = 6$ Skyrmion.

| $I$ | $J^P$ | Wavefunction | $E \times 10^4$ | $E$ (MeV) |
|---|---|---|---|---|
| 0 | $1^+$ | $\|1, 0\rangle \otimes \|0, 0\rangle$ | 6.5 | 1.6 |
|   | $3^+$ | $\|3, 0\rangle \otimes \|0, 0\rangle$ | 38.9 | 9.7 |
|   | $4^-$ | $(\|4, 4\rangle - \|4, -4\rangle) \otimes \|0, 0\rangle$ | 73.5 | 18.3 |
|   | $5^+$ | $\|5, 0\rangle \otimes \|0, 0\rangle$ | 97.3 | 24.2 |
|   | $5^-$ | $(\|5, 4\rangle + \|5, -4\rangle) \otimes \|0, 0\rangle$ | 105.9 | 26.4 |
| 1 | $0^+$ | $\|0, 0\rangle \otimes \|1, 0\rangle$ | 47.4 | 11.8 |
|   | $2^+$ | $\|2, 0\rangle \otimes \|1, 0\rangle$ | 66.8 | 16.7 |
|   |       | $\|2, 2\rangle \otimes \|1, 1\rangle + \|2, -2\rangle \otimes \|1, -1\rangle$ | 62.6 | 15.6 |
|   | $2^-$ | $\|2, 2\rangle \otimes \|1, -1\rangle + \|2, -2\rangle \otimes \|1, 1\rangle$ | 73.1 | 18.2 |
|   | $3^+$ | $\|3, 2\rangle \otimes \|1, 1\rangle - \|3, -2\rangle \otimes \|1, -1\rangle$ | 82.0 | 20.4 |
|   | $3^-$ | $\|3, 2\rangle \otimes \|1, -1\rangle - \|3, -2\rangle \otimes \|1, 1\rangle$ | 92.5 | 23.1 |
|   | $4^+$ | $\|4, 0\rangle \otimes \|1, 0\rangle$ | 112.2 | 28.0 |
|   |       | $\|4, 2\rangle \otimes \|1, 1\rangle + \|4, -2\rangle \otimes \|1, -1\rangle$ | 108.0 | 26.9 |
|   | $4^-$ | $\|4, 2\rangle \otimes \|1, -1\rangle + \|4, -2\rangle \otimes \|1, 1\rangle$ | 118.5 | 29.5 |
|   |       | $(\|4, 4\rangle + \|4, -4\rangle) \otimes \|1, 0\rangle$ | 120.9 | 30.1 |
| 2 | $0^-$ | $\|0, 0\rangle \otimes (\|2, 2\rangle - \|2, -2\rangle)$ | 137.4 | 34.2 |
|   | $1^+$ | $\|1, 0\rangle \otimes \|2, 0\rangle$ | 148.6 | 37.0 |
|   | $1^-$ | $\|1, 0\rangle \otimes (\|2, 2\rangle + \|2, -2\rangle)$ | 143.9 | 35.9 |
|   | $2^+$ | $\|2, 2\rangle \otimes \|2, 1\rangle - \|2, -2\rangle \otimes \|2, -1\rangle$ | 157.3 | 39.2 |
|   | $2^-$ | $\|2, 0\rangle \otimes (\|2, 2\rangle - \|2, -2\rangle)$ | 156.9 | 39.1 |
|   |       | $\|2, 2\rangle \otimes \|2, -1\rangle - \|2, -2\rangle \otimes \|2, 1\rangle$ | 167.8 | 41.8 |

weakly coupled symmetric tops,

$$H = \frac{1}{2V_{11}}\mathbf{J}^2 + \frac{1}{2U_{11}}\mathbf{I}^2 + \left(\frac{U_{33}}{2\Delta_{33}} - \frac{1}{2V_{11}}\right)L_3^2$$
$$+ \left(\frac{V_{33}}{2\Delta_{33}} - \frac{1}{2U_{11}}\right)K_3^2 + \frac{W_{33}}{\Delta_{33}}L_3K_3, \qquad (9.4)$$

where $\Delta_{33} = U_{33}V_{33} - W_{33}^2$. Its allowed quantum states are listed in Table 9.5, together with their energy levels computed using the inertia tensors in the final column of Table 9.4 and the conversion factor $e^3 F_\pi = 2492$ MeV from Table 9.1. The small $L_3 K_3$ term splits the energies between the isospin 1 states where $L_3$ and $K_3$ have, respectively, the same and opposite signs.

The theory qualitatively reproduces part of the experimental spectrum for nuclei with baryon number 6, shown in Fig. 9.1. For the isospin 0 states of Lithium-6, the ground state and first-excited state are correctly predicted as having spin/parities $J^P = 1^+$ and $3^+$, but the excitation energy

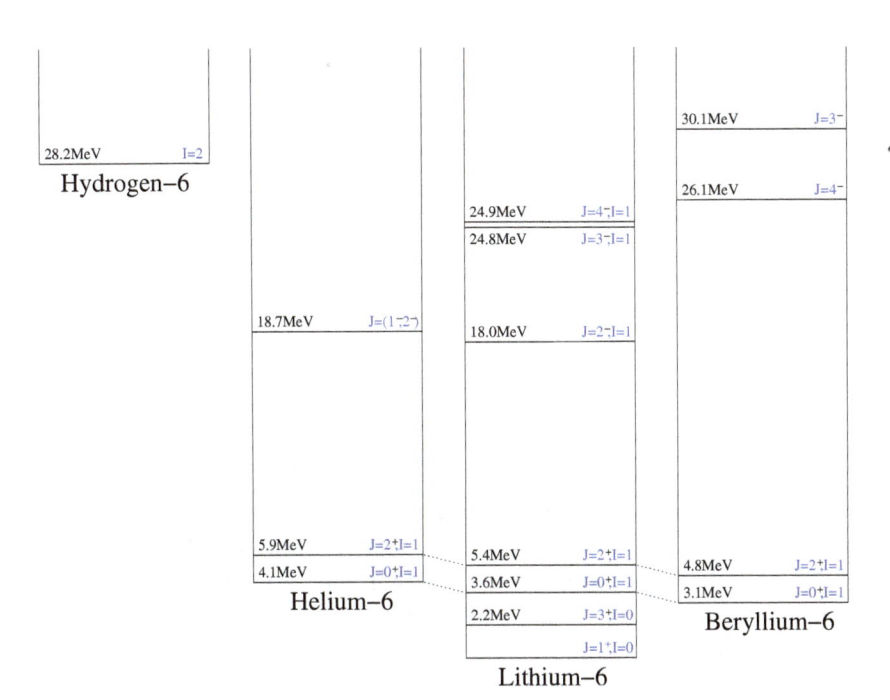

Fig. 9.1: Energy level diagram for nuclei with baryon number 6. Data are from ref.[226]. $J^P$ and $I$ are shown for each state, where known, and brackets denote uncertainty.

of the $3^+$ state is overpredicted by roughly a factor of five. Clearly, this nucleus tends to break up into $B = 4$ and $B = 2$ clusters, so the effective moment of inertia again gets larger with spin, something not taken into account in rigid-body quantization. Experimentally, the breakup threshold for Lithium-6 into an $\alpha$-particle and a deuteron is just 1.474 MeV which is less than 2.2 MeV, the excitation energy of the $3^+$ state. The theory misses the higher-energy $2^+$ and $1^+$ states of Lithium-6 that are observed, but these are broad resonances and deformations of the Skyrmion are probably needed to model them.

The lowest predicted state with isospin 1 is a $0^+$ state, which is seen experimentally as an isotriplet comprising the Beryllium-6 and Helium-6 ground states and an excited state of Lithium-6. An excited $2^+$ state of this isotriplet is also predicted and observed. However, its excitation energy is again overestimated here. The lowest predicted negative parity state with isospin 1 is a $2^-$ state with excitation energy 18.2 MeV. Lithium-6 has an

Table 9.6: $B = 8$ twisted 2-cube Skyrmion properties.

| $m$ | 0.5 | 1 | 1.5 | 0.832 |
|---|---|---|---|---|
| $U_{11}$ | 403 | 299 | 243 | 329 |
| $U_{22}$ | 374 | 291 | 242 | 315 |
| $U_{33}$ | 418 | 326 | 271 | 353 |
| $V_{11}$ | 4740 | 4052 | 3490 | 4269 |
| $V_{33}$ | 1990 | 1390 | 1109 | 1556 |
| $\langle r^2 \rangle^{\frac{1}{2}}$ | 2.316 | 2.017 | 1.787 | 2.109 |
| $\mathcal{M}_8$ | 1106 | 1213 | 1323 | 1177 |

observed $2^-$ resonance state with isospin 1 at 18.0 MeV, seen in Helium-3/Hydrogen-3 collisions, although its Beryllium-6 and Helium-6 partners are not yet confirmed. Further isospin 1 states of both Beryllium-6 and Lithium-6 with $J^P = 3^-$ and $4^-$ have also been observed, and are predicted with roughly the correct excitation energies, although rigid-body quantization is probably not justified at these high energies. A model allowing the splitting of the $B = 6$ Skyrmion into two $B = 3$ clusters would be preferable.

The ground state of Hydrogen-6 (which is again a resonance) has isospin 2 and is observed with energy 28.2 MeV above the Lithium-6 ground state, with an undetermined spin/parity. The lowest predicted state with isospin 2 is a $0^-$ state with excitation energy 34.2 MeV, and it would be interesting to check this spin/parity prediction. Higher-spin excited isospin 2 states are also expected, but have not been observed.

## 9.4 $B = 8$

Recall that when $m \approx 1$, the stable $B = 8$ Skyrmion has $D_{4h}$ symmetry, and resembles two touching $B = 4$ cubes with a relative 90° twist (see Fig. 8.2), matching the known physics that Beryllium-8 is an almost bound configuration of two $\alpha$-particles. The quantum Hamiltonian is the uncoupled sum of a symmetric top in space and an asymmetric top in isospace [29],

$$H = \frac{1}{2V_{11}}\mathbf{J}^2 + \left(\frac{1}{2V_{33}} - \frac{1}{2V_{11}}\right) L_3^2 + \frac{1}{2U_{11}}K_1^2 + \frac{1}{2U_{22}}K_2^2 + \frac{1}{2U_{33}}K_3^2, \quad (9.5)$$

with the inertia tensors given in Table 9.6. As anticipated on symmetry grounds, the inertia tensor $U_{ij}$ has three distinct eigenvalues (although they are rather close), whereas $V_{ij}$ has only two.

Table 9.7: Energy levels of the quantized $B = 8$ Skyrmion. There is more than one allowed state for several of the $J^P$ values.

| $I$ | $J^P$ | $E \times 10^4$ | $E$ (MeV) |
|---|---|---|---|
| 0 | $0^+$ | 0.0 | 0.0 |
|   | $2^+$ | 7.0 | 3.0 |
|   | $4^+$ | 23.4, 56.1 | 10.2, 24.3 |
| 1 | $0^-$ | 30.0 | 13.0 |
|   | $2^+$ | 44.6 | 19.3 |
|   | $2^-$ | 37.1, 46.3 | 16.1, 20.1 |
|   | $3^+$ | 51.6 | 22.4 |
|   | $3^-$ | 53.3 | 23.1 |
|   | $4^+$ | 61.0 | 26.5 |
|   | $4^-$ | 53.5, 62.7, 86.2 | 23.2, 27.2, 37.4 |
| 2 | $0^+$ | 87.6, 93.5 | 38.0, 40.6 |
|   | $0^-$ | 90.9 | 39.4 |
|   | $2^+$ | 94.6, 100.5, 108.1 | 41.0, 43.6, 46.9 |
|   | $2^-$ | 98.0, 103.0 | 42.5, 44.7 |

A rational map with the same symmetry as the $B = 8$ Skyrmion is

$$R(z) = \frac{z^8 - az^4 + 1}{z^2(z^4 + 1)}, \tag{9.6}$$

and using this one finds that the FR constraints on quantum states are [174]

$$e^{i\frac{\pi}{2}L_3} e^{i\pi K_3} |\Psi\rangle = |\Psi\rangle,$$
$$e^{i\pi L_1} |\Psi\rangle = |\Psi\rangle. \tag{9.7}$$

For states with isospin 0 it is easy to see that the allowed spin/parities are $0^+$, $2^+$ and $4^+$ with $L_3 = 0$, and there is a further $4^+$ state with $|L_3| = 4$ having much higher energy because the $B = 8$ Skyrmion is substantially prolate and $V_{33} \ll V_{11}$. The three lower-lying states form a rotational band where the Skyrmion performs an end-over-end rotation, whereas in the $|L_3| = 4$ state the Skyrmion spins about its long axis.

The allowed states and their energies are listed in Table 9.7. The energies are calculated using the standard results for symmetric and asymmetric tops [160].

Fig. 9.2 is a simplified experimental energy-level diagram for nuclei with baryon number 8. The Skyrme theory predictions for positive parity states agree well with experiment. The isospin 0 ground state of Beryllium-8 is correctly determined to be a $0^+$ state, and the ground state rotational band

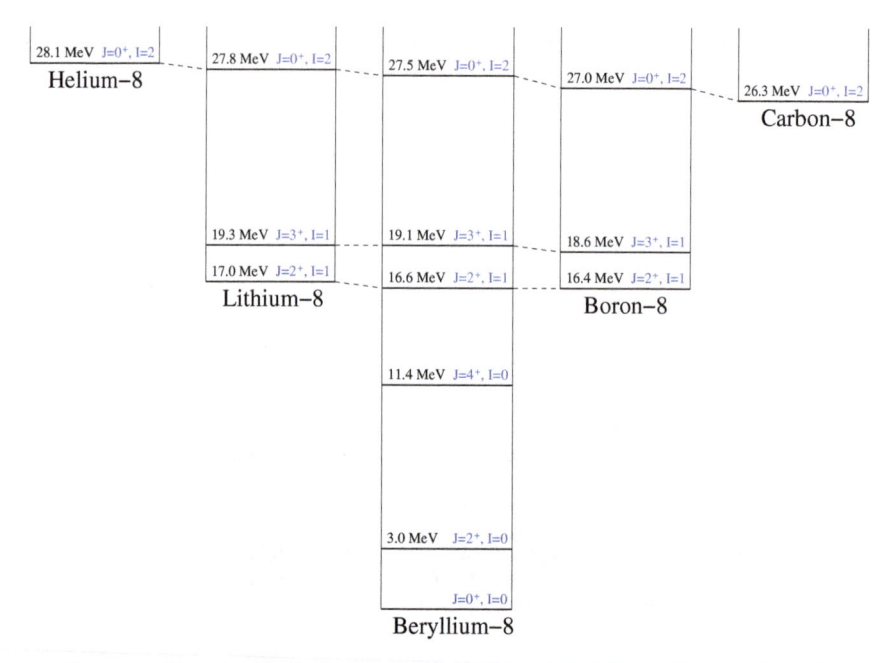

Fig. 9.2: Energy level diagram for $B = 8$ nuclei. Data from ref.[227].

is remarkably well reproduced; there are $2^+$ and $4^+$ states predicted at 3.0 MeV and 10.2 MeV, close to the experimental values of 3.0 MeV and 11.4 MeV. The higher-energy $4^+$ state with isospin 0 is predicted at 24.3 MeV, and such states are seen experimentally at 19.9 MeV and 25.5 MeV.

Less successful is that the lowest allowed state with isospin 1 is predicted to be a $J^P = 0^-$ state with excitation energy 13.0 MeV above the Beryllium-8 ground state, and above this, states with $J^P = 2^+$ and $3^+$ are predicted, together with $2^-$ and $3^-$ states. These should all be observed in the spectrum of Boron-8 and Lithium-8. However, the ground states of Boron-8 and Lithium-8 have $J^P = 2^+$, and the second-excited states have $J^P = 3^+$. These states are joined by excited states of Beryllium-8 to form isotriplets. A $0^-$ state is not seen, and the observed first-excited $1^+$ state is missed by the theory. The predictions for the excitation energies of the $2^+$ and $3^+$ isotriplets are 19.3 MeV and 22.4 MeV, matching the experimental mean energies of 16.7 MeV and 19.0 MeV quite well. Perhaps the $0^-$ isotriplet is still to be discovered, for it is known that low-lying $0^-$ states are difficult to observe, as was experienced in the search for the bottomonium and charmonium ground state mesons $\eta_b$ and $\eta_c$ [18, 194]. We will see later that the lowest $0^-$ state of Oxygen-16 is also problematic. The experimental situation for negative parity states with baryon number 8 and isospin 1 is altogether rather confused, because the few states that have been observed do not all form complete isotriplets.

The lowest predicted states with isospin 2 have $J^P = 0^+$ and excitation energy 38.0 MeV. A complete isospin 2 quintet that includes the $0^+$ ground states of Carbon-8 and Helium-8 is observed with mean energy 27.3 MeV.

In summary, the rigid-body quantization of the $B = 8$ Skyrmion gives quite a good spectrum of states. The most serious problem is the prediction that the lowest states with isospin 1 have spin/parity $0^-$. Possibly the experimental observations are incomplete, although this is not very likely, because Lithium-8 is a well-studied nucleus with a lifetime of nearly 1 s. More likely, the rigid-body quantization needs extending. It may be theoretically important to allow for dynamical twisting motion between the two cubes in the $B = 8$ Skyrmion. This could affect the isospin 1 states more than the isospin 0 states.

## 9.5  $B = 10$

The $B = 10$ Skyrmion, shown in Fig. 9.3, has $D_{2h}$ symmetry and can be interpreted as two $B = 4$ cubes bound together by a pair of $B = 1$ Skyrmions.

Fig. 9.3: $B = 10$ Skyrmion

Let us quantize the $B = 10$ Skyrmion in detail, as this example was not considered in Chapter 7. We use the rational map ansatz to determine its

Table 9.8: $B = 10$ Skyrmion properties.

| $m$ | 0.5 | 1 | 1.5 | 0.830 |
|---|---|---|---|---|
| $U_{11}$ | 511 | 383 | 303 | 421 |
| $U_{22}$ | 508 | 380 | 298 | 418 |
| $U_{33}$ | 459 | 351 | 285 | 383 |
| $V_{11}$ | 4250 | 3120 | 2360 | 3463 |
| $V_{22}$ | 5860 | 4520 | 3700 | 4917 |
| $V_{33}$ | 5730 | 4400 | 3590 | 4794 |
| $W_{33}$ | $-10.4$ | $-4.8$ | 0.7 | $-6.7$ |
| $\langle r^2 \rangle^{\frac{1}{2}}$ | 2.455 | 2.047 | 1.745 | 2.174 |
| $\mathcal{M}_{10}$ | 1373 | 1516 | 1657 | 1468 |

FR constraints. A suitable rational map is

$$R(z) = \frac{z^{10} + az^8 + bz^6 + cz^4 + dz^2 + e}{ez^{10} + dz^8 + cz^6 + bz^4 + az^2 + 1}, \quad (9.8)$$

with $a = 3.02$, $b = 4.98$, $c = 14.83$, $d = -9.37$ and $e = 0.28$. The $D_2$ rotation group is generated by the symmetries

$$R(-z) = R(z), \quad R\left(\frac{1}{z}\right) = \frac{1}{R(z)}, \quad (9.9)$$

and the associated FR constraints are determined using Krusch's formula (7.24) to be

$$e^{i\pi L_3}|\Psi\rangle = |\Psi\rangle, \quad e^{i\pi L_1}e^{i\pi K_1}|\Psi\rangle = -|\Psi\rangle. \quad (9.10)$$

The signs $\chi_{\text{FR}}$ form one of the non-trivial one-dimensional representations of $D_2$. The rational map (9.8) also has the spatial inversion symmetry

$$R\left(-\frac{1}{\bar{z}}\right) = \frac{1}{\overline{R(z)}}. \quad (9.11)$$

The right hand side is a pure reflection in isospace, so the inversion operation $\mathscr{I}$ is equivalent to a 180° rotation in isospace. The parity operator is therefore $\mathcal{P} = e^{i\pi K_3}$, as in other examples.

These symmetries of the $B = 10$ Skyrmion imply that the inertia tensors $U_{ij}$ and $V_{ij}$ are diagonal, and the only non-zero component of the mixed inertia tensor is $W_{33}$. The quantum Hamiltonian is that of a system of coupled asymmetric tops,

$$H = \frac{1}{2V_{11}}L_1^2 + \frac{1}{2V_{22}}L_2^2 + \frac{U_{33}}{2\Delta_{33}}L_3^2$$
$$+ \frac{1}{2U_{11}}K_1^2 + \frac{1}{2U_{22}}K_2^2 + \frac{V_{33}}{2\Delta_{33}}K_3^2 + \frac{W_{33}}{\Delta_{33}}L_3 K_3, \quad (9.12)$$

Table 9.9: Energy levels of the quantized $B = 10$ Skyrmion.

| $I$ | $J^P$ | $E \times 10^4$ | $E$ (MeV) |
|---|---|---|---|
| 0 | $1^+$ | 2.5 | 1.1 |
|   | $2^+$ | 6.6 | 2.9 |
|   | $3^+$ | 12.7, 16.1 | 5.5, 7.0 |
|   | $4^+$ | 21.1, 24.3 | 9.2, 10.6 |
| 1 | $0^+$ | 23.8 | 10.4 |
|   | $0^-$ | 24.9 | 10.8 |
|   | $1^-$ | 27.5 | 11.9 |
|   | $2^+$ | 30.0, 31.7 | 13.1, 13.8 |
|   | $2^-$ | 31.1, 31.6, 32.8 | 13.5, 13.8, 14.3 |
|   | $3^+$ | 37.8 | 16.5 |
|   | $3^-$ | 37.6, 38.9, 41.1 | 16.4, 16.9, 17.9 |
|   | $4^+$ | 44.4, 46.1, 51.1 | 19.3, 20.1, 22.2 |
|   | $4^-$ | 45.5, 46.2, 47.2, 49.3, 52.2 | 19.8, 20.1, 20.5, 21.5, 22.7 |
| 2 | $0^+$ | 76.0 | 33.1 |
|   | $0^-$ | 72.7 | 31.6 |
|   | $1^+$ | 73.9, 78.5 | 32.2, 34.2 |
|   | $1^-$ | 74.0 | 32.2 |
|   | $2^+$ | 78.1, 82.2, 82.7, 83.9 | 34.0, 35.8, 36.0, 36.5 |
|   | $2^-$ | 78.9, 79.2, 80.6 | 34.3, 34.4, 35.0 |
| 3 | $0^+$ | 142.9, 147.5 | 62.2, 64.2 |
|   | $0^-$ | 144.1, 153.2 | 62.7, 66.7 |
|   | $1^+$ | 150.0 | 65.2 |
|   | $1^-$ | 146.8, 155.7 | 63.9, 67.7 |

where $\Delta_{33} = U_{33}V_{33} - W_{33}^2$ as before. The calculation of energy levels is similar to the case of a general asymmetric top. However, the final term in (9.12) mixes states of the form $|J, L_3\rangle + |J, -L_3\rangle$ and $|J, L_3\rangle - |J, -L_3\rangle$ (and similarly for isospin basis states). In Table 9.9 we list the allowed states, together with their energy eigenvalues. The energies are calculated by diagonalising the Hamiltonian in matrix form, separately for each combination of $J^P$ and $I$, and using the moments of inertia in the final column of Table 9.8 [29,254]. The precise forms of the energy eigenstates are omitted as they add little to the discussion.

The experimental energy level diagram for $B = 10$ nuclei is shown in Fig. 9.4. Boron-10 has a $J^P = 3^+$ ground state, a $1^+$ state at 0.7 MeV, a $2^+$ state at 3.6 MeV and an excited $3^+$ state at 4.8 MeV, all with isospin 0. Skyrme theory predicts a $1^+$ ground state, a $2^+$ state and two $3^+$ states –

the expected states but in the wrong order. This problem with the spin of the ground state arises in other models of Boron-10, for example in models involving nucleon-nucleon potentials derived using chiral perturbation theory [191]. Rigid-body Skyrmion quantization only allows positive parity states with isospin 0, but at least one negative parity state is observed. The excitation of further degrees of freedom of the Skyrmion is needed to account for negative parity states.

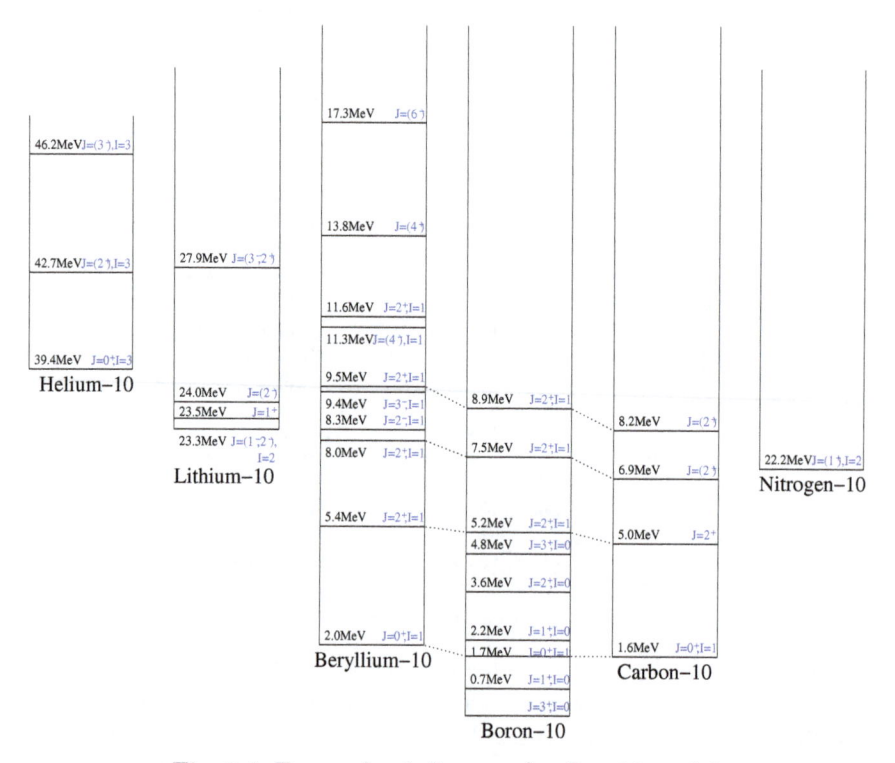

Fig. 9.4: Energy level diagram for $B = 10$ nuclei

The lowest theoretically allowed state with isospin 1 has $J^P = 0^+$. This matches the observed $0^+$ isotriplet that includes the Carbon-10 and Beryllium-10 ground states. More than two $2^+$ excitations of this isotriplet have been observed, although only two are predicted. The second-excited $3^+$ state in Boron-10 at 7.0 MeV may have isospin 1, matching the predicted state. Additionally, $1^-$, $2^-$ and $3^-$ states with isospin 1 have been observed in the spectrum of Beryllium-10, and the Skyrme theory predictions for their excitation energies are close to the experimental values. These states do not yet lie in complete isotriplets. $J^P = 1^+$ states with isospin 1 are theoretically disallowed, and none have been observed. A low-lying $0^-$ state is predicted, but has not been observed.

For isospins 2 and 3, there are just a few observed states, and little certainty about their spins or parities. The lowest predicted states with isospin 2 have spin/parities $0^-$ and $1^-$. The Lithium-10 ground state (an isospin 2 resonance) may have spin/parity $1^-$. The Helium-10 ground state has isospin 3 and $J^P = 0^+$, in agreement with the theory. This nucleus is interesting because it is doubly magic, with 2 protons and 8 neutrons, but is still a broad resonance because it is very unstable to two-neutron emission. Theoretically, the energy of isospin excitations should increase in proportion to $I(I+1)$. This quadratic behaviour is approximately realised in the nuclear energy level diagram, provided one subtracts the relatively large spin energy of about 5 MeV from the Boron-10 ground state energy level.

## 9.6 $B = 12$

Among the nuclei with baryon number 12, the most interesting and important is Carbon-12. This is the smallest nucleus that is a genuine bound state of $\alpha$-particles. Experimenters have devoted much effort to this nucleus, and its spectrum is known rather well, up to states with spin 5. Within Skyrme theory, there has been particular attention given to this nucleus too.

There are two Skyrmions with baryon number 12, having $D_{3h}$ and $D_{4h}$ symmetries (see Figs. 8.3 and 8.4). The first has an equilateral triangular shape and the second an extended linear shape, analogous to the triangle and linear chain structures of three $\alpha$-particles. We first consider the rigid-body quantization of the $D_{3h}$-symmetric Skyrmion. The spectrum of excited states of Carbon-12 that emerges can then be usefully extended by also quantizing the $D_{4h}$-symmetric Skyrmion [162]. Using the moments of inertia of both these Skyrmions, the energies of their quantized rotational excitations can be calculated. For the triangular $D_{3h}$-symmetric Skyrmion, the isospin 0 states include the $0^+$ ground state of Carbon-12 and its rotational excitations, and there are also states with isospin 1 and 2, matching the ground states of Oxygen-12, Nitrogen-12, Boron-12 and Beryllium-12. The isospin 0 states of the $D_{4h}$-symmetric chain Skyrmion give a good match to the rather recently established rotational band of the Hoyle state. The Hoyle state is the first-excited $0^+$ state of Carbon-12, and is of great

Table 9.10: $B = 12$ $D_{3h}$-symmetric Skyrmion properties.

| $m$ | 0.5 | 1 | 1.5 | 0.685 |
|---|---|---|---|---|
| $U_{11}$ | 588 | 444 | 364 | 527 |
| $U_{33}$ | 653 | 500 | 396 | 590 |
| $V_{11}$ | 6487 | 5037 | 4087 | 5891 |
| $V_{33}$ | 9743 | 7684 | 6240 | 8909 |
| $W_{11}$ | −49 | −40 | −37 | −45 |
| $W_{33}$ | −42 | −35 | −40 | −38 |
| $\langle r^2 \rangle^{\frac{1}{2}}$ | 2.674 | 2.265 | 1.952 | 2.511 |
| $\mathcal{M}_{12}$ | 1653 | 1816 | 1982 | 1713 |

astrophysical importance for the synthesis of heavy elements [131]. The ratio of the matter radii of the Hoyle state and ground state, calculated using the Skyrmions, also matches the experimental value.

In Chapter 10 we will discuss a more unified model where these two Skyrmions are connected by a deformation mode, but since they are both minima of the energy, treating them independently is quite a good approximation.

### 9.6.1 Quantizing the $D_{3h}$-symmetric Skyrmion

The $D_{3h}$-symmetric $B = 12$ Skyrmion can be obtained using the double rational map ansatz, and we use this to determine its FR constraints. The ansatz uses the outer map $R_2$ of degree 11 and the standard inner map $R_1$ of degree 1, given in eqs.(8.5) and (8.6). Both maps satisfy

$$R\left(e^{i\frac{2\pi}{3}}z\right) = e^{i\frac{2\pi}{3}} R(z), \quad R\left(\frac{1}{z}\right) = \frac{1}{R(z)}. \tag{9.13}$$

As the baryon number is a multiple of four, the FR signs form the trivial representation of $D_3$, so the FR constraints on quantum states are

$$e^{i\frac{2\pi}{3}L_3} e^{i\frac{2\pi}{3}K_3} |\Psi\rangle = |\Psi\rangle, \quad e^{i\pi L_1} e^{i\pi K_1} |\Psi\rangle = |\Psi\rangle. \tag{9.14}$$

The maps also have the combined reflection symmetry

$$R\left(\frac{1}{\bar{z}}\right) = \frac{1}{\overline{R(z)}}, \tag{9.15}$$

so the parity operator is equivalent to $\mathcal{P} = e^{i\pi L_3} e^{i\pi K_3}$. The $D_{3h}$ symmetry implies that the inertia tensors are diagonal, with $U_{11} = U_{22}$, $V_{11} = V_{22}$ and $W_{11} = W_{22}$; the quantum Hamiltonian is that of a system of coupled symmetric tops and is rather complicated. The Hamiltonian and its energy levels have been discussed in detail by Wood [254].

Here we make a simplification. The elements of the $W$-tensor are much smaller than those of the $U$- and $V$-tensors, as shown in Table 9.10, so we set the $W$-tensor to zero. This approximation introduces errors in the excitation energies of about 2%, which is less than various other errors due to the approximate nature of Skyrme theory, the neglect of vibrational degrees of freedom, etc. The quantum Hamiltonian now becomes the uncoupled sum of a symmetric top in space and a symmetric top in isospace,

$$H = \frac{1}{2V_{11}}\mathbf{J}^2 + \frac{1}{2U_{11}}\mathbf{I}^2 + \left(\frac{1}{2V_{33}} - \frac{1}{2V_{11}}\right)L_3^2 + \left(\frac{1}{2U_{33}} - \frac{1}{2U_{11}}\right)K_3^2. \tag{9.16}$$

$|L_3|$ and $|K_3|$ become good quantum numbers, and the energy of a quantum state is

$$E_{J^P,I,|L_3|,|K_3|} = \frac{1}{2V_{11}}J(J+1) + \frac{1}{2U_{11}}I(I+1) + \left(\frac{1}{2V_{33}} - \frac{1}{2V_{11}}\right)|L_3|^2$$
$$+ \left(\frac{1}{2U_{33}} - \frac{1}{2U_{11}}\right)|K_3|^2. \tag{9.17}$$

The parity is positive (negative) if $|L_3| + |K_3|$ is even (odd).

For isospin 0, the FR constraints (9.14) require that $|L_3| = 0 \mod 3$ and $J$ is even if $L_3 = 0$. The parity is positive (negative) if $|L_3|$ is even (odd). The allowed states are therefore exactly the same as those resulting from the rigid-body quantization of an equilateral triangle of bosonic $\alpha$-particles. Up to spin 6 the allowed states have spin/parities $J^P = 0^+, 2^+, 3^-, 4^-, 4^+, 5^-, 6^+, 6^-, 6^+$, where the negative parity states have $|L_3| = 3$ and the second $6^+$ state has $|L_3| = 6$. These are the predicted states of Carbon-12 in the ground state rotational band, arising from the rigid-body quantization of the $D_{3h}$-symmetric Skyrmion.

The spin/parities and energies (9.17) for all the allowed states up to spin 6 for isospin 0, spin 2 for isospin 1, and spin 1 for isospin 2 are

$$E_{0^+,0,0,0} = 0,$$
$$E_{2^+,0,0,0} = 3/V_{11},$$
$$E_{3^-,0,3,0} = 3/2V_{11} + 9/2V_{33},$$
$$E_{4^-,0,3,0} = 11/2V_{11} + 9/2V_{33},$$
$$E_{4^+,0,0,0} = 10/V_{11},$$
$$E_{5^-,0,3,0} = 21/2V_{11} + 9/2V_{33},$$
$$E_{6^-,0,3,0} = 33/2V_{11} + 9/2V_{33},$$
$$E_{6^+,0,0,0} = 21/V_{11},$$
$$E_{6^+,0,6,0} = 3/V_{11} + 18/V_{33},$$
$$E_{1^+,1,1,1} = 1/V_{11} + 1/U_{11},$$
$$E_{1^+,1,0,0} = 1/2V_{11} + 1/2V_{33} + 1/2U_{11} + 1/2U_{33},$$
$$E_{2^-,1,2,1} = 1/V_{11} + 2/V_{33} + 1/2U_{11} + 1/2U_{33},$$
$$E_{2^+,1,1,1} = 5/2V_{11} + 1/2V_{33} + 1/2U_{11} + 1/2U_{33},$$
$$E_{0^+,2,0,0} = 3/U_{11},$$
$$E_{1^-,2,1,2} = 1/2V_{11} + 1/2V_{33} + 1/U_{11} + 2/U_{33},$$
$$E_{1^+,2,1,1} = 1/2V_{11} + 1/2V_{33} + 5/2U_{11} + 1/2U_{33}. \quad (9.18)$$

An approximate spectrum was obtained from these formulae in ref.[29], using estimated inertia tensor values. Better is to use the exact energy formulae allowing for a non-zero $W$-tensor, together with the inertia tensor values in the final column of Table 9.10. The results are shown in Table 9.11. The table also gives the energies in physical units, obtained using the conversion factor $e^3 F_\pi = 6216$ MeV from Table 9.1.

### 9.6.2 Comparison with experimental data

The isospin 0 states in the ground state rotational band of Carbon-12 up to $J^P = 5^-$ have all been observed [94,151,185], although for some time the $4^-$ state was misidentified as a $2^-$ state [11]. The evidence for an intrinsic $D_{3h}$ symmetry is clear from the spin/parities occurring in this band. However, because $V_{33}/V_{11} = 1.51$ for the Skyrmion, the states with $|L_3| = 3$ are predicted to be about 2 MeV below those having the same spin with $|L_3| = 0$, and the data are hardly consistent with this. In Section 10.4, we will consider this issue further. The three spin 6 states are characteristic of any model involving rigid rotations of an equilateral triangle, although their

Table 9.11: Energy levels of the $D_{3h}$-symmetric $B = 12$ Skyrmion. To each quantum state there correspond dominant values of $|L_3|$ and $|K_3|$.

| $I$ | $J^P$ | $|L_3|$ | $|K_3|$ | $E \times 10^4$ | $E$ (MeV) |
|---|---|---|---|---|---|
| 0 | $0^+$ | 0 | 0 | 0.0 | 0.0 |
|  | $2^+$ | 0 | 0 | 5.1 | 3.2 |
|  | $3^-$ | 3 | 0 | 7.6 | 4.7 |
|  | $4^-$ | 3 | 0 | 14.4 | 8.9 |
|  | $4^+$ | 0 | 0 | 17.0 | 10.6 |
|  | $5^-$ | 3 | 0 | 22.9 | 14.2 |
|  | $6^-$ | 3 | 0 | 33.1 | 20.6 |
|  | $6^+$ | 0 | 0 | 35.7 | 22.2 |
|  |  | 6 | 0 | 25.3 | 15.7 |
| 1 | $1^+$ | 1 | 1 | 19.4 | 12.1 |
|  |  | 0 | 0 | 20.7 | 12.9 |
|  | $2^-$ | 2 | 1 | 21.8 | 13.5 |
|  | $2^+$ | 1 | 1 | 22.8 | 14.2 |
| 2 | $0^+$ | 0 | 0 | 56.9 | 35.4 |
|  | $1^-$ | 1 | 2 | 54.1 | 33.6 |
|  | $1^+$ | 1 | 1 | 57.4 | 35.7 |
|  | $2^-$ | 1 | 2 | 57.5 | 35.7 |
|  |  | 2 | 1 | 59.7 | 37.1 |
|  | $2^+$ | 2 | 2 | 57.1 | 35.5 |
|  |  | 1 | 1 | 60.6 | 37.7 |
|  |  | 0 | 0 | 62.3 | 38.7 |

precise energy ratios depend on $V_{33}/V_{11}$. No observed states of Carbon-12 with isospin 0 have yet been confirmed as having spin 6 [247]. A new search for spin 6 states would be worthwhile.

Isospin excitations of the $B = 12$ triangular Skyrmion give further states. Two isospin 1 triplets are predicted with $J^P = 1^+$. One such isotriplet is observed, and includes the ground states of Nitrogen-12 and Boron-12 with mean excitation energy 15.1 MeV, to be compared to the predicted value of 12.1 MeV. These states may be an unresolved doublet of isotriplets. Also predicted are a $2^+$ and a $2^-$ isotriplet, with the $2^-$ isotriplet lying just below the $2^+$ isotriplet. Experimentally, these states are seen but in the opposite energy order. Higher-energy $1^-$ and $0^+$ isotriplets are observed too, but they are not predicted as rigid-body excitations of the $D_{3h}$-symmetric Skyrmion. An incomplete $3^-$ isotriplet is observed at 18.5 MeV, and predicted at 16.4 MeV. An isolated $4^-$ state of Boron-12 is observed at 19.7 MeV, and predicted at 19.0 MeV.

A nearly complete isospin 2 quintet with $J^P = 0^+$, which includes the ground states of Oxygen-12 and Beryllium-12, is observed with an average excitation energy of 27.7 MeV. Such an isoquintet is predicted at 35.4 MeV. However, the theory also predicts the existence of a $1^-$ isoquintet with an excitation energy less than that of the $0^+$ isoquintet, and such states are only seen at higher energy. An incomplete $2^+$ isoquintet is experimentally observed with an average excitation energy roughly 2 MeV above the $0^+$ isoquintet. Theoretically, three $2^+$ isoquintets are expected, just above the $0^+$ isoquintet.

### 9.6.3 The chain Skyrmion and the Hoyle band

The ground state rotational band of Carbon-12 doesn't account for all the isospin 0 states of this nucleus. To model more states it is useful to combine the rigid-body quantum states of the triangular $B = 12$ Skyrmion with those of the chain Skyrmion, as this is one way to obtain an excited $J^P = 0^+$ state – the Hoyle state. Here, we set $m = 1$. Both the $D_{3h}$-symmetric and $D_{4h}$-symmetric Skyrmions have spatial moment of inertia tensors of symmetric-top type, with two distinct eigenvalues $V_{33}$ and $V_{11} = V_{22}$. $V_{33}$ is larger than $V_{11}$ for the oblate, triangular Skyrmion, and smaller for the prolate, chain Skyrmion.

In both cases, the quantum Hamiltonian for purely rotational motion is

$$H = \frac{1}{2V_{11}}\mathbf{J}^2 + \left(\frac{1}{2V_{33}} - \frac{1}{2V_{11}}\right) L_3^2, \qquad (9.19)$$

and the energy eigenvalues are simply

$$E_{J,|L_3|} = \frac{1}{2V_{11}} J(J+1) + \left(\frac{1}{2V_{33}} - \frac{1}{2V_{11}}\right) |L_3|^2. \qquad (9.20)$$

The moments of inertia for the triangular Skyrmion are $V_{11} = 5037$ and $V_{33} = 7684$ when $m = 1$, as given in Table 9.10. The moments of inertia for the chain Skyrmion are $V_{11} = 12699$ and $V_{33} = 2106$ [162]. These values are similar to what one finds using the single moment of inertia of the $B = 4$ Skyrmion and the parallel axis theorem [254].

However, the FR constraints are different for the chain and triangular Skyrmions. For the chain Skyrmion, oriented with its long axis along the $x^3$-axis, the constraints are

$$e^{i\frac{\pi}{2} L_3} e^{i\pi K_1} |\Psi\rangle = |\Psi\rangle, \quad e^{i\pi L_1} e^{i\pi K_1} |\Psi\rangle = |\Psi\rangle, \qquad (9.21)$$

and the parity operator is $\mathcal{P} = e^{i\pi K_3}$. The states with isospin 0 therefore all have positive parity. The lowest-energy states have $L_3 = 0$ and $J$ even.

The resulting $0^+$, $2^+$ and $4^+$ states are identified as the rotational band of the Hoyle state (the Hoyle band). There is a further $4^+$ state with $|L_3| = 4$ but this has much higher energy, and higher still is a $5^+$ state. There is therefore no ambiguity assigning the observed $5^-$ state to the ground state band of the triangular Skyrmion.

Because the ground state and Hoyle state of Carbon-12 are both $0^+$ states, their energies are simply the classical Skyrmion energies in the rigid-body approach. The energies of the triangular and chain Skyrmions are estimated as, respectively, 1816 and 1812 in Skyrme units, which is unfortunately not the preferred order, but the difference of these large numbers is not very certain. Figs. 8.3 and 8.4 show that the chain Skyrmion has two strong bonds between the $B = 4$ cubes, whereas the triangular Skyrmion has three weaker bonds, so the total energies are similar. The classical energy of a strong bond is about 14.2 in Skyrme units, corresponding to about 93 MeV, but this gives no useful prediction for the 7.65 MeV energy difference between the Hoyle state and the Carbon-12 ground state, which is just 0.07% of the total mass of Carbon-12.

We simply accept the observed energy difference between the $0^+$ states, and add the rotational energy (9.20) for the states of higher spin. The resulting predictions for the energies of the states in the ground state band and the Hoyle band are shown in Fig. 9.5, together with the observed energies. The Skyrmions allow us to confidently predict the slopes of these rotational bands, because they only depend on the moments of inertia. For each of the $L_3 = 0$ bands there are observed $J^P = 0^+, 2^+, 4^+$ states, and additionally there are $3^-$ and $5^-$ states in the ground state band. Skyrme theory predicts the ratio of the slopes to be just the ratio of the $V_{11}$ values for the chain and triangular Skyrmions, which is $12699/5037 = 2.52$. One can estimate the ratio of the experimental slopes from the energy differences between the $4^+$ and $0^+$ states. For the ground state band the energy difference is 14.1 MeV, and for the Hoyle band it is $13.3 - 7.65 = 5.65$ MeV. The ratio is 2.50, agreeing with the prediction [162].

The $V_{11}$ ratio 2.52 reflects the extended structure and separation of the $B = 4$ Skyrmion subunits. For three ideal point $\alpha$-particles with a fixed bond length separating them, arranged as an equilateral triangle and as a linear chain, the ratio of the $V_{11}$ values is 4. For Skyrmions the ratio is smaller, partly because the bond is less tight in the triangular Skyrmion, but mainly because of the extended form of the $B = 4$ cubes.

The Hoyle state excitations with $|L_3| = 4$, starting with a $4^+$ state, and including $5^+$ and $6^+$ states, are of considerably higher energy because of the

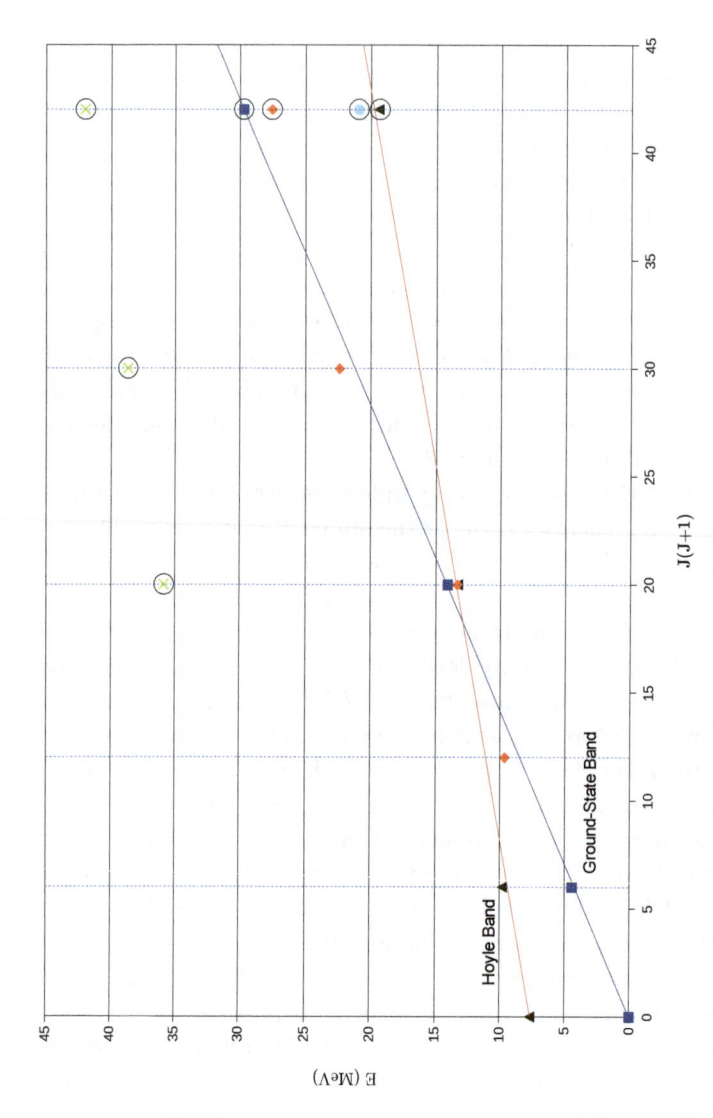

Fig. 9.5: Carbon-12 states in the ground state band and Hoyle band. The Skyrme theory predictions for the band slopes are shown as the blue and red lines. Observed states assigned to the ground state band are shown as blue squares ($L_3 = 0$) and red diamonds ($|L_3| = 3$); observed states in the Hoyle band are shown as black triangles ($L_3 = 0$). Further predicted states are encircled (light blue disc denotes $|L_3| = 6$ in the ground state band, and green cross denotes $|L_3| = 4$ in the Hoyle band).

strongly prolate nature of the chain Skyrmion, and have not been observed. In total, five states with spin 6 are predicted, three in the ground state band (with $|L_3| = 0$, 3 and 6) and two in the Hoyle band (with $|L_3| = 0$ and 4). Of these, the two lowest are the $|L_3| = 6$ state in the ground state band and the $L_3 = 0$ state in the Hoyle band, with predicted energies close to 20 MeV. Only the $|L_3| = 3$ state has negative parity.

The chain Skyrmion also has isospin excitations, which supplement the states derived using the triangular Skyrmion. The allowed states and their excitation energies have been investigated by Wood [254]. There are no off-diagonal terms or spin/isospin cross terms here. The lowest-energy isospin 1 excitation of the Hoyle state is predicted to have $J^P = 0^-$, with energy a few MeV above the four isospin 1 states of the triangular Skyrmion tabulated above.

Some states of Carbon-12 with isospin 0 are not modelled by rigid-body excitations of either of the Skyrmions we have discussed. The first of these is the second-excited $0^+$ state at 9.9 MeV and the next two are the $1^-$ and $2^-$ states below 12 MeV. In the next chapter we will consider models allowing for Skyrmion deformations, which can accommodate these states.

### 9.6.4 Matter radii

The root-mean-square matter radius

$$\langle r^2 \rangle^{\frac{1}{2}}_{\text{matter}} = \left( \frac{\int r^2 \, \mathcal{E}(\mathbf{x}) \, d^3x}{\int \mathcal{E}(\mathbf{x}) \, d^3x} \right)^{\frac{1}{2}} \qquad (9.22)$$

provides a further test of the Skyrme theory picture for Carbon-12. $\mathcal{E}(\mathbf{x})$ is the static energy density, interpreted as a matter density. The triangular and chain Skyrmions have $\langle r^2 \rangle^{\frac{1}{2}}_{\text{matter}}$ values 2.28 and 2.80 respectively, in Skyrme units, whose ratio is 1.23. The experimental matter radius for the Carbon-12 ground state is 2.43 fm [224] and for the Hoyle state it is inferred to be 2.89 fm [68], giving a ratio of 1.19. The prediction of Skyrme theory is good here.

Replacing $\mathcal{E}$ by the baryon density $\mathcal{B}$ gives radii 1% smaller, as the baryon density has a more compact tail than the energy density. These radii are the Skyrme theory predictions for the charge radii of the Carbon-12 ground state and Hoyle state, because the charge density is half the baryon density for isospin 0 states. The ratio is still 1.23.

In the point $\alpha$-particle model the ratio of matter radii is $\sqrt{2}$, and other models predict larger ratios [65]. The smaller experimental ratio has been a reason to prefer models where the Hoyle state is not a linear chain of α-particles, but rather an obtuse triangular structure [79]. Skyrme theory seems to favour the linear chain, with its large $D_{4h}$ symmetry group. A linear chain admits far fewer low-lying rotational states than an obtuse triangle, just the $J^P = 0^+, 2^+, 4^+$ states for which there is clear experimental evidence. The expectation from Skyrme theory is that rotational excitations with $J^P = 3^-$ and $4^-$, suggested in [185], will not be seen in the Hoyle band. States with these spin/parities require a deformation of the linear chain and therefore have higher energy.

# Chapter 10

# Skyrmion Deformations and Vibrations

## 10.1 The Need to Consider Vibrations

In addition to collective rotational excitations, it is well known that nuclei can vibrate. The vibrational frequencies are such that the excitation energies of single vibrational phonons are comparable to the excitation energies associated with rotations. In other words, the phonon energy $\hbar\omega$, where $\omega$ is a typical vibrational frequency, is comparable to the rotational energy gap $\hbar^2/V$ between a spin 0 and spin 1 state, where $V$ is a typical moment of inertia. This makes nuclei different from small molecules, whose vibrational energies are much greater than the rotational energies. In practice, it can become difficult to interpret energy spectra of nuclei and distinguish vibrational from rotational excitations. It is necessary to consider the fine detail of the spin/parities and energies in order to find a convincing interpretation.

The importance of deformations and vibrations of Skyrmions was recognised early on [238, 239]. To model the deuteron, with its small binding energy, the toroidal $B = 2$ Skyrmion must be allowed to deform towards a configuration with two separated $B = 1$ Skyrmions [166], as discussed in Section 7.4. Selected vibrational modes of the $B = 4$ Skyrmion were considered in Walhout's model of the $\alpha$-particle and its excitations [240]. Recall that a purely rotational excitation of the $B = 4$ Skyrmion has spin 4 or more, and a high energy. A model that allows for vibrational excitations coupled to rotations can give an improved spectrum for Helium-4, including states of lower spin.

Vibrational modes of the $B = 2$, $B = 3$ and $B = 4$ Skyrmions were studied by Barnes, Baskerville and Turok [21, 24] (making some approximations), and this study was extended by Baskerville to $B = 7$ [23]. Small random perturbations of the underlying stable Skyrmions were evolved dynamically, and the frequencies of the vibrational modes extracted by Fourier analysis. The degeneracies of the vibrational modes, which depend on the intrinsic symmetry of each Skyrmion, were noted. The shapes of the distorted, vibrating Skyrmions were also illustrated for each mode.

A more broad-ranging and systematic study of Skyrmion vibrations has been carried out by Gudnason and Halcrow [108], and the vibrational modes are illustrated with beautiful animations on their website [109]. Their results are for all Skyrmions (with pion mass parameter $m = 1$) up to $B = 8$, and they confirm the earlier results for particular Skyrmions but go much further in classifying the vibrational modes, their symmetry character, and their frequencies. All the frequencies are between $\omega = 0$ and $\omega = 1.5$ in Skyrme units. Since $m = 1$, the modes with $\omega < 1$ are genuine normal modes of oscillation, but for $\omega > 1$ they are quasi-normal modes that possess long-range tails and slowly radiate away. These modes would become genuine normal modes if $m$ were increased. In the numerical simulations carried out in a finite box, it is not easy to distinguish normal from quasi-normal modes.

Quite generally, the vibrations can be classified in the following way. The lowest modes have zero frequency and are called *zero modes*. They correspond to the motions induced by the action of the symmetry generators of Skyrme theory. For Skyrmions with $B = 3$ or more, there are nine zero modes, associated with translations, rotations and isorotations of the Skyrmion. For the $B = 1$ hedgehog there are only six zero modes, because rotations are equivalent to isorotations, and for the toroidal $B = 2$ Skyrmion there are eight zero modes, because a rotation about the main symmetry axis is equivalent to an isorotation. The next-lowest modes, with positive frequency, are associated with deformations of the Skyrmion that tend to break it up into clusters with smaller baryon numbers. These are called *monopole modes* because non-abelian monopoles have analogous breakup modes. Higher than these, and separated by a significant frequency gap, is the non-degenerate *breather mode*, where the Skyrmion's overall size oscillates but the Skyrmion's symmetry is unbroken. Above the breather are more complicated modes, where individual parts of the Skyrmion have oscillating sizes. Some of these higher-frequency modes can be interpreted in terms of relative isorotations between different parts.

The zero modes and monopole modes can together be interpreted as resulting from deformations of the rational map describing the Skyrmion. The numerator and denominator of the rational map for a Skyrmion with baryon number $B$ are complex polynomials of algebraic degree $B$. If the coefficients are all freely varying, the map has $2B+1$ complex parameters. (The numerator and denominator each have $B+1$ parameters, but one factor cancels between them.) It is more convenient to discuss real parameters, so there are $4B+2$ of these[1]. The results of Gudnason and Halcrow confirm that the Skyrmions from $B=3$ up to $B=7$, and the most stable Skyrmion with $B=8$, have $4B+2$ vibrational modes below what is clearly a non-degenerate breather mode.

Lin and Piette estimated the frequencies of the monopole modes for $B=2$ and $B=4$ by explicitly varying the parameters of the Skyrmion's rational map [169]. Qualitatively, they found the correct modes, with their degeneracies determined by the Skyrmion's symmetry. However, the frequencies were substantially wrong and not in the right order, because the rational map ansatz is too rigidly defined. It doesn't properly allow Skyrmions to separate into subclusters. It would be interesting to find a generalisation of the rational map ansatz that produces Skyrme fields more closely resembling non-abelian monopole field configurations. Monopoles tend to properly separate into subclusters as the rational map parameters vary.

Note that for the $B=1$ Skyrmion, a total of six zero modes and monopole modes are expected, and these are all in fact zero modes. The first positive-frequency mode is the breather, an example of a quasi-normal mode. Excitation of the breather mode, combined with rotational quantization, gives a model for the Roper resonance states of nucleons, which have spin $\frac{1}{2}$ and isospin $\frac{1}{2}$ [1]. The Roper resonances have an energy exceeding the nucleon masses by approximately 500 MeV, so they are more energetic than the delta resonances, and broad. This kind of excitation energy is far larger than what is usually considered in nuclear physics and suggests that vibrational modes of higher-$B$ Skyrmions above the breather mode, where parts of the Skyrmion change size, can be assumed to be unexcited. This includes all the quasi-normal modes of Skyrmions.

The total number of vibrational modes of a Skyrmion is potentially unlimited, because the field can deform locally at arbitrary short length scales. However, there have been a few attempts to find a natural cutoff. We have mentioned the $4B+2$ zero and monopole modes as one candidate. However, this constrains the relative orientations if the initial Skyrmion separates entirely into $B=1$ Skyrmions. Instead, since a $B=1$ Skyrmion

---

[1]Non-abelian monopoles have just $4B-1$ parameters arising from the same rational maps, because the three parameters associated with isorotations of a Skyrmion correspond to gauge transformations of a monopole.

has six modes, it is tempting to seek $6B$ modes for a Skyrmion with baryon number $B$. For example, to discuss two-nucleon scattering within Skyrme theory, it would be desirable to have a set of Skyrme fields with 12 degrees of freedom. A candidate for a 12-dimensional manifold of Skyrme field configurations has been proposed [16], but its details are hard to calculate, and it has not yet been possible to use this manifold in practice. Even more ambitiously, Gudnason and Halcrow have argued, based of their study of higher-frequency Skyrmion vibrations, that $7B$ modes is a more natural choice [108]. This allows each $B = 1$ Skyrmion constituent to have an independent position, orientation and also size.

One reason for wishing to find the complete spectrum of vibrations of a Skyrmion is to calculate the total zero-point energy of the vibrational modes. In principle this should be added to the classical Skyrmion energy (plus any contribution from the spin and isospin) to obtain an estimate of the total energy of a nucleus, treating Skyrme theory as a true quantum field theory. However, this type of calculation is not feasible for general $B$. The number of modes is really infinite, and even if truncated, the sum of the zero-point energies is very large. It may be possible to renormalise this energy by subtracting some large constant, for example $B$ times the sum of the zero-point energies of the $B = 1$ Skyrmion. A calculation like this may work, because the number of modes (after truncation) is essentially proportional to $B$, but it seems not to have been attempted. The effect of all the vibrational modes, even in their ground states, is to produce a fuzzier version of the Skyrmion, and probably increase its size. This could help to bring the rather precisely shaped Skyrmion closer to the more uniform density that is observed in the interior of nuclei, and reduce the Skyrmion binding energies. However, the zero-point motion of Skyrmion vibrations is usually ignored, and instead the parameters of Skyrme theory are adjusted to match the sizes and masses of real nuclei.

In practice, to model the excitation spectra of larger nuclei, one needs to consider far fewer than $7B$ or even $4B$ modes. The lowest modes of positive frequency invariably correspond to Skyrmion motions involving relatively large subclusters. Some of these clearly contribute to nuclear spectra, but the choice of modes for larger $B$ has to be selective. Rather than consider many modes, it is more effective to treat just a few modes in an anharmonic way, in order to model the motion and rearrangement of key subclusters of the Skyrmion.

For example, for $B = 12$ the best current model allows three $B = 4$ subclusters ($\alpha$-particles) to move freely in space. This requires just nine degrees of freedom, of which six are zero modes. In equilibrium, the $B = 12$ Skyrmion has an equilateral triangular shape. The positive-frequency vibrational modes are a degenerate pair that change the angular shape of the triangle, and a breather mode that changes its size. Even this description can be simplified. Rawlinson found a useful model for the spectrum of Carbon-12 where just one degree of freedom of the triangle is excited [203], allowing the shape to deform from an equilateral triangle via isosceles triangles into a straight chain. This model is discussed below. It improves on the model for Carbon-12 that was described in Section 9.6, where three $B = 4$ Skyrmions could form either an equilateral triangle or a straight chain. Rawlinson's model interpolates between these, and leads to additional states in the quantized dynamics. Less symmetric triangular deformations can account for a few more states, giving a complete low-energy spectrum for Carbon-12.

Another example where it helps to carefully consider vibrational excitations is the $B = 7$ Skyrmion [113]. In its unexcited state this Skyrmion has icosahedral symmetry, and its lowest allowed spin/parity is $J^P = \frac{7}{2}^-$. In a 1-phonon vibrational state the spin can be lower, as required phenomenologically. A final example we mention here is the $B = 32$ Skyrmion. An estimate for its vibrational excitation spectrum has been found by Feist, by modelling the Skyrmion as a cluster of eight $\alpha$-particles at the vertices of a cube, connected by springs [83].

In the remainder of this chapter we will discuss quantized vibrational models for the Skyrmions with baryon numbers $B = 4$, 7 and 12, examples that have been studied in some detail. When the vibrations are treated using a harmonic oscillator approximation, their coupling to rotations results in *rovibrational states*. The coupling introduces Coriolis terms to the Hamiltonian in some cases. The example of $B = 16$ and Oxygen-16 is postponed until Chapter 11, because here the $\alpha$-particle dynamics is modelled nonlinearly, and the spectrum is more complicated. A rovibrational model for the excited states of Calcium-40, using a tetrahedrally-symmetric $B = 40$ Skyrmion, is discussed in Chapter 12. This just involves harmonic vibrational excitations, but it is quite elaborate because there are so many observed states to consider.

## 10.2 The $\alpha$-Particle and its Vibrational Excitations

The ground state of Helium-4 (the $\alpha$-particle) has spin/parity $0^+$. Being a small doubly-magic nucleus, it has no low-lying excited states. The first-excited state has energy more than 20 MeV, just above the breakup threshold for Helium-4. So analysis of excited states in terms of vibrational harmonic oscillator states is an approximation, and nonlinear effects should be considered. Walhout [240] and (much more recently) Rawlinson [203] have been able to understand the Helium-4 spectrum in terms of vibrations and rotations of the cubic $B = 4$ Skyrmion quite well, and their modelling takes some account of the anharmonic aspects.

The quadrupole vibrations have lowest frequency. For a spherical body, such vibrations are 5-fold degenerate and the 1-phonon excited states have spin/parity $2^+$. However, the $B = 4$ Skyrmion has only octahedral symmetry $O_h$, and the quadrupole vibrations split into a doublet and a triplet with rather similar frequencies. This is analogous to the splitting of quadratic Cartesian polynomials into the doublet $2z^2 - x^2 - y^2$ and $2x^2 - y^2 - z^2$ (with $2y^2 - z^2 - x^2$ linearly related to these), and the triplet $xy, yz$ and $zx$. A further low-frequency mode is the non-degenerate mode that breaks octahedral but preserves tetrahedral symmetry. In one direction four separate Skyrmions are pulled out from alternating vertices of the $B = 4$ cube, forming a tetrahedron; in the opposite direction they form the dual tetrahedron. This mode is analogous to the Cartesian monomial $xyz$. The final modes below the breather are a degenerate triplet of modes involving relative size changes of the edges of the cube.

The breather mode is next, but occurs above a significant gap in the frequencies, so it makes sense to limit the dynamical analysis to the lower-frequency modes below this. Rawlinson has modelled the rovibrational excited states of the $\alpha$-particle in the range 20 – 30 MeV in terms of these four groups of vibrational modes, which transform respectively as irreps $E^+, F_2^+, A_2^-$ and $F_2^-$ of $O_h$. Their frequencies are in the range 0.46 – 0.62 in Skyrme units [108]. Phenomenologically, it appears that the observed spectrum can be fitted using rovibrational states of just the 1-phonon and 2-phonon excitations of the $F_2^+$-, $A_2^-$- and $F_2^-$-modes. The Hamiltonian incorporates a Coriolis coupling between the vibrational and rotational motion that occurs for vectorial F-modes. The vibrational frequencies are treated as free parameters in Rawlinson's model – an improvement on the calculated Skyrmion frequencies that can account for the anharmonic effects of large vibrational amplitudes and Skyrmion subclustering. In total, the model incorporates six parameters – three vibrational frequencies, a rotational moment of inertia, and two Coriolis couplings. The physical motivation for ignoring the quadrupole E-mode is that $\alpha$-particle breakup into two deuterons via this mode requires more energy than breakup into

Table 10.1: Spectrum of α-particle excitations. The spin/parity, number of phonons of each type, and energy of each state in MeV are shown. $E$ is the energy in Rawlinson's model, $E_{\text{exp}}$ is the observed energy of the associated state, and $E_{\eta=0}$ is the model energy if the Coriolis couplings are neglected.

| $J^P$ | $N_{F_2^-}$ | $N_{F_2^+}$ | $N_{A_2^-}$ | $E$ | $E_{\text{exp}}$ | $E_{\eta=0}$ |
|---|---|---|---|---|---|---|
| $0^+$ | 0 | 0 | 0 | 0 | 0 | 0 |
| $0^+$ | 2 | 0 | 0 | 19.4 | 20.2 | 19.4 |
| $0^-$ | 1 | 1 | 0 | 21.9 | 21.0 | 21.4 |
| $2^-$ | 1 | 0 | 0 | 22.2 | 21.8 | 21.7 |
| $0^+$ | 0 | 2 | 0 | 23.4 |  | 23.4 |
| $1^-$ | 1 | 1 | 0 | 24.2 | 24.3 | 25.4 |
| $1^+$ | 1 | 0 | 1 | 28.3 | 28.3 | 28.8 |
| $1^-$ | 0 | 1 | 1 | 28.5 | 28.4 | 30.8 |
| $2^-$ | 1 | 1 | 0 | 28.9 | 28.4 | 33.4 |
| $0^-$ |  |  |  |  | 28.6 |  |
| $2^+$ | 0 | 2 | 0 | 28.4 | 28.7 | 35.4 |
| $2^+$ | 2 | 0 | 0 | 29.9 | 29.9 | 31.4 |

Hydrogen-3 and a proton (or Helium-3 and a neutron) via an F-mode, so the E-mode excitations (whose minimal spin is 2) are of high energy. The spectrum of α-particle excitations in the model is tabulated in Table 10.1, and compared with the observed spectrum of isospin 0 states of Helium-4. It is shown again in Fig. 10.1.

In the interpretation of Walhout, and others, the relatively low-lying $J^P = 0^+$ state at 20.2 MeV is a 1-phonon excitation of the breathing mode. In Rawlinson's model, by contrast, this state is interpreted as a 2-phonon excitation of the $F_2^-$-mode, so a single $F_2^-$-phonon has a fairly low energy near 10 MeV. The $F_2^+$-phonon has a similar energy. The lowest allowed spin/parity for a 1-phonon $F_2^-$ excitation is $2^-$, with an associated rotational energy of about 10 MeV and hence the total energy of the $2^-$ state is above that of the 20.2 MeV $0^+$ state, despite having one fewer phonon. This is an illustration of vibrational and rotational energies being of similar magnitude in nuclei. In all, Rawlinson's model accommodates 11 states up to 30 MeV, all with spins 0, 1 or 2. They match the observed states of Helium-4 rather well, in terms of spin/parity and energy. A $0^+$ state is predicted at about 23 MeV that has not been observed, and a $0^-$ state is observed at 28.6 MeV that is not captured by the model. This $0^-$ state cannot be interpreted as a 1-phonon excitation of the $A_2^-$ tetrahedral vibration, as that leads to states with minimal spin 3 when coupled to rotations (the vibrational wavefunction changes sign under a 90° rotation about each Cartesian axis). More likely, it is a 3-phonon state of F-mode excitations. The purely rotational $4^+$ excitation, mentioned in Chapter 7, is predicted to have energy 40 – 50 MeV, well above the energy of all the excitations with smaller spin for which there is clear experimental evidence.

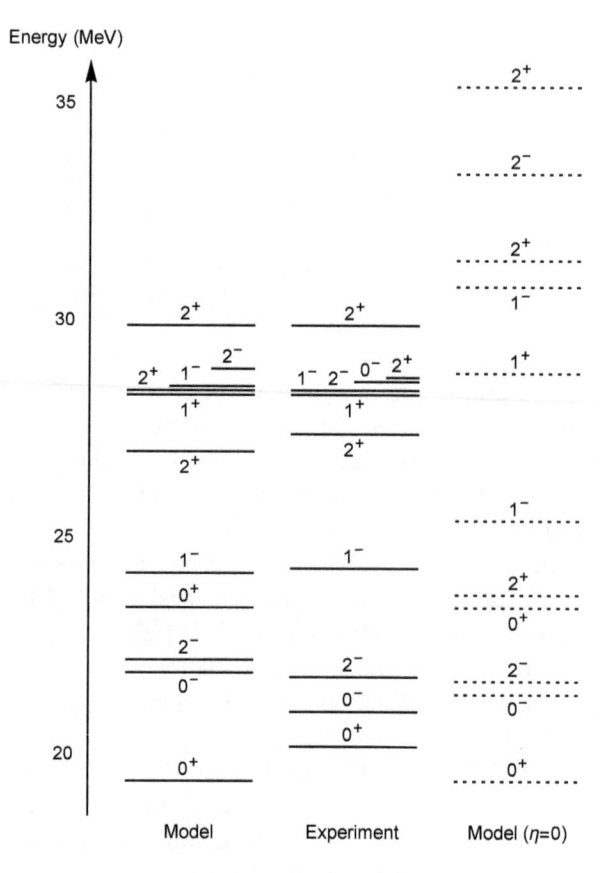

Fig. 10.1: Spectrum of $\alpha$-particle excitations

We will see later that low-lying, excited $0^+$ states of some larger nuclei can be interpreted as 2-phonon states of a shape-deforming vibrational mode, rather than as a 1-phonon state of a breather mode.

## 10.3 Deformations of the $B = 7$ Skyrmion

Recall that the $B = 7$ Skyrmion has $Y_h$ symmetry and a dodecahedral shape with 12 holes in the baryon density, as shown in Fig. 7.4. Rigid-body quantization requires states with half-integer spin and isospin because $B$ is odd. The FR constraints are quite tricky to find and solve, because of the subtle geometry of the dodecahedron. As usual, the states of minimal energy are those with isospin $\frac{1}{2}$, as the isospin moments of inertia are substantially less than the spin moments of inertia. These should model the ground state and some of the excited states of the isodoublet nuclei Beryllium-7 and Lithium-7. In rigid-body quantization, the lowest allowed spin/parity is $\frac{7}{2}^-$ [135], as we explained in Section 7.7. Spin $\frac{7}{2}$ is large for small nuclei like Beryllium-7/Lithium-7. It is only slightly less than the predicted spin 4 for the rotationally-excited $\alpha$-particle – a state that has not been seen. The estimated spin energy is about 20 MeV. Beryllium-7/Lithium-7 both have a $\frac{7}{2}^-$ state, but it is not their ground state, which has spin/parity $\frac{3}{2}^-$. The mean energy of the $\frac{7}{2}^-$ state is, however, only 4.6 MeV above the ground state and is less than the excitation energy of the lowest state with spin $\frac{5}{2}$ – a feature unique among small nuclei with odd baryon numbers. This suggests that the ground state is a rotational excitation of the $Y_h$-symmetric Skyrmion deformed to have lower symmetry than $Y_h$, but at some energy cost. The easiest way to realise this deformation is to assume that the Skyrmion has been excited to a 1-phonon state of a suitable vibrational mode. The 1-phonon quantum state has its maximal amplitude at some deformed shape, and the amplitude vanishes at the Skyrmion itself.

Gudnason and Halcrow's calculations show that the $B = 7$ Skyrmion has 16 vibrational modes below the breather. These have only four distinct frequencies because the $Y_h$ symmetry leads to quite large frequency degeneracies. The lowest-frequency, 5-fold degenerate modes transform under the irrep $H_g$ of $Y_h$. The preferred 1-phonon state is aligned so that the Skyrmion is stretched symmetrically between a pair of dodecahedral vertices. The advantage of this alignment is that a slight further stretch, breaking the symmetry, deforms the Skyrmion into $B = 3$ and $B = 4$ subclusters, as shown in Fig. 7.4. This is the nonlinear deformation that requires least energy. It is along a rather flat direction in the landscape of $B = 7$ Skyrme field configurations.

Halcrow has exploited this to understand Beryllium-7/Lithium-7 better [113]. The allowed states of the deformed Skyrmion can be found by restricting the FR constraints to those of the residual symmetry group $C_{3v}$. The lowest state now has spin/parity $\frac{3}{2}^-$, correctly modelling Beryllium-7/Lithium-7. Its energy is the sum of the phonon energy and the spin energy, whose total is estimated to be a few MeV less than the pure spin energy of the $\frac{7}{2}^-$ state, although this depends on the calibration and requires some modelling of the anharmonic character of the vibration and of the way the phonon changes the moment of inertia. A significant prediction is that the vibrationally unexcited $\frac{7}{2}^-$ state is spatially more compact than all the states of lower spin requiring a deformed Skyrmion. Its matter radius is estimated to be 7% smaller than that of the $\frac{3}{2}^-$ ground state.

Further Beryllium-7/Lithium-7 states are predicted in the 1-phonon sector of the $H_g$-mode. One has spin $\frac{1}{2}$ and two have spin $\frac{5}{2}$. This fits the observed spectrum of states quite well, although the energy ordering is not right. Rovibrational states of the $B = 7$ Skyrmion have not been so systematically investigated as those of the $B = 4$ Skyrmion, and further improvement of the modelling is desirable. The calculations are trickier, because of the non-zero isospin and the need to work with rotational double groups. In summary, the main conclusion of Halcrow's study is that through deformations of the $B = 7$ Skyrmion, the long-standing, incorrect Skyrme theory prediction of a spin $\frac{7}{2}$ ground state for Beryllium-7/Lithium-7 can be circumvented.

## 10.4  $B = 12$ Skyrmion Deformations and Carbon-12

The existence of two $B = 12$ Skyrmions, one having the form of an equilateral triangle of cubic $B = 4$ subunits and the other a straight chain, has enabled an understanding of several states of Carbon-12 – those in the ground state band and those in the Hoyle band – as we saw in Section 9.6. Further states arise through deformations of these two basic Skyrmions. The $B = 4$ Skyrmion itself is hard to excite, and below 20 MeV it behaves like an $\alpha$-particle with no substructure. Excited states of Carbon-12 therefore arise mainly from the relative motion of the three $B = 4$ constituents of the $B = 12$ Skyrmions.

Rawlinson has developed a model with just one shape degree of freedom, allowing for a deformation of the equilateral triangle into the straight chain via isoceles triangles [203]. We will discuss this here, and in the next subsection discuss in a more qualitative way two overlapping models that allow for harmonic shape vibrations of either the equilateral triangle or the straight chain, and how they mesh together. These models have three

$$\mathbf{x}_1 = (0, s)$$

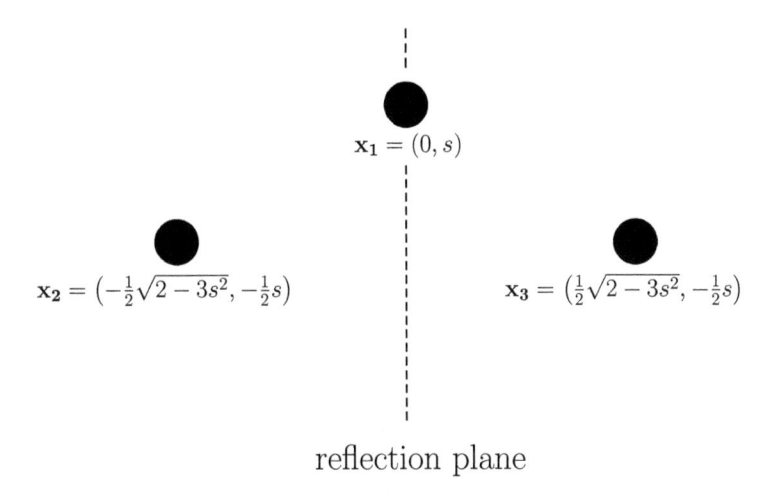

$$\mathbf{x}_2 = \left(-\tfrac{1}{2}\sqrt{2-3s^2}, -\tfrac{1}{2}s\right) \qquad \mathbf{x}_3 = \left(\tfrac{1}{2}\sqrt{2-3s^2}, -\tfrac{1}{2}s\right)$$

reflection plane

Fig. 10.2: Cartesian coordinates of an isosceles triangle of three $B = 4$ Skyrmions. $s$ runs from 0 (chain) to $1/\sqrt{3}$ (equilateral triangle).

vibrational degrees of freedom, allowing for all possible triangular shapes, and can accommodate some states missed in Rawlinson's model.

In detail, Rawlinson's model allows for isosceles triangle configurations with summit angle between 60° and 180°, and it therefore incorporates the intermediate, symmetric bent-arm configuration, and the L-shaped solution. Fig. 10.2 shows the geometry of these triangles. There is a potential energy barrier between the equilateral triangle and the chain, whose height is fitted using experimental data. One interesting aspect of this model is that there are three routes for deforming an equilateral triangle into a chain, because any of the three $B = 4$ Skyrmions can become the central one in the chain. The shape dynamics is therefore quantum mechanics on a graph with three edges, shown in Fig. 10.3, and wavefunctions are functions on this graph. One boundary condition is the continuity condition that the wavefunction has a unique value at the vertex where the edges meet. The derivative of the wavefunction on each edge must also satisfy the condition that there is no net probability current into or out of the vertex. In practice, by bosonic symmetry, the wavefunction is the same along each edge, and its derivative must therefore be zero at the vertex. In addition to the shape dynamics, the 3-dimensional rotational degrees of freedom need to be

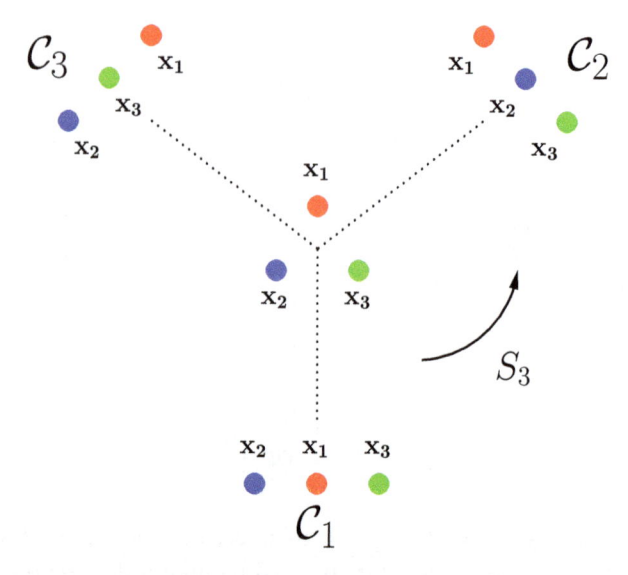

Fig. 10.3: Graph on which Carbon-12 wavefunctions are defined. $\mathcal{C}_1, \mathcal{C}_2$ and $\mathcal{C}_3$ are chain configurations related by permutations of the three $B = 4$ Skyrmion constituents. $S_3$ is the permutation group.

quantized. The spin of a state then couples to the shape dynamics through centrifugal effects. For details of the Hamiltonian, see ref.[203].

This model supports the ground state rotational band and the Hoyle band of Carbon-12, with varying shape wavefunctions along the graph edges. The $0^+$ ground state has large probability close to the equilateral triangle, whereas the $0^+$ Hoyle state has large probability close to the chain. Higher-spin states in the ground state band and Hoyle band are similarly concentrated at one or other of these special configurations. The wavefunctions of the $3^-$, $4^-$ and $5^-$ states in the ground state band have to vanish at the chain. This suggests, in agreement with observations, that they should have slightly higher energies than those obtained by extrapolating the energies of the $0^+$, $2^+$ and $4^+$ states in the same band, although curiously this is not what has emerged from the calculations so far, and some adjustment to the model seems desirable. However, the real strength of the model is that it also admits $1^-$ and $2^-$ states, matching the states seen experimentally. These have nodeless wavefunctions that vanish at both ends of the graph edges, i.e. at the equilateral triangle and the chain, so their maximum probability occurs at a bent arm. The squared magnitudes

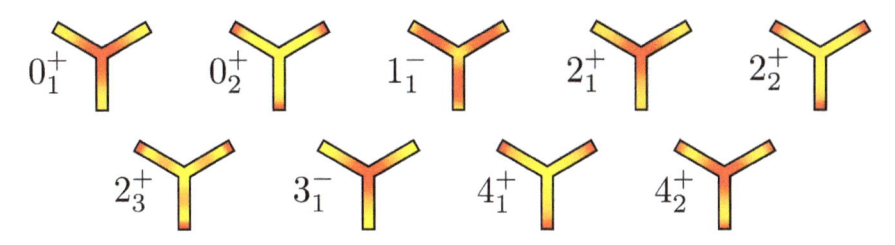

Fig. 10.4: The shape wavefunctions for several Carbon-12 states in Rawlinson's graph model. Red indicates a high squared magnitude. States are labelled by their spin/parity and energy ordering.

of the wavefunctions are shown in Fig. 10.4 for several states.

It is worth noting that no triangle of three bosonic $\alpha$-particles can be in a $J^P = 0^-$ state. This is because a triangle always has an $h$ symmetry (reflection in the plane of the triangle), which implies that the parity operator is equivalent to a 180° rotation in that plane. Such a rotation has eigenvalue $+1$ acting on a spin 0 state, so a $0^-$ state is forbidden. Consistent with this, no $0^-$ state with isospin 0 is seen in the Carbon-12 spectrum.

The energy spectrum of Rawlinson's model is shown in Fig. 10.5, and it matches quite well the experimental spectrum of Carbon-12 up to 15 MeV excitation energy, and beyond. Partly, it is similar to the spectrum in Fig. 9.5, obtained by rigid-body quantization of the equilateral triangular Skyrmion and straight chain Skyrmion, although the calculated energies are a little different. It extends that spectrum because it incorporates previously excluded states, in particular, states with $J^P = 1^-$ and $2^-$.

Still missing are the second-excited $0^+$ state at 9.9 MeV, and any state with $J^P = 1^+$. To understand these we require a model with further degrees of freedom. The lowest observed $1^+$ state is at 12.7 MeV, and can be interpreted as a rotational excitation of a (non-isosceles) scalene triangle, as we will see next.

## 10.4.1  *A multiphonon model for Carbon-12*

A stable equilateral triangle of three, dynamical point particles has nine degrees of freedom, of which three are associated with genuine vibrations of the triangle and six with translations and rotations. The symmetry group

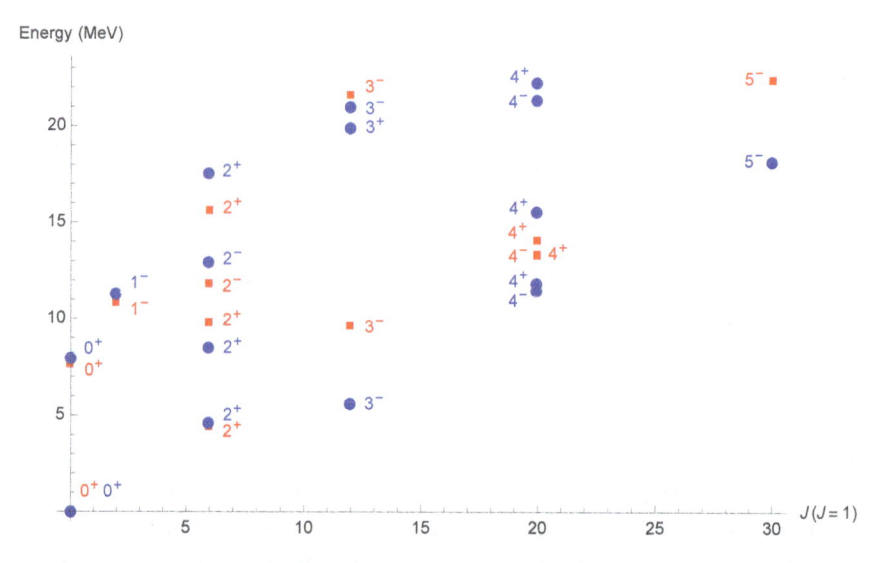

Fig. 10.5: Spectrum of Carbon-12, comparing the calculated energies in Rawlinson's model (blue circles) with observed energies (red squares)

of the equilateral triangle is $D_{3h}$ but the $h$ element acts trivially on the vibrational modes, so we may classify the modes by the irreps (irreducible representations) of $D_3$ – the trivial irrep $A_1$, the nontrivial 1-dimensional irrep $A_2$, and the 2-dimensional irrep E. The three vibrational modes decompose, in fact, just into the irreps E and $A_1$ under $D_3$. The E-modes change the shape of the triangle whereas the $A_1$-mode is the breather mode, changing the size. The 1-phonon excitation of the breather can be identified with the 9.9 MeV second-excited $0^+$ state of Carbon-12, above the ground state and Hoyle state. This was missed in Rawlinson's model. The 2-dimensional linear space of E-mode vibrations can be extended into the 2-dimensional shape space of triangles [150], with nonlinear dynamics, but a full quantum mechanics on shape space (coupled to rotations) has not yet been developed. For the moment we just consider the vibrational E-modes in the harmonic approximation.

The lowest excited states of the E-mode have one E-phonon; higher excited states are multiphonon states. The states with $n$ E-phonons transform under $D_3$ as the $n$-th symmetric power of the irrep E, denoted by $E^n$, whose decompositions into $D_3$ irreps are as follows:

$$E^0 = A_1\,, \tag{10.1}$$

$$E^1 = E\,, \tag{10.2}$$

$$E^2 = A_1 \oplus E\,, \tag{10.3}$$

$$E^3 = A_1 \oplus A_2 \oplus E\,, \tag{10.4}$$

$$E^4 = A_1 \oplus 2E\,. \tag{10.5}$$

The appearance of $A_2$ in $E^3$ is significant.

The allowed spin/parities for rotational excitations of each vibrational state (rovibrational states) are known [127]. They depend on the irrep of $D_3$ classifying the vibrational state, so we just need to list the allowed spin/parities $J^P$ for each of the irreps $A_1$, $A_2$ and E. These are

$$A_1 \longrightarrow 0^+, 2^+, 3^-, 4^\pm, 5^-, 6^\pm, \ldots, \tag{10.6}$$

$$A_2 \longrightarrow 1^+, 2^+, 3^+, \ldots, \tag{10.7}$$

$$E \longrightarrow 1^-, 2^\pm, 3^\pm, 4^\pm, \ldots. \tag{10.8}$$

The spin/parities for the trivial $A_1$ irrep are those familiar from the Carbon-12 ground state band.

The rovibrational states classified above can occur in Rawlinson's three-edged graph model only if the vibrational wavefunction is non-vanishing on the graph. The remaining states are exotic, and will tend to have higher energies. The $1^+$ state is exotic because it arises from a 3-phonon $A_2$ state whose wavefunction vanishes for all isosceles triangles and therefore vanishes on the graph. This can be seen by explicit construction of the relevant harmonic oscillator wavefunction in the amplitude plane for the 2-fold degenerate E-modes. In suitable polar coordinates $(\rho, \theta)$, Rawlinson's graph is along the half-lines $\theta = 0, \frac{2\pi}{3}, \frac{4\pi}{3}$, and the wavefunctions of $n$-phonon states are polynomials in $\rho\cos\theta$ and $\rho\sin\theta$ times a Gaussian factor. There is a 3-phonon wavefunction proportional to $\sin 3\theta$ that vanishes on the graph, transforming under the $A_2$ irrep, and this gives the $1^+$ state. Its wavefunction is peaked on a scalene triangle, whose only symmetry is the $h$ reflection in the plane of the triangle. One may also verify directly that rigid-body quantization of a scalene triangle of particles allows a $1^+$ state. (In fact, all $J^P$ combinations are allowed except $0^-$.)

The straight chain Skyrmion has an energy very similar to that of the equilateral triangle, and is also an energy minimum. It therefore makes sense to reinterpret quantum states on Rawlinson's graph in a different way, as rovibrational states of the chain. A straight chain of three pointlike particles has five translational and rotational degrees of freedom, out of nine degrees of freedom overall, so it has four true vibrational modes. These are classified by irreps $\Sigma_g^+$, $\Sigma_u^+$ and $\Pi_u$ of the symmetry group of the chain, $D_{\infty h}$. The first two are 1-dimensional ($\Sigma_g^+$ being the trivial irrep) and the last is 2-dimensional. The first vibrational mode, transforming trivially, is the breather. The second is a linear vibration with the central particle moving oppositely to the end particles, a motion excluded from Rawlinson's graph. The last mode has a 2-fold degeneracy and corresponds to bending deformations of the chain into an isosceles triangle (in any direction orthogonal to the chain), so it matches motion along the graph.

Rovibrational states for excitations of a chain with $D_{\infty h}$ symmetry are described by Herzberg [127] (p.372 and Fig. 99). The ground state band of the chain, having no phonons, matches the Hoyle band with $J^P = 0^+, 2^+, 4^+, \ldots$, in agreement with the states on Rawlinson's graph that have maxima at the chain. The rovibrational states with one phonon in the (bending) $\Pi_u$-mode have $J^P = 1^-, 2^-, 3^-, 4^-, \ldots$. It is clear that the $1^-$ and $2^-$ states match those with the same spin/parity on the graph, because their wavefunctions vanish linearly at the chain, being 1-phonon states, and must also vanish at the equilateral triangle. The $3^-$ state and $4^-$ state similarly match states on the graph. They vanish linearly at the chain, but can extend to be non-vanishing at the equilateral triangle, so they are simultaneously a 1-phonon state around the chain, and in the ground state band of the equilateral triangle. Their energies are therefore expected to be lower than would be obtained by simple extrapolation from the $1^-$ and $2^-$ states, and this matches what is experimentally observed.

Further states with one $\Sigma_u^+$-phonon or two $\Pi_u$-phonons can be regarded as higher-energy excitations of the Hoyle state, which have not yet been identified experimentally. None of these states have spin/parity $1^+$. The 1-phonon state of the $\Sigma_g^+$-mode (simultaneously the 1-phonon breather of the chain and equilateral triangle) is observed as the second-excited $0^+$ state of Carbon-12 at 9.9 MeV, as mentioned earlier.

The $1^+$ state is still to be accounted for. To obtain a rovibrational state with $J^P = 1^+$, by exciting the straight chain, one needs a vibrational tensor product state with one $\Sigma_u^+$-phonon and one $\Pi_u$-phonon. The tensor product has the decomposition

$$\Sigma_u^+ \otimes \Pi_u = \Pi_g \tag{10.9}$$

and the $\Pi_g$ irrep allows spin/parities $1^+, 2^+, 3^+, \ldots$. Because of the $\Sigma_u^+$-phonon, such a vibrational state vanishes along Rawlinson's graph. It is non-vanishing only for scalene triangles, and is therefore identified with the exotic $1^+$ state we found earlier as a 3-phonon excitation of the equilateral triangle.

In summary, the rovibrational states of the straight chain, with their $J^P$ values, overlap the rovibrational states of the equilateral triangle, although their classification is by a different symmetry group. These states, which include the 1-phonon breather excitation, qualitatively account for all the observed states of Carbon-12 with isospin 0 below about 15 MeV, and a few with higher energy.

We have not given energies for these multiphonon rovibrational states, because the harmonic oscillator approximation that is being used here gives different results depending on whether the equilateral triangle or the chain is excited, so the results would be of limited value. Instead, finding the energies by solving a Schrödinger equation on Rawlinson's graph is more illuminating and consistent. However, this approach misses the $1^+$ state. Since the $1^+$ state arises from a 3-phonon excitation of the equilateral triangle, or as a combination of two 1-phonon excitations of the chain, one can predict a higher energy for this state than for the lowest $1^-$ state, which is a 1-phonon state. The observed energies are 10.85 MeV for the $1^-$ state and 12.71 MeV for the $1^+$ state, which is the right ordering, though substantially different from what a harmonic oscillator calculation would give.

A fully consistent model for all the Carbon-12 states with isospin 0 requires the development of a Hamiltonian for motion on the entire shape space of triangles, coupled to rotations and the breather mode.

Chapter 11

# Modelling Oxygen-16

## 11.1 Introduction

The energy spectrum of the Oxygen-16 nucleus has posed a challenge to nuclear physicists for decades. Wefelmeier [241] and Wheeler [246] suggested in the 1930s that Oxygen-16 can be modelled as a cluster of four $\alpha$-particles, with the ground state having these particles in a tetrahedral arrangement. Early work on the vibrational excitations of a tetrahedral structure, following Wefelmeier and Wheeler, was that of Dennison [70], who applied to the Oxygen-16 nucleus many insights gained from studying the spectra of tetrahedrally-symmetric molecules like methane ($CH_4$). Dennison's work was followed up by Kameny [144] and then by Robson [206].

The binding energy of Oxygen-16 and of several other small nuclei that contain a whole number of $\alpha$-particles can be interpreted in terms of the number of short $\alpha - \alpha$ bonds [112, 241]; for Oxygen-16 there are six. This picture of the $0^+$ ground state has been verified in many different models such as the shell [56, 195], lattice *ab initio* [80] and AMD [146] models, giving credence to the old idea; some clustering is also confirmed in many experiments [234]. For a review of $\alpha$-particle clustering in light nuclei, see [97], and for recent discussions of the cluster structure in Oxygen-16, see [95, 98, 145].

Of course, as the ground state of Oxygen-16 has $J^P = 0^+$, the mean particle density in this state is spherical in any model. However, a conceptual difference arises for the low-lying $3^-$ state at 6.13 MeV, which is known through its E3 (octupole) decay strength to be a highly collective excitation. If the ground-state intrinsic shape is spherical, then this state is a vibrational excitation, perhaps with a tetrahedral character to account for the spin and parity. If the intrinsic shape is tetrahedral then this state is simply a rotational excitation.

Despite the general agreement about the structure of the ground state, there is no consensus on the structure of the excited states. For example, the first-excited state, the $0^+$ state at 6.05 MeV, has been described as a four-particle-four-hole (4p4h) state within the shell model [25], an excited breather mode of the tetrahedron [40, 70], or correlated with a bent-square (bent-rhomb) or flat-square configuration of $\alpha$-particles [35, 55, 80].

The first suggestion has been put in doubt by more extensive studies [134], and surprisingly, there appears no need to identify any states purely as shell-model excitations. All appear to be collective. Nevertheless, a classification of the excited states of Oxygen-16 is available entirely in shell-model terms, see e.g. [228] (p.7), so the conclusion seems to be that collective excitations and shell-model excitations are alternative ways of modelling the same nuclear states.

A tetrahedral structure of four $\alpha$-particles has three vibrational frequencies, associated with the irreps A, E and F of the tetrahedral group[1]. These irreps have, respectively, dimensions 1, 2 and 3, and the A-mode is the breather. Each vibrational state is classified by the number of phonons of each mode that are excited, and it has an associated rotational band of rovibrational excitations, whose allowed spin/parities are controlled by the representation theory of the tetrahedral group. In the rotational band of states with one or more vectorial F-phonons, it is important to take into account the (quantum) Coriolis effect.

The breathing-mode interpretation of the first-excited $0^+$ state depends on the A-mode frequency being sufficiently low relative to other vibrational modes. Bijker and Iachello considered an algebraic vibrational model incorporating all tetrahedral vibrations, but not including any square configuration explicitly [40]. They explored the consequences of large symmetry algebras that determine the relative frequencies of the various tetrahedral modes. In particular, they considered the possibility that these frequencies are all equal at about 6 MeV. However, there are limitations to the success of their overall fit of the Oxygen-16 spectrum. Bauhoff, Schultheis and Schultheis [35] didn't systematically study vibrations at all, but instead found rotational bands for a number of distinct configurations of four $\alpha$-particles, including tetrahedral and bent-square configurations. This is similar to the model of Carbon-12 based on distinct triangular and chain configurations of three $\alpha$-particles.

---

[1] We shall be more careful later to distinguish the $A_1$, $A_2$, E, $F_1$ and $F_2$ irreps of $T_d$.

Fig. 11.1: Path from $B = 16$ tetrahedron to dual tetrahedron via flat and bent squares

These descriptions of Oxygen-16 do not necessarily contradict each other, as quantum states of dynamical $\alpha$-particles can overlap tetrahedral, bent-square and flat-square configurations. However, to understand the first-excited $0^+$ state and other states in detail, it seems clear that what is needed is a model of $\alpha$-particles that includes large amplitude, anharmonic vibrations extending from the tetrahedron as far as the bent- and flat-square configurations, and beyond. In this chapter we analyse the spectrum of Oxygen-16 using such a dynamical model, where the four $\alpha$-particles are represented by four $B = 4$ Skyrmions [116, 117]. This is a further development in the long history of modelling the excitation spectrum of Oxygen-16 in terms of vibrational excitations of a tetrahedral intrinsic structure.

## 11.2 The Vibrational E-Manifold

In Skyrme theory there is a dynamical trajectory, shown in Fig. 11.1, connecting tetrahedral and flat-square configurations of four $B = 4$ cubes, via a bent square. Two $B = 8$ Skyrmions (pairs of $B = 4$ cubes) approach each other and open out to form a tetrahedron, which flattens out into a square, before deforming into the dual (inverted) tetrahedron and then breaking into two $B = 8$ Skyrmions again, having picked up a 90° twist. There are three of these trajectories passing through each tetrahedron, corresponding to the three pairs of opposing edges. A small motion along any one of them excites a small-amplitude vibrational mode of the tetrahedron, and these modes can be linearly superposed. However, these modes are degenerate, and when the three modes are excited equally they cancel out, so together they only generate a 2-dimensional space of small-amplitude vibrations of the tetrahedron – the E-vibrations – where E is the 2-dimensional irrep of the tetrahedral group. This space of small-amplitude E-vibrations can be nonlinearly extended into a 2-dimensional curved surface $\mathcal{M}_\text{E}$ that includes the complete trajectories passing through the tetrahedron and its dual, as shown in Fig. 11.1. We refer to $\mathcal{M}_\text{E}$ as the vibrational *E-manifold* of the $B = 16$ Skyrmion.

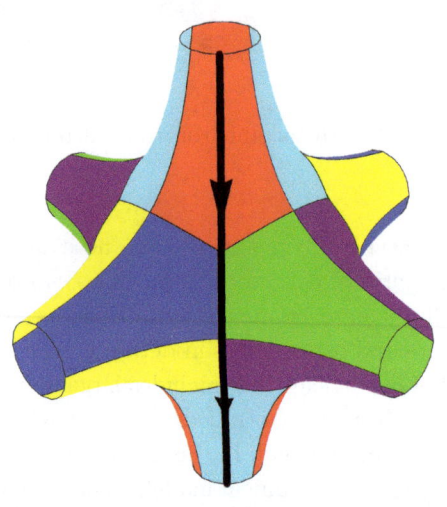

Fig. 11.2: The E-manifold $\mathcal{M}_\text{E}$ – a surface with six punctures, constructed from eight (curved) equilateral triangles. Regions with the same colouring are related by $D_2$ symmetry. The scattering path in Fig. 11.1 is represented by the thick black line.

Each point on $\mathcal{M}_\text{E}$ corresponds to a configuration of four spatial points – the centres of the $B = 4$ Skyrmions – and also the associated Skyrme field configuration itself, both having $D_2$ symmetry relative to the standard, fixed frame of Cartesian axes in space. $D_2$ is the group generated by 180° rotations about the Cartesian axes. The degrees of freedom on $\mathcal{M}_\text{E}$ are the correlated positions of the four points, all of which lie on a single, fixed surface in $\mathbb{R}^3$. Because of the $D_2$ symmetry, if one is at $\mathbf{x} = (x, y, z)$, the others are at $(x, -y, -z), (-x, y, -z)$ and $(-x, -y, z)$. These are alternating vertices of a cuboid, whose scale size is fixed using a potential that disfavours the $B = 4$ Skyrmions being too close together or too far apart.

The E-manifold is therefore topologically a 2-sphere, parametrised by the direction of $(x, y, z)$. To account for the asymptotics seen in Fig. 11.1, this surface must stretch out to infinity in six directions as in Fig. 11.2, so the 2-sphere has six punctures.

### 11.2.1 The Hamiltonian and quantum states

The total space of vibrational E-manifold field configurations is $\mathcal{M}_\mathrm{E} \times SO(3)$, which allows for rigid rotations of each configuration in $\mathcal{M}_\mathrm{E}$. Quantum states are therefore coupled vibrational and rotational excitations, i.e. rovibrational states. Let us first consider the vibrational states. The rescaled Schrödinger equation for the vibrational wavefunction $\psi$ is

$$-\Delta_\mathrm{vib}\,\psi + V\psi = E_\mathrm{vib}\,\psi\,, \qquad (11.1)$$

where $E_\mathrm{vib}$ is the vibrational energy. $V$ is the static energy of a configuration on $\mathcal{M}_\mathrm{E}$, and the kinetic operator is proportional to the Laplace–Beltrami operator

$$-\Delta_\mathrm{vib} = -(\det g)^{-\frac{1}{2}}\partial_\alpha \left((\det g)^{\frac{1}{2}} g^{\alpha\beta} \partial_\beta\right)\,, \qquad (11.2)$$

where $g_{\alpha\beta}$ is the metric on $\mathcal{M}_\mathrm{E}$, $\det g$ its determinant, and $g^{\alpha\beta}$ its inverse, expressed in terms of two suitable coordinates.

In principle, both the metric and potential can be calculated from the Skyrme Lagrangian, but this is computationally challenging, and in [116] a rather simple model was constructed, having the right symmetries and overall geometry. $\mathcal{M}_\mathrm{E}$ is modelled as the surface of a curved, regular octahedron – a topological 2-sphere with six punctures at the octahedral vertices. The punctures represent the asymptotic configurations with two infinitely-separated $B = 8$ Skyrmions, so the metric is complete, and is not that of the usual round 2-sphere with punctures. The surface can be tesselated into eight (curved) equilateral triangles, where each triangle centre corresponds to a tetrahedral configuration, and the three edge mid-points correspond to flat-square configurations. The choice of potential $V$ is motivated by $\alpha$-particle models which find that the tetrahedral configuration has the lowest energy [35], the flat square is a saddle point with slightly higher energy, and the separated $B = 8$ clusters have higher energy still.

$\mathcal{M}_\mathrm{E}$ has octahedral symmetry $O_h$ and points related by $D_2$ must be identified, with the wavefunction having equal values at identified points. This is because of Finkelstein–Rubinstein symmetry constraints, or more physically because the $D_2$ symmetry permutes the $\alpha$-particles, and $\alpha$-particles are bosons. A non-trivial $O_h/D_2 \cong S_3 \times \mathbb{Z}_2$ symmetry group still acts on $\mathcal{M}_\mathrm{E}$, where $S_3$ is the permutation group of the $x$-, $y$- and $z$-axes, and inversion acts on $\mathcal{M}_\mathrm{E}$ as the $\mathbb{Z}_2$ symmetry

$$\mathbf{x} = (x, y, z) \mapsto (-x, -y, -z). \tag{11.3}$$

Each vibrational energy eigenstate therefore has an $S_3$-species and a parity eigenvalue.

The inclusion of paths from the tetrahedron to its dual, tunnelling through flat squares, enables a significant energy difference to be created between vibrational quantum states with opposite parities, which would not be possible if only local vibrations around the tetrahedron were considered. Positive parity wavefunctions are equal at the tetrahedron and its dual, whereas negative parity wavefunctions flip sign. Therefore, a positive parity wavefunction can be non-zero at a flat square mid-way between the tetrahedron and its dual, whereas a negative parity wavefunction has to be zero, raising the energy.

## 11.3 E-Manifold States

We now present in detail the vibrational E-manifold states, their energies, and the spin/parities allowed for their rotational excitations. We also clarify how the vibrational states can be classified in terms of E-phonon counting, analogous to the way Carbon-12 excitations were classified as multiphonon states around the equilateral triangular $B = 12$ Skyrmion.

A planar projection of one quarter of $\mathcal{M}_\mathrm{E}$ is shown in Fig. 11.3. The remaining three quarters are related by the action of $D_2$. Fig. 11.4 illustrates the tetrahedral and square configurations at points of enhanced symmetry on the projected surface. The full symmetry group of the E-manifold and the quantized dynamics on it is $O_h$, as we have mentioned. As wavefunctions are $D_2$-invariant, they are classified by a representation of $S_3$ and a parity eigenvalue $\pm$. Note that $S_3$ permutes the three curves emerging at 120° angular separation from each tetrahedral point.

$S_3$ has three irreducible representations – the 1-dimensional trivial irrep T, the 1-dimensional sign irrep S, and the 2-dimensional standard irrep St. E-manifold wavefunctions are therefore labelled by species/parity $T^+$, $S^+$, $St^+$ or $T^-$, $S^-$, $St^-$ with an additional subscript $n$ to denote the number of phonons in the state (which is described carefully below). The energy increases with the phonon number. The wavefunctions of species T and S are denoted by $\psi$, but those of species St have a 2-fold degeneracy, and two selected orthogonal wavefunctions are denoted $u$ and $v$.

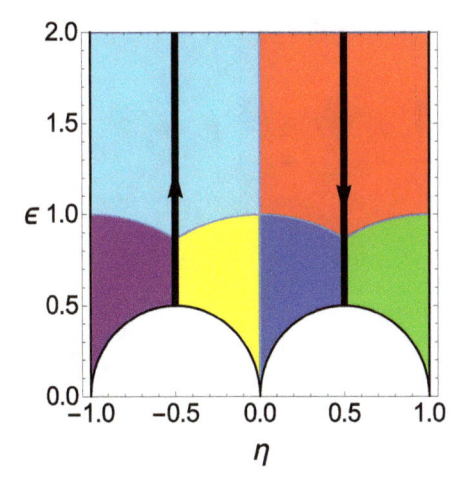

Fig. 11.3: The correspondence between the sphere and projected plane for one quarter of the E-manifold, $\mathcal{M}_{\mathrm{E}}$. The coloured regions correspond to those in Fig. 11.2.

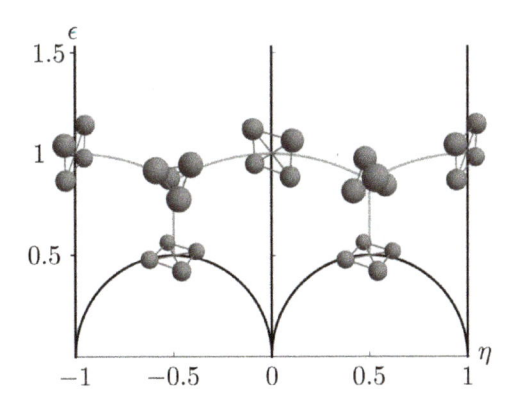

Fig. 11.4: Tetrahedral and flat-square configurations on the projected plane. The balls represent $B = 4$ Skyrmions or $\alpha$-particles.

The low-energy wavefunctions on the E-manifold have been calculated numerically [117] and are illustrated in Fig. 11.5, alongside their energies $E_{\mathrm{vib}}$. The two lowest states $\psi_{\mathrm{T0}}^{+}$ and $\psi_{\mathrm{S0}}^{-}$ have wavefunctions concentrated around the minima of the potential energy at the tetrahedron $\mathbf{x} = (1, 1, 1)$ and its dual $\mathbf{x} = (-1, -1, -1)$, but have opposite parities. Tunnelling between these is via the flat-square saddle point. Wavefunctions of the species S are constrained to vanish at the square while those of species T are not.

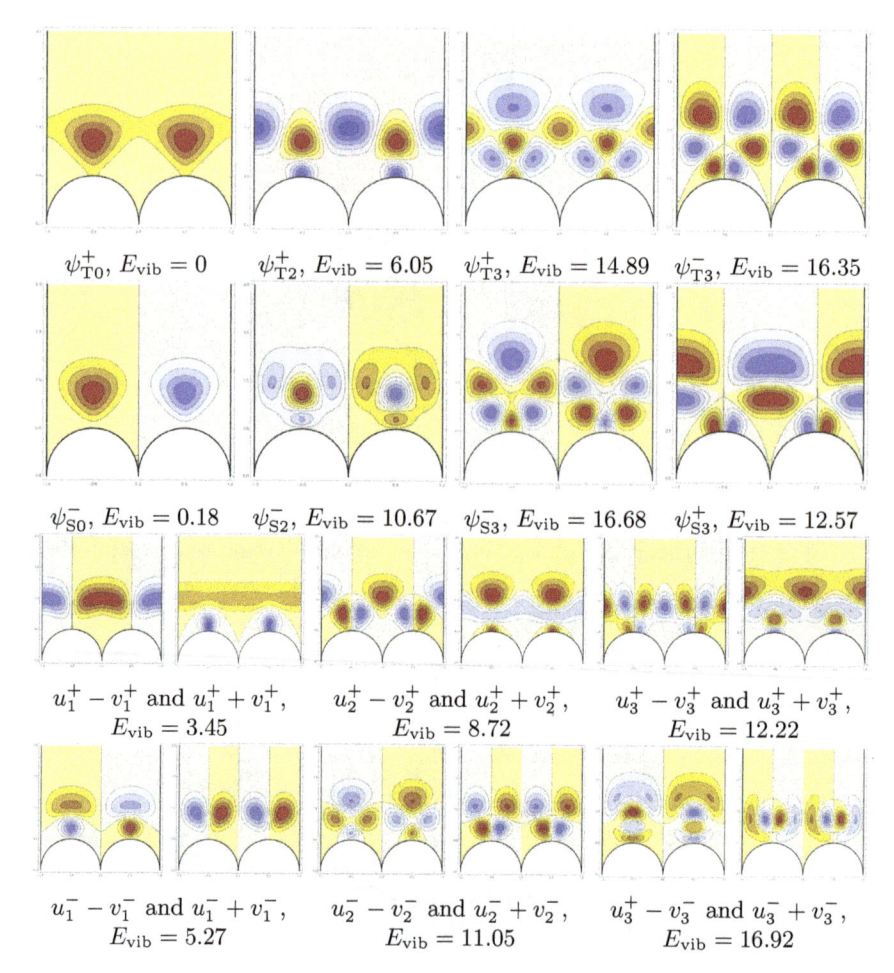

$\psi_{T0}^+$, $E_{\text{vib}} = 0$  $\psi_{T2}^+$, $E_{\text{vib}} = 6.05$  $\psi_{T3}^+$, $E_{\text{vib}} = 14.89$  $\psi_{T3}^-$, $E_{\text{vib}} = 16.35$

$\psi_{S0}^-$, $E_{\text{vib}} = 0.18$  $\psi_{S2}^-$, $E_{\text{vib}} = 10.67$  $\psi_{S3}^-$, $E_{\text{vib}} = 16.68$  $\psi_{S3}^+$, $E_{\text{vib}} = 12.57$

$u_1^+ - v_1^+$ and $u_1^+ + v_1^+$,  $u_2^+ - v_2^+$ and $u_2^+ + v_2^+$,  $u_3^+ - v_3^+$ and $u_3^+ + v_3^+$,
$E_{\text{vib}} = 3.45$  $E_{\text{vib}} = 8.72$  $E_{\text{vib}} = 12.22$

$u_1^- - v_1^-$ and $u_1^- + v_1^-$,  $u_2^- - v_2^-$ and $u_2^- + v_2^-$,  $u_3^- - v_3^-$ and $u_3^- + v_3^-$,
$E_{\text{vib}} = 5.27$  $E_{\text{vib}} = 11.05$  $E_{\text{vib}} = 16.92$

Fig. 11.5: Vibrational wavefunctions on the E-manifold, with contours from −1 (blue) to +1 (red). Those in the first (second) row transform as the trivial irrep T (sign irrep S) of $S_3$; those in the third (fourth) row transform as the standard irrep St with positive (negative) parity. The numerical subscript indicates the number of E-phonons. Wavefunctions and axes are scaled for clarity. Energies are in MeV.

At low energy, the tunnelling amplitude is small so the energy gap between $\psi_{T0}^+$ and $\psi_{S0}^-$ is small. Rotational excitations of these two lowest vibrational states jointly form the tetrahedral ground-state band. As the phonon number and energy increases, the tunnelling amplitude increases, creating a larger energy gap between states that would otherwise be degenerate parity doubles.

The state $\psi_{T2}^+$ is identified with the low-lying $0^+$ state of Oxygen-16 at 6.05 MeV. It is concentrated around both the tetrahedron and the square configurations. The lowest-energy states of species St are a positive-parity pair $u_1^+ - v_1^+$ and $u_1^+ + v_1^+$, and a negative-parity pair $u_1^- - v_1^-$ and $u_1^- + v_1^-$, all of which have to vanish at the tetrahedra. Their lowest-allowed spin/parities are $2^+$ and $2^-$, respectively. The positive-parity pair are concentrated around the flat-square configurations. The negative-parity pair have to vanish at the flat squares, and instead are concentrated around bent-square configurations. The presence of wavefunction nodes at the flat squares significantly raises the energy of the $2^-$ state compared to the $2^+$ state.

### 11.3.1 E-phonons

Here, we will clarify how E-manifold states are interpreted in terms of E-phonon excitations around the tetrahedral $B = 16$ Skyrmion. The Hamiltonian is not simply that of a harmonic oscillator in the neighbourhood of the Skyrmion, and wavefunctions are influenced by anharmonic interactions of vibrational E-phonons, but it is close enough to harmonic that multiphonon states can easily be recognised from the wavefunctions illustrated in Fig. 11.5.

This analysis uses Table 11.1, the character table of the tetrahedral group $T_d$ [160]. $T_d$ is the 24-element subgroup of $O_h$ that preserves the configuration of four particles at the vertices of a regular tetrahedron, and it acts non-trivially but linearly on small deformations away from the tetrahedron. The 24 elements lie in five conjugacy classes, denoted $Id$, $C_3$, $C_2$, $\sigma_d$ and $S_4$. These consist, respectively, of rotations by $0°$, $120°$ and $180°$, and rotations by $180°$ and $90°$ combined with an inversion. The group $T_d$ has five irreps, the trivial and non-trivial 1-dimensional irreps $A_1$ and $A_2$, the 2-dimensional irrep E, and the two 3-dimensional irreps $F_1$ and $F_2$. (The 3-dimensional irreps are sometimes written as $T_1$ and $T_2$.)

As $T_d$ is a subgroup of $O_h$, each E-manifold state classified by an irrep of $O_h$ is also classified by a $T_d$ irrep. The states of species $T^+$ and $S^-$ are in the $A_1$ irrep of $T_d$, the states of species $T^-$ and $S^+$ are in the $A_2$ irrep, and the states of species $St^+$ and $St^-$ are in the E irrep. No E-manifold states are classified by $F_1$ or $F_2$. The $T_d$ classification misses the parity label. This can be read off from the behaviour of the wavefunction under

Table 11.1: Character table of $T_d$.

|   | Id | $C_3$ | $C_2$ | $\sigma_d$ | $S_4$ |
|---|---|---|---|---|---|
| $A_1$ | 1 | 1 | 1 | 1 | 1 |
| $A_2$ | 1 | 1 | 1 | -1 | -1 |
| E | 2 | -1 | 2 | 0 | 0 |
| $F_2$ | 3 | 0 | -1 | 1 | -1 |
| $F_1$ | 3 | 0 | -1 | -1 | 1 |

reflection in the vertical line splitting in half each subfigure of Fig. 11.5. If the wavefunction changes sign, then the parity is negative. In particular, the states $\psi_{T0}^+$ and $\psi_{S0}^-$ are invariant under $T_d$ and are interpreted as states with no E-phonons, but with opposite parities. The pair of states $u_1^+ \pm v_1^+$ transform under the irrep E, and are therefore interpreted as degenerate states with one E-phonon, as are the pair $u_1^- \pm v_1^-$, and these pairs have opposite parities.

Multiphonon states also occur. As E-phonons are bosons, multiphonon states transform as a symmetrised tensor product of 1-phonon irreps. For a representation d with character $\chi$, the symmetric square $d^2$ (we drop the subscript "symm.") has character $[\chi^2](g) = \frac{1}{2}\{\chi(g)^2 + \chi(g^2)\}$ for each group element $g$. Applied to the irrep E, this formula gives the characters of $E^2$, the space of states with two E-phonons, which is a reducible representation of $T_d$ with decomposition $E^2 = A_1 \oplus E$. Decompositions of higher symmetrised powers of E, denoted $E^n$, are given in [127] (p.127), and are identical to the decompositions for the irreps of $D_3$ used in Section 10.4.1. They are

$$E^0 = A_1, \tag{11.4}$$

$$E^1 = E, \tag{11.5}$$

$$E^2 = A_1 \oplus E, \tag{11.6}$$

$$E^3 = A_1 \oplus A_2 \oplus E, \tag{11.7}$$

$$E^4 = A_1 \oplus 2E. \tag{11.8}$$

The dimension of $E^n$ is $n+1$, as expected for $n$-phonon states of a 2-dimensional harmonic oscillator. Combining the above decompositions with the estimate that an $n$-phonon state has approximately $n$ times the energy of a 1-phonon state, and inspecting the shape of the wavefunctions in Fig. 11.5 close to the tetrahedral configuration, we can classify these wavefunctions as follows. The wavefunctions $\psi_{T0}^+$, $\psi_{S0}^-$ are 0-phonon states, $u_1^+ \pm v_1^+$, $u_1^- \pm v_1^-$ are 1-phonon states, $\psi_{T2}^+$, $\psi_{S2}^-$ are 2-phonon $A_1$ states and $u_2^+ \pm v_2^+$, $u_2^- \pm v_2^-$ are 2-phonon E states. The wavefunctions $\psi_{T3}^+$, $\psi_{S3}^+$ are 3-phonon $A_1$ states, $\psi_{T3}^-$, $\psi_{S3}^-$ are 3-phonon $A_2$ states and $u_3^+ \pm v_3^+$, $u_3^- \pm v_3^-$ are 3-phonon E states. In all cases, there is a pair of states distinguished by the parity label. This classification is verified by looking at the nodes of the wavefunctions, and comparing with harmonic oscillator states near the tetrahedral point expressed in (plane) polar coordinates. For example, 1-phonon states have one linear node, and 2-phonon $A_1$ states have one radial node.

Classifying the E-manifold states in terms of E-phonons helps compare our classification of states with the classifications created by others, in particular those of Bijker and Iachello [40]. However, for the energy of an $n$-phonon state we use the E-manifold energies $E_{\text{vib}}$ shown in Fig. 11.5, rather than the harmonic estimate $n\omega_E$, where $\omega_E$ is the 1-phonon energy. These are noticeably different, showing the importance of tunnelling through the flat-square configurations.

### 11.3.2 Rovibrational states

We now consider the rovibrational states, i.e. the bands of rotational excitations of each vibrational E-manifold state we have found. In each band, the allowed spin/parities $J^P$ are those for which the $O_h$-decomposition of the $O(3)$ representation with spin/parity $J^P$ contains the irrep of $O_h$ classifying the E-manifold state. These are essentially the same as for vibrational states around the tetrahedron, classified by irreps of the $T_d$ subgroup, so we work with these.

To find these spin/parities, we start with the characters $\chi$ of the irrep $J^P$ of $O(3)$. For a pure rotation by angle $\theta$, the character is

$$\chi_{J^P}(\theta) = \frac{\sin(J+\frac{1}{2})\theta}{\sin\frac{1}{2}\theta}, \tag{11.9}$$

for either parity. For a rotation by $\theta$ combined with inversion $\mathscr{I}$ it is

$$\chi_{J^P}(\theta, \mathscr{I}) = \pm \frac{\sin(J+\frac{1}{2})\theta}{\sin\frac{1}{2}\theta}, \tag{11.10}$$

with the sign matching the parity $P$. Next, these characters need to be restricted to the conjugacy classes in the $T_d$ subgroup – rotations by $0°$, $120°$ and $180°$, and rotations by $120°$ and $90°$ combined with inversion. They then become characters of the (generally reducible) $J^P$ representation of $T_d$, whose decomposition into irreps can be found using the $T_d$ characters in Table 11.1. If an irrep d occurs here and simultaneously as an irrep of a multiphonon excitation (i.e. on the right hand side of eqs.(11.4) – (11.8)), then that multiphonon excitation can have a rovibrational $J^P$ state. If d has multiplicity $\nu$ then the $J^P$ state occurs with multiplicity $\nu$.

For example, the $J^P = 2^+$ representation of $O(3)$ is 5-dimensional and its characters for the five $T_d$ conjugacy classes are, respectively, $\chi_{2^+} = 5, -1, 1, 1, -1$. Its decomposition into $T_d$ irreps is therefore

$$(J^P = 2^+)\Big|_{T_d} = \mathrm{E} \oplus \mathrm{F}_2, \qquad (11.11)$$

so $2^+$ states occur in the rovibrational E-band (E$^1$-band), E$^2$-band and E$^3$-band, and with multiplicity 2 in the E$^4$-band.

More systematically, by using the formulae (11.9) and (11.10) and the $T_d$ character table, one finds that the rovibrational bands associated with the five $T_d$ irreps are [127] (p.450),

$$\mathrm{A}_1 \longrightarrow 0^+, 3^-, 4^+, 6^\pm, 7^-, 8^+, \ldots, \qquad (11.12)$$

$$\mathrm{A}_2 \longrightarrow 0^-, 3^+, 4^-, 6^\pm, 7^+, 8^-, \ldots, \qquad (11.13)$$

$$\mathrm{E} \longrightarrow 2^\pm, 4^\pm, 5^\pm, 6^\pm, 7^\pm, 8^\pm, 8^\pm, \ldots, \qquad (11.14)$$

$$\mathrm{F}_1 \longrightarrow 1^+, 2^-, 3^\pm, 4^\pm, 5^\pm, 5^+, 6^\pm, 6^-, 7^\pm, 7^\pm, 8^\pm, 8^\pm, \ldots, \qquad (11.15)$$

$$\mathrm{F}_2 \longrightarrow 1^-, 2^+, 3^\pm, 4^\pm, 5^\pm, 5^-, 6^\pm, 6^+, 7^\pm, 7^\pm, 8^\pm, 8^\pm, \ldots. \qquad (11.16)$$

E-manifold states only occur in the irreps $A_1$, $A_2$ and E. The extra parity label for E-manifold states means that the positive and negative parity states within a band can be assigned different vibrational energies.

So, for example, the rovibrational excitations of $\psi_{T0}^+, \psi_{T2}^+$ and $\psi_{T3}^+$ are the positive parity $A_1$ states $0^+, 4^+, 6^+, \ldots$, whereas the excitations of $\psi_{S0}^-, \psi_{S2}^-$ and $\psi_{S3}^-$ are the negative parity $A_1$ states $3^-, 6^-, 7^-, \ldots$, and these have different vibrational energies. Among the more exotic states, the 3-phonon wavefunction $\psi_{T3}^-$ allows a rovibrational $A_2$ state $0^-$, and the 3-phonon wavefunction $\psi_{S3}^+$ allows a rovibrational $A_2$ state $3^+$. Note also that $0^+$ states only arise from E-manifold states in the irrep $A_1$. Above the ground state with no E-phonons, there is one $0^+$ state with each of two, three and four E-phonons.

We will see shortly that irreps $F_1$ and $F_2$ occur when additional F-phonons are considered. Equations (11.11) and (11.16) show that further $2^+$ states, for example, occur in the rotational bands associated with the $F_2$ irrep.

## 11.4 Beyond E-Vibrations

The more complete energy spectrum of the Oxygen-16 nucleus up to 20 MeV and beyond requires combining E-vibrational states with vibrational A- and F-phonons. The E-vibrations are treated anharmonically using the 2-dimensional E-manifold of configurations, as above, but can also be interpreted in terms of E-phonons, as we have explained. The A- and F-phonons are treated in the harmonic approximation.

Energies of the rotational excitations of each vibrational state are modelled as in the classic work of Dennison, Robson and others, with centrifugal corrections included. States with F-phonons require Coriolis corrections too, and the Coriolis parameter $\zeta$ is chosen positive to ensure the right splitting of the $3^+$ and $3^-$ states near 11 MeV. Altogether, about 80 states with isospin 0 are predicted below 20 MeV, and these match more than 60 experimentally tabulated states quite well. Several high-spin states are predicted, up to spin 9, and these match some of the observed high-spin states in the energy range 20 – 30 MeV (all of which have natural parity, i.e. $P = (-1)^J$). The model is mainly phenomenological but it receives input from the properties of the $B = 16$ Skyrmions.

Let us describe this in more detail. It is assumed that an A-phonon has energy about 12 MeV, an F-phonon has energy about 6.5 MeV, and an E-phonon has the lowest energy of about 3.5 MeV. The above ordering of frequencies has more than one motivation. First, in a simple tetrahedral model of four equal-mass particles connected by six equal springs, the frequencies of the A-, F- and E-modes are in the proportions 2, $\sqrt{2}$ and 1, respectively [246]. A breather state, arising from the excited A-mode, therefore has rather high energy. We can calibrate the A-mode frequency in Oxygen-16 by using the spring model and comparing with Carbon-12. The ratio of breather frequencies between a tetrahedron with six springs and a triangle with three springs is $2/\sqrt{3} \simeq 1.15$, so the 9.9 MeV breather state in Carbon-12 implies an 11.4 MeV breather state in Oxygen-16. The A-phonon energy is close to this. The F-phonon energy is fixed by identifying the lowest $1^-$ state at 7.12 MeV as a 1-phonon F-excitation of the E-manifold ground state. The total energy of this state has a contribution of about 0.7 MeV from the rotational and Coriolis energies.

In previous rovibrational models of Oxygen-16, the first-excited $0^+$ state at 6.05 MeV was usually identified with the 1-phonon breather state, requiring the A-phonon energy to be 6.05 MeV. But in the model here this state is a 2-phonon E-excitation, so the E-phonon energy has to be near

3 MeV. This energy allows a good fit of the lowest $2^+$ and $2^-$ states as excitations with one E-phonon. The rotational energies of these states are comparable with their vibrational E-phonon energy.

The ground state rotational band of Oxygen-16 has the states associated with a rotating tetrahedron, having spin/parities $0^+, 3^-, 4^+, \ldots$ as in the list (11.12). These are based on the $\psi_{T0}^+$ and $\psi_{S0}^-$ E-manifold states of opposite parity, with their small energy difference of 0.18 MeV. Energies are relative to the Oxygen-16 ground state energy here, so the $0^+$ state has zero energy. The rotational energy in a spin $J$ state, including centrifugal corrections, is taken to be

$$E_{\text{rot}} = \text{B}J(J+1) - \text{C}(J(J+1))^2 + \text{D}(J(J+1))^3. \quad (11.17)$$

Here, B is Dennison's shorthand for $\frac{1}{2V}$, with $V$ the moment of inertia; it is called the *rotational constant* of the band. To fit the $3^-$ state at 6.13 MeV, B is set to be 0.56 MeV. States with spin 4 and beyond have a significant centrifugal energy correction. The parameters C and D are chosen to have the constant values $\text{C} = 4.5 \times 10^{-3}$ MeV and $\text{D} = 2.8 \times 10^{-5}$ MeV. This ensures that the rotational energy increases approximately linearly with $J$ between $J = 3$ and $J = 8$. The formula (11.17) is a simplified version of one proposed by Sood [217, 253].

For all E-manifold states having at least one E-phonon, the rotational constant B is reduced to 0.45 MeV. This can be interpreted physically; the 0-phonon wavefunctions are concentrated at the tetrahedral configuration while the 1-phonon E-wavefunctions are concentrated at the flat and bent squares. The squares have a larger moment of inertia and hence a smaller B. The same value of B is used for states with one F-phonon or one A-phonon, and no E-phonons. For states with two F-phonons, combined E- and F-phonons, or combined E- and A-phonons, $\text{B} = 0.4$. This steady decrease of B as the number of phonons increases resembles the pattern used in molecular physics. Crossing of rotational bands in the energy *versus* spin diagram is possible as the spin increases, because of the differences in B between bands.

The final ingredient is the Coriolis correction to the energies. This has been studied in depth by theoretical molecular chemists since the 1930s. Herzberg gives an illuminating review [127], but the key original paper discussing Coriolis effects in tetrahedral molecules is by Johnston and Dennison [143]. The Coriolis effect arises because vibrational motion can carry an internal angular momentum, and this influences the total rotational energy. Vibrational excitations only involving A- and E-phonons have no Coriolis energy correction, but excitations involving F-phonons do. We will focus on the F-band states – states in the rotational band with just one F-phonon. Similar corrections occur when one F-phonon is combined with any number of A- and E-phonons.

The correction depends on a parameter $\zeta$. It was assumed by Dennison [70] and others [40, 206] that $\zeta = -0.5$ for four pointlike $\alpha$-particles, by analogy with a calculation for the four hydrogen atoms in methane when the central carbon atom is decoupled. However, for extended structures as in the Skyrmion picture, where four $B = 4$ Skyrmions partly merge and are not vibrating in the same way as hydrogen atoms vibrate in methane, $\zeta$ can be different. Robson has mentioned this possibility [207]. Values of $\zeta$ in the range $-1$ to $1$ are common for molecules with various geometries, and it is found that a best fit for Oxygen-16 occurs for $\zeta$ close to 0.2.

The F-mode vibrations are triply-degenerate and transform under the $F_2$ irrep of the group $T_d$. This irrep is the restriction to $T_d$ of the vectorial $J^P = 1^-$ representation of $O(3)$, which explains why an F-phonon has one unit of vibrational angular momentum, and negative parity. Each F-band state has as its underlying E-manifold state the ground state with no E-phonons. The underlying state can be rotationally excited, and its spin/parity $0^+, 3^-, 4^+, 6^\pm, \ldots$ needs to be combined with the vibrational angular momentum of the F-phonon using the usual Clebsch–Gordon angular momentum rules. Let us denote the spin operators of the underlying E-manifold state by $\mathbf{R}$, and those of the vibrational angular momentum of the F-phonon by $\mathbf{l}$. $\mathbf{R}$ and $\mathbf{l}$ commute, and the total angular momentum is

$$\mathbf{J} = \mathbf{R} + \mathbf{l}. \tag{11.18}$$

The parity of the combined state is the opposite of the parity of the underlying state, because the F-phonon has negative parity. The allowed F-band states therefore have spin/parities (up to $R = 8$)

$$1_0^-, 2_3^+, 3_3^+, 4_3^+, 3_4^-, 4_4^-, 5_4^-, 5_6^\pm, 6_6^\pm, 7_6^\pm, 6_7^+, 7_7^+, 8_7^+, 7_8^-, 8_8^-, 9_8^-, \tag{11.19}$$

where the usual spin/parity label $J^P$ is supplemented by a subscript $R$ to denote the underlying rotational angular momentum $R = J+1$, $J$ or $J-1$. Note that the $J^P$ values occurring here are exactly the same as those in the list (11.16), but reinterpreted.

The total rotational energy, including the Coriolis and centrifugal corrections, is derived from the Hamiltonian [127, 143]

$$H_{\rm rot} = {\rm B}(\mathbf{J} - \zeta\mathbf{l})^2 - {\rm C}(J(J+1))^2 + {\rm D}(J(J+1))^3. \tag{11.20}$$

Expanding out, this is

$$H_{\rm rot} = {\rm B}J(J+1) - 2{\rm B}\zeta\, {\bf J}\cdot{\bf l} + 2{\rm B}\zeta^2 - {\rm C}(J(J+1))^2 + {\rm D}(J(J+1))^3\,, \quad (11.21)$$

where we have set $l(l+1) = 2$ for internal angular momentum $l = 1$. (This expansion is valid even though $\bf J$ and $\bf l$ do not commute, because the component pairs $J_i$ and $l_i$ do commute.) By squaring eq.(11.18) we find $2{\bf J}\cdot{\bf l} = J(J+1) - R(R+1) + 2$, so the energy eigenvalues of $H_{\rm rot}$ are

$$E_{\rm rot} = {\rm B}J(J+1) - 2{\rm B}\zeta(1-\zeta) - {\rm C}(J(J+1))^2 + {\rm D}(J(J+1))^3$$

$$+ 2{\rm B}\zeta \begin{cases} J+1 & \text{if } R = J+1\,, \\ 0 & \text{if } R = J\,, \\ -J & \text{if } R = J-1\,. \end{cases} \quad (11.22)$$

As mentioned earlier, $\zeta$ is calibrated using the energies of the $3_3^+$ and $3_4^-$ states in the F-band (where the subscript again denotes $R$). The lowest F-band state should clearly be identified with the experimental $1^-$ state at 7.12 MeV, and spin 3 states are 4 MeV to 5 MeV above this. There is just one experimentally confirmed $3^+$ state, at 11.08 MeV, and this is identified as the $3_3^+$ state in the F-band; the $3_4^-$ state in the F-band is identified with the experimental $3^-$ state at 11.60 MeV, the first-excited $3^-$ state. In the model, there are two sources for the 0.52 MeV energy splitting between them. First, the underlying E-manifold states have an energy splitting of $-0.18$ MeV, because the $R = 4$ state with positive parity has underlying state $\psi_{\rm T0}^+$ and the $R = 3$ state with negative parity has underlying state $\psi_{\rm S0}^-$. The additional 0.70 MeV is from the Coriolis splitting between the $J = 3$ states with $R = 4$ and $R = 3$, which is $8{\rm B}\zeta$. As $2{\rm B} = 0.9$ for the F-band, the calibration requires $\zeta = 0.194$. This positive value for $\zeta$ gives reasonable energy splittings for several other states, up to spin/parity $5^\pm$. With $\zeta$ fixed, the term $-2{\rm B}\zeta(1-\zeta)$ in $E_{\rm rot}$ is constant for the entire F-band and has value $-0.14$ MeV.

It is known from molecular physics that $\zeta$ is unchanged for all bands of species $E^n \otimes F_2$. The rotational energy formula (11.22) therefore extends to combined bands with a single F-phonon, and only the vibrational energy changes. In the $F^2$-band, with two F-phonons, the Coriois corrections are a little different. For the details of this, see ref.[117].

## 11.5 The Complete Oxygen-16 Energy Spectrum

These assumptions for the model and its parameters lead to a rovibrational energy spectrum for the Oxygen-16 nucleus where essentially all states up to 20 MeV excitation energy are fitted moderately well. There are just over 60 such states with isospin 0 in the experimental tables [77, 228], although some uncertainties remain in the spin/parity and isospin assignments for several of the higher-energy states. The model predicts one new 4⁻ state below 15 MeV. There are no confirmed 6⁻ states in the tables, but the model predicts a few of these above 17 MeV. It also predicts several higher-spin states between 20 MeV and 30 MeV, of which some have been observed.

In detail, each vibrational state is classified by its numbers of A-, F- and E-phonons, and has an associated rovibrational band. We will discuss all states with total energy less than 20 MeV, and states with spin 6 and higher up to 30 MeV. This means there are at most one A-phonon, two F-phonons or four E-phonons. Combined vibrational states also occur, including those with one A- or F-phonon and one or two E-phonons. Recall that (multi-)E-phonon states are an interpretation for what are really E-manifold states.

The energy spectrum of the model is plotted in Fig. 11.6, where states from each rotational band are displayed in a different colour. Energy $E$ is plotted against spin $J$, up to $J = 6$. The total energy of each state is

$$E = n_A \omega_A + n_F \omega_F + E_{\text{vib}} + E_{\text{rot}}, \qquad (11.23)$$

where $n_A$ and $n_F$ are the numbers of A- and F-phonons, with $\omega_A = 12.05$ MeV and $\omega_F = 6.55$ MeV, and $E_{\text{vib}}$ is the underlying E-manifold energy. $E_{\text{rot}}$ is the total rotational energy (11.17) if there is no F-phonon, or (11.22) if there is one F-phonon. (For states with two F-phonons see [117].) The spectrum is rather dense and is plotted again in Fig. 11.7; here, each spin/parity is considered separately and it is shown how the model states can be matched with experimental states.

The lowest-lying $0^+, 3^-$ and $4^+$ states lie in the tetrahedral ground state rotational band. Extrapolating this band to higher spins gives $6^\pm, 7^-, 8^+$ and $9^\pm$ states. Of these, the natural parity states have been observed close to the predicted energies, but these energies are not the lowest for those spins, because of band crossing. This was noted earlier by Robson, who identified the $6^+$ state at 16.27 MeV as belonging to the tetrahedral ground state band, although there is a $6^+$ *yrast state*[2] at 14.82 MeV.

Next is the E-band (E¹-band) with one E-phonon. The E-manifold Hamiltonian predicts a significant lifting of the parity doubling characteristic of this band, produced by the non-negligible tunnelling probability from the tetrahedron to its dual via the flat square. There is a good fit to the low-lying 6.92 MeV $2^+$ and 8.87 MeV $2^-$ states of Oxygen-16. The

---

[2]The lowest-energy state for a given spin/parity is known as the yrast state.

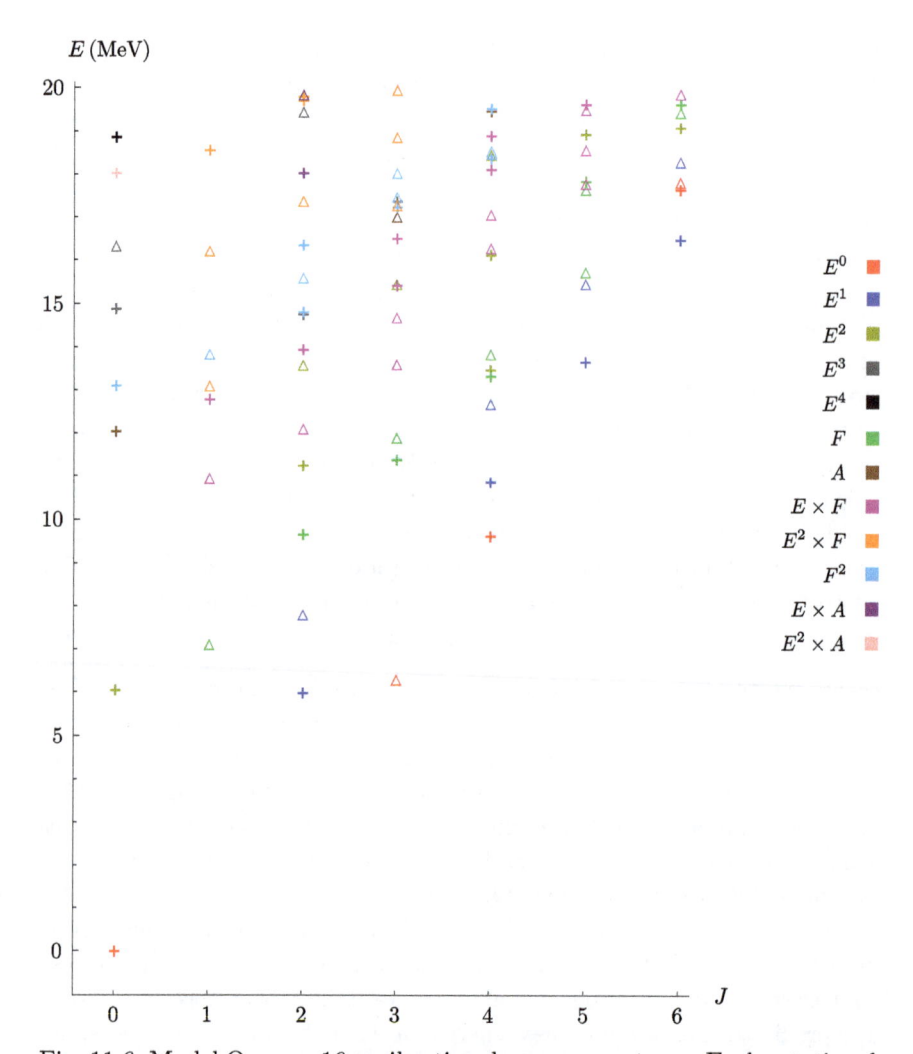

Fig. 11.6: Model Oxygen-16 rovibrational energy spectrum. Each rotational band is coloured differently. Positive (negative) parity states are displayed as pluses (triangles).

observed energy splitting is 1.95 MeV, compared with 1.82 MeV predicted in the model. The $2^-$ state has higher energy, because its wavefunction has a node at the flat square, as mentioned earlier. The $E^1$-band has the first-excited $4^+$ state and the lowest predicted $4^-$ state, and the lowest $5^+$ and $5^-$ states, always with the same splitting between the parity doubles.

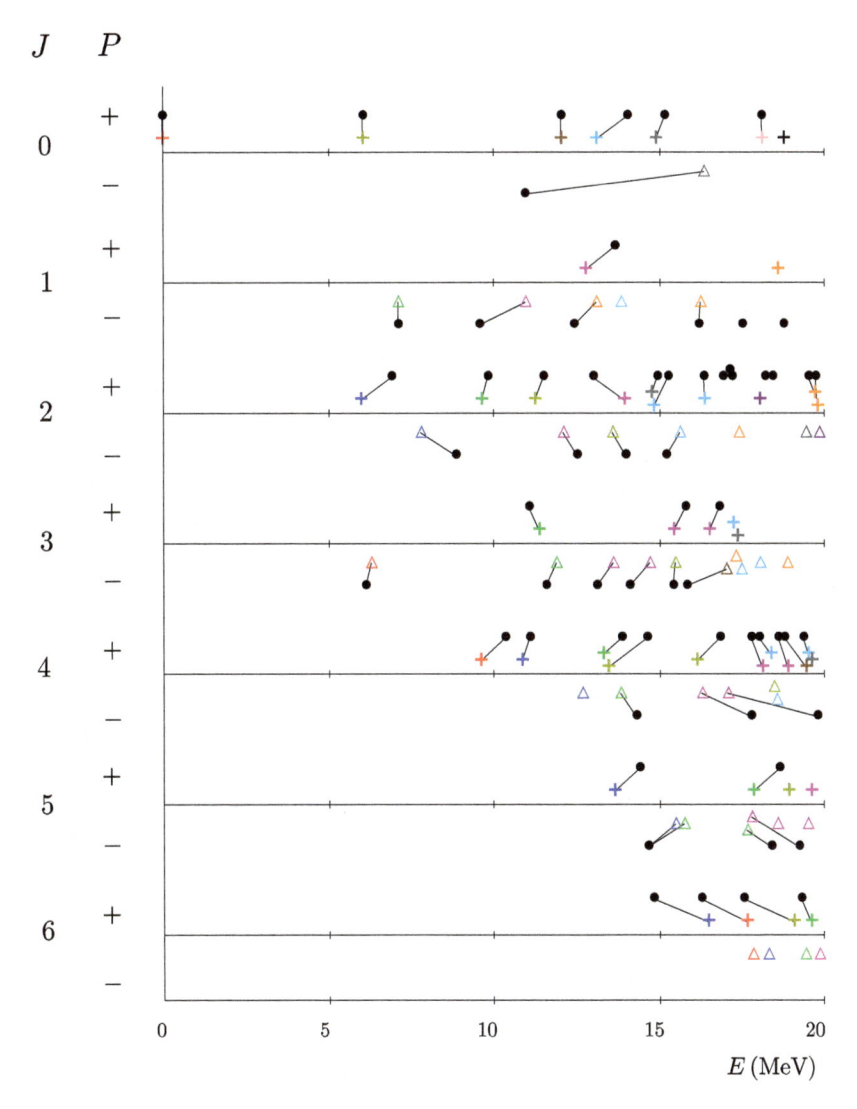

Fig. 11.7: Comparison between model and experimental Oxygen-16 energies. Positive (negative) parity states are displayed as pluses (triangles), and coloured as in Fig. 11.6 according to their rotational band. Experimental states are displayed as black dots. Proposed identifications between model and experimental states are shown by lines.

The $6^+$ yrast state also appears to be in this band, although its observed energy of 14.82 MeV is less than that predicted. The $8^+$ yrast state is in this band too. Beyond spin 5, no unnatural parity states have yet been observed, but the model predicts several $6^-$, $7^+$ etc. states above 17 MeV.

The $E^2$-band is interesting, as the first-excited $0^+$ state at 6.05 MeV is interpreted as belonging to this band. Because of the decomposition $E^2 = A_1 \oplus E$, this band combines spin/parities in the ground state band and the E-band, so it has states with spin/parities $0^+$, $3^-$ and $4^+$, and also $2^\pm$ and $4^\pm$. The predicted $3^-$ state at 15.47 MeV can be identified with the fourth-excited $3^-$ state at 15.41 MeV. This unexpectedly high energy occurs because of the relatively large vibrational energy of the underlying $\psi_{S2}^-$ state, $E_{\text{vib}} = 10.67$ MeV. Models which treat this $3^-$ state as a simple rotational excitation of the $0^+$ state at 6.05 MeV give it an energy of 12 MeV or less. However, such an approach leads to too many $3^-$ states with low energy. The $2^\pm$ states in the $E^2$-band match excited states with these spin/parities, and the two $4^+$ states can be matched with the observed states either side of 15 MeV. The single $4^-$ state is predicted to lie above 18 MeV, close to where a couple of such states are observed.

The states in the F-band, which have significant Coriolis energy, match experimental states quite well. The band has the $1^-$ yrast state and the first-excited $2^+$ state, then close together the $3^+$ yrast state observed at 11.08 MeV and the first-excited $3^-$ state at 11.60 MeV, discussed earlier in the context of calibrating the Coriolis coupling parameter $\zeta$. The $4^+$ state matches an observed state, and the F-band also has a nearby $4^-$ state. So from the E-band and F-band two $4^-$ states are predicted below 15 MeV, at 12.69 MeV and 13.84 MeV respectively, but just one is observed at 14.30 MeV. This is the first serious difference between the predictions of the model and what is observed; however, there is some experimental uncertainty here [228]. The model of Bijker and Iachello makes a similar prediction [40]. In the F-band there are two $5^-$ states and one $5^+$ state. Further spin 5 states are predicted below 20 MeV, arising from the E × F band and $E^2$-band. The F-band gives rise to the $7^-$ yrast state just above 20 MeV.

Let us now consider the higher-energy $0^+$ states. It seems agreed that there is no $0^+$ state at 11.26 MeV. (It is tabulated, but was observed with a weak signal just once [42].) There are clearly-observed $0^+$ states at 12.05, 14.03 and 15.10 MeV. In the rovibrational model these can be matched to states in the $F^2$-, $E^3$- and A-bands. The predicted energies for the first two of these are 13.11 and 14.89 MeV, so it is most likely that the A-band state is at 12.05 MeV. This is the reason for calibrating the A-phonon energy to be 12.05 MeV, close to the value of 11.4 MeV estimated using the breather state of Carbon-12, as mentioned earlier.

The $E^3$-band also contains a $0^-$ state (because of the $A_2$ irrep in the decomposition of $E^3$) whose predicted energy is 16.35 MeV, far higher than the observed energy of 10.96 MeV for the $0^-$ yrast state. This is the model's worst prediction. The state's vibrational wavefunction $\psi_{T3}^-$ is shown in Fig. 11.5, and it is clear that its energy is rather sensitive to the form of the model potential on the E-manifold, away from the tetrahedron and square configurations where this wavefunction has to vanish. Softening the potential could help.

The $E \times F$ band decomposes into $F_1$ and $F_2$ subbands. The states in the $F_2$ subband have the same spin/parities as those in the F-band but somewhat higher energies; the states in the $F_1$ subband have reversed parities.

$3^+$ states provide a stringent test of the model as only three are observed below 20 MeV, at 11.08, 15.78 and 16.82 MeV. The third of these has uncertain spin and isospin. In the model, these states arise from the F-band (one) and $E \times F$ band (two), with predicted energies 11.39, 15.43 and 16.52 MeV. This seems to confirm that the observed state at 16.82 MeV is definitely a $3^+$ state with isospin 0. The $E \times F$ band also contains a single $1^+$ state at 12.79 MeV, fairly close in energy to the observed state at 13.66 MeV, the only $1^+$ state known. These rare spin/parity states are especially important for comparing models.

The most interesting states with energy greater than 20 MeV are those with spin 6 or more, as there has been significant debate about high-spin states of Oxygen-16. Some of these have been discovered recently [94], and do not appear in the tables [77, 228]. Freer et al. found at least three $8^+$ states between 23 MeV and 30 MeV by studying Beryllium-8 decay channels [93], and also discovered a $6^+$ state at 21.2 MeV, close in energy to $6^+$ states previously observed at 21.4 MeV and 21.6 MeV [12, 92]. These three states may all be from a single broad resonance. There are two established $7^-$ states at 20.86 MeV and 21.62 MeV, and probably more at higher energy. There is also evidence of a $9^-$ state at around 30 MeV [200]. Experimental work at these energies is difficult and more states should be discovered as experimental techniques continue to improve. The observed high-spin spectrum and the predicted spectrum in the rovibrational model are considered in more detail in [117]. Both are dense above 20 MeV.

Overall, the rovibrational model of a cluster of four $B = 4$ Skyrmions, discussed here, successfully describes approximately 70 states in the observed spectrum of Oxygen-16, with most of the energy predictions matching measured energies to within about 1 MeV. The model incorporates tetrahedral, flat- and bent-square configurations, and configurations interpolating between them on the 2-dimensional E-manifold. The quantized dynamics on the E-manifold accounts well for the energy splitting between the parity-doubled states with $J^P = 2^\pm, 4^\pm, 5^\pm, \ldots$ in the E-band. Exceptionally, the predicted energy for the lowest $0^-$ state is about 5 MeV too high. This state vanishes at the tetrahedral and square configurations, so it is sensitive to details of the model which are little explored by other states. A number of further $3^-$ and $4^-$ states just below 20 MeV are predicted, and numerous unnatural parity states with high spin should exist, starting with a $6^-$ state in the ground state band between 17 MeV and 18 MeV. Perhaps the most robust prediction is the existence of a further $4^-$ state around 13 MeV. Finding this state would help confirm the model.

Chapter 12

# Modelling Calcium-40

## 12.1 Tetrahedral Structure of Calcium-40

The principal evidence that the Oxygen-16 nucleus is a tetrahedral cluster of four $\alpha$-particles is the presence of low-lying $J^P = 3^-$ and $J^P = 4^+$ states and the absence of a $2^+$ state below the $3^-$ state. The ratio of the excitation energies of the $4^+$ and $3^-$ states is approximately $\frac{5}{3}$, as expected for a tetrahedral rotational band whose collective rotational energy scales as $J(J+1)$. The strong transitions between the $3^-$ state and the $0^+$ ground state, and between the $4^+$ state and the ground state provide further evidence for collectivity. Interpreting the $3^-$ state as a 1-phonon, octahedral vibration of an intrinsically spherical nucleus is less convincing, because no obvious multiplet of close-to-degenerate 2-phonon states with spin/parities $6^+, 4^+, 2^+$ and $0^+$ is observed.

Lezuo in the 1970s drew attention to similar features in the energy spectrum of the two larger doubly-magic nuclei Calcium-40 and Lead-208 [168]. Calcium-40 has a low-lying $3^-$ state, and then three $4^+$ states in the required energy range for a ratio $\frac{5}{3}$, but one has energy 15% too low and the other two have energy 5% too high. Lezuo mentioned that a fuller understanding of the spectrum would require a modelling of vibrational modes.

In Oxygen-16, the E-mode has lowest frequency, and in the last chapter we explained how the first-excited $0^+$ state at 6.05 MeV is interpreted as a state with two E-phonons. Quantum mechanics on the E-manifold of $B = 16$ Skyrme field configurations replaces the harmonic oscillator quantization of the E-mode, but states can still be classified in terms of E-phonons (by examining the wavefunctions close to the tetrahedral Skyrmion). By combining E-manifold states with A- and F-phonons, and rotations, we saw that the spectrum of states can be matched rather well with the known isospin 0 states of Oxygen-16. Here we show that a similar approach can be applied to Calcium-40.

Finding a Skyrmion with baryon number 40 to model Calcium-40 has been difficult, but by using a suitable multi-layer rational map ansatz as a starting point, a tetrahedrally-symmetric solution has been found by numerical relaxation [161]. This solution is shown in Fig. 12.1. It resembles a tetrahedral Skyrmion of baryon number 56 [163] (56 is a tetrahedral number), with four clusters of baryon number 4 cut off from the vertices. The final structure is closer to spherical, and because it doesn't have the pointedness of an ideal tetrahedron, it is more stable.

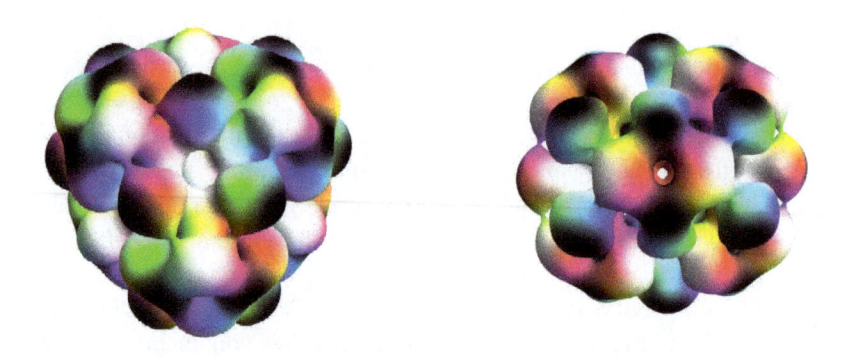

Fig. 12.1: Two views of the $B = 40$ Skyrmion

This Skyrmion solution also emerges by stacking 40 $B = 1$ Skyrmions as a tetrahedrally-symmetric subcluster of the FCC lattice, with all nearest neighbours in the attractive channel. A model of the relevant cluster is shown in Fig. 12.2, in the same two orientations as in Fig. 12.1. The steel balls correspond to the cores of the $B = 1$ Skyrmions, and the magnetic rods connecting them correspond to the nearest-neighbour attractive forces – these are not materially present in the Skyrmion solution. Physically, such a cluster appears as a static solution in a variant of Skyrme theory (to be discussed further in Section 14.1.1) called the lightly-bound model, where the basic Skyrmions attract but hardly merge [101, 102]. It is particularly clear in the lightly-bound model how to orient the basic Skyrmions to achieve maximal attraction and greatest stability; the $B = 4$ Skyrmion is tetrahedral in this variant model. The $B = 40$ Skyrmion in Fig. 12.1, which is a solution of the standard Skyrme theory with pion mass $m = 1$, is obtained by relaxation of its analogue in the lightly-bound model.

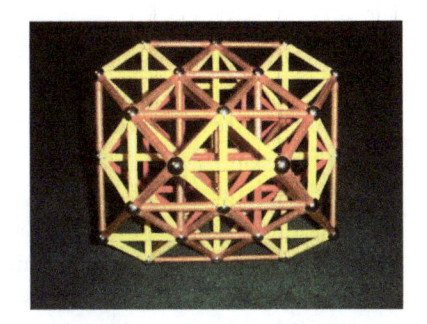

Fig. 12.2: Two views of the cluster with baryon number 40

Notice that the $B = 40$ Skyrmion can be identified as a bound cluster of ten distinct, tetrahedral $B = 4$ Skyrmions ($\alpha$-particles). These tetrahedral subclusters are highlighted by yellow magnetic rods in Fig. 12.2. It is encouraging that a tetrahedrally-symmetric, quasi-spherical $B = 40$ Skyrmion exists. From this, one can develop a phenomenological model of states of Calcium-40, based on the rovibrational spectrum of a tetrahedral structure [179]. This again involves A-, F- and E-phonons, combined with rotations. The model is less detailed than that for Oxygen-16, but is qualitatively similar.

We will see that more than 100 excited states of the Calcium-40 nucleus, with isospin 0 and spins up to 16, can be classified into tetrahedral rovibrational bands. Almost all observed states below 8 MeV can be accommodated, as well as many of the high-spin states above 8 MeV. The rotational bands for Calcium-40 have some similarity to the bands for Oxygen-16, but the A-mode vibrational frequency is lower relative to the E-mode and F-mode frequencies.

For Calcium-40, states up to spin 16 have been observed, whereas for Oxygen-16 the highest confirmed spin is about 9. Such high spins have not been measured directly, but are inferred from gamma ray spectra. The model here makes substantial use of the interesting and detailed spectra of Calcium-40 states with positive parity, up to $J^P = 16^+$, that have been tabulated and gathered into proposed rotational bands in [132]. There, no use is made of the theory of tetrahedral vibrations, and the bands have the typical structure associated with ellipsoidally deformed nuclei [46]. The negative parity states are less completely classified, but several known states are tabulated and gathered into bands in [230]. Again, a tetrahedral structure is not discussed, but it is argued that some type of permanent octupole deformation needs to be considered. We find here that just a few tetrahedral rovibrational bands can unify several of the simpler rotational bands identified in refs.[132] and [230]. For the complete set of experimentally determined states, we use the standard compilation [77].

Perhaps the most important observation, providing evidence for the tetrahedral structure of Calcium-40, is that the number of states in one of the unified bands jumps by one starting at $J^P = 8^+$ and jumps by one again starting at $14^+$. Such jumps are expected in the rotational band of a tetrahedron, vibrationally excited by one E-phonon. In contrast, the rotational bands of an ellipsoidally deformed nucleus have no such jumps, and it needs to be assumed that new bands start for some reason at $8^+$ and $14^+$.

An interesting issue is whether the entire spectrum of isospin 0 states of Calcium-40 can be interpreted using a rovibrational model, or whether there are independent shell-model excitations. Four states of Calcium-40 can be clearly identified as shell-model, particle-hole (1p1h) excitations. The particle states are from the unfilled $f_{\frac{7}{2}}$ shell and the holes are $d_{\frac{3}{2}}$ states from the filled sd-shell [221]. Combined, they give states with spin/parities $5^-, 4^-, 3^-$ and $2^-$. Their energies, and even their ordering, are not easily or precisely predicted by shell-model calculations, but the trend is for the energy to decrease as the spin increases. In particular, the $5^-$ state, identified with the observed $5^-$ yrast state at 4.491 MeV, is strikingly low in energy. These four 1p1h states of Calcium-40 are not easily fitted into collective rotational bands. However, our rovibrational model for Oxygen-16 suggests that they do not need to be treated as supplementary to the collective states. It seems best to identify these four states as belonging to rotational bands of Calcium-40, but with significantly displaced energies. As collective states, they have wavefunctions and energies that are significantly affected by their overlap (superposition) with pure 1p1h excitations, but it is not necessary to postulate that they are totally distinct. The effect is largest for the $5^-$ state. In the rovibrational model it lies in the E-band, with one E-phonon, but its energy is shifted down by more than 2 MeV compared to what would be expected by extrapolation from other states in the E-band.

## 12.2 Rovibrational Bands

Here we list the expected spin/parities of rovibrational states of the tetrahedral $B = 40$ Skyrmion. They are obtained using Table 11.1, the character table of the tetrahedral group $T_d$. We need to extend the lists in eqs.(11.12) – (11.16) up to spin 16. Spins up to 12 are tabulated in ref.[127] (p.450). As for Oxygen-16, the vibrational modes of the Skyrmion modelling Calcium-40 are A-, F- and E-modes, transforming according to the $A_1$, $F_2$ and E irreps.

The tetrahedral ground state band consists of the rotational excitations with no phonons. The vibrational state transforms under the trivial $A_1$ irrep of $T_d$, and has rotational excitations with spin/parities

ground state (g.s.) band :
$$0^+, 3^-, 4^+, 6^+, 6^-, 7^-, 8^+, 9^+, 9^-, 10^+, 10^-, 11^-, 12^+, 12^+, 12^-,$$
$$13^+, 13^-, 14^+, 14^-, 15^+, 15^-, 15^-, 16^+, 16^+, 16^-, \ldots. \qquad (12.1)$$

(Note that multiplicities greater than 1 start at spin 12.) There is some periodic structure here. Whenever $J$ increases by 6, the number of allowed states (ignoring the parity label) increases by 1. So, for example, there are no spin 2 states, one spin 8 state and two spin 14 states. This is because the characters (11.9) of $O(3)$ for rotations by 120° and 180° repeat when $J$ increases by 6, whereas the character $2J+1$ for the rotation by 0° increases linearly with $J$. Similarly, the difference between the number of positive parity and negative parity states has period 4, being $1, 0, 0, -1$ for $J = 0, 1, 2, 3$ mod 4. This is because the characters (11.10) for the rotations by 180° and 90° combined with inversion repeat when $J$ increases by 4.

The next important band is the E-band, consisting of rovibrational states where the Skyrmion is excited by one E-phonon. Here the states all occur symmetrically as parity doubles. The allowed spin/parities are

E-band :
$$2^\pm, 4^\pm, 5^\pm, 6^\pm, 7^\pm, 8^\pm, 8^\pm, 9^\pm, 10^\pm, 10^\pm, 11^\pm, 11^\pm, 12^\pm, 12^\pm,$$
$$13^\pm, 13^\pm, 14^\pm, 14^\pm, 14^\pm, 15^\pm, 15^\pm, 16^\pm, 16^\pm, 16^\pm, \ldots. \qquad (12.2)$$

In the E-band, the number of states of each parity separately increases by 1 when the spin $J$ increases by 6. In particular, there is an upward step at spin 8 and spin 14, and these steps are observed experimentally [132]. For low spins, the energies are higher in the E-band than in the ground state band because of the E-phonon energy, but the bands cross over at higher spin, because the E-phonon stretches the Skyrmion, increasing its effective moment of inertia and decreasing the band slope.

A further band is the 2-phonon, $E^2$-band. As noted before in (11.6), $E^2 = A_1 \oplus E$. The 2-phonon state transforming under the trivial $A_1$ irrep has rotational excitations with the spin/parities of the ground state band (12.1), and the doublet transforming under the E irrep have rotational excitations with spin/parities as in the E-band (12.2). The energies are slightly higher for the 2-phonon than 1-phonon states, and the E-band and $E^2$-band appear not to cross over.

The triplet of vibrational states with one F-phonon have a vectorial character, and carry a vibrational spin/parity $1^-$. Their rotational excitations combine an underlying spin/parity $R = 0^+, 3^-, 4^+, 6^\pm, \ldots$ from the ground state band with the vibrational spin/parity $1^-$ using the Clebsch–Gordon rules, as explained in the last chapter. This gives the F-band spin/parities

F-band :
$$1_0^-; 2_3^+, 3_3^+, 4_3^+; 3_4^-, 4_4^-, 5_4^-; 5_6^\pm, 6_6^\pm, 7_6^\pm; \ldots, \qquad (12.3)$$

where the states are grouped according to the underlying spin $R$ (indicated by the subscript). Because of the relatively high energy of the F-phonon, we do not need to consider higher spins here.

## 12.3 Interpreting the Calcium-40 Spectrum

The primary data for the spectrum of excited states for Calcium-40 are the ENSDF adopted levels [77]. Although several states have an uncertain spin/parity, it was shown in ref.[179] that 104 of the states can be assigned with some confidence to tetrahedral rovibrational bands. The experimental energies and spin/parities are reproduced here in Fig. 12.3. Filled circles (triangles) indicate positive (negative) parity states, and the colouring distinguishes the various rovibrational bands to which the states are assigned. States with a given spin, in one band, do not all have the same energy. In Oxygen-16, part of the energy splitting between positive and negative parity states arose from the tunnelling between the tetrahedral Skyrmion and its dual, which could be calculated. The splittings have not been calculated in Calcium-40, as there is no precise dynamical model for the anharmonic tetrahedral deformations.

From the adopted levels, almost all isospin 0 states below 8 MeV are retained. The states at 6.160, 6.422, 7.421 and 7.481 MeV are dropped as they are either doubtful or have unknown (low) spins. Above 8 MeV, only

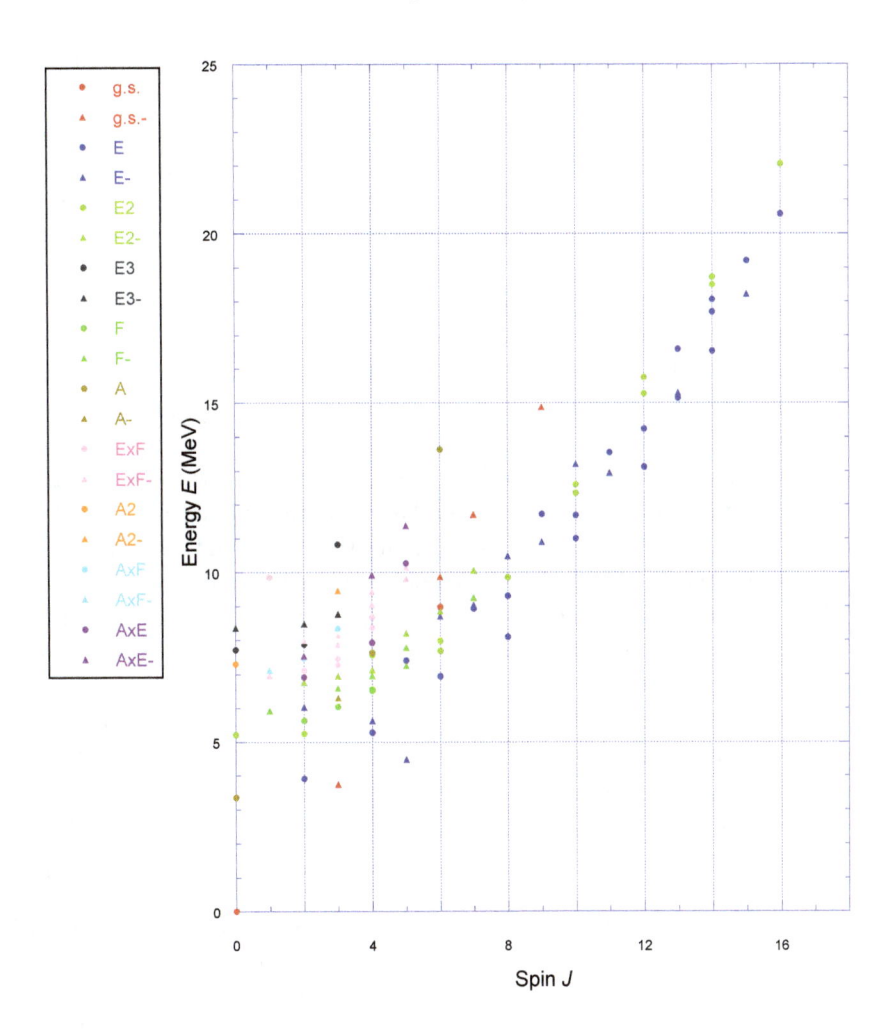

Fig. 12.3: Observed Calcium-40 energy spectrum, with the rovibrational band assignments. Positive parity states are displayed as filled circles, negative parity states as filled triangles. (E2– denotes the negative parity part of the $E^2$-band, etc.)

selected states are retained, mainly of spin 5 and higher. States of lower spin are very numerous and would lie in various multiphonon bands; only a few of these can be classified, even tentatively.

In the range between approximately 7 and 10 MeV, the experimental data only partially constrain the possible spins and parities. Above 9 MeV, high-spin states have their energy and spin mainly inferred from the pattern of gamma decays, and all states that have been assigned to bands in [132] and [230] are retained. The highest of these is a 16$^+$ state at 22.106 MeV.

The band assignment that one can be most confident about is the E-band, and we discuss this first. 31 observed states up to spin 16, of positive and negative parity, are identified as being in this band. They are shown in blue in Fig. 12.3. Many are from the positive parity bands 2, 3 and 4 in [132], and from the negative parity bands in the rightmost two columns of Fig. 5 in [230]. Just a few expected states with positive parity are missing from this E-band – a second 11$^+$ state, a second 15$^+$ state, and a second and third 16$^+$ state are expected, according to the list (12.2), but have not yet been identified. A greater number of negative parity states are missing. This is because there is just one identified state with each of the spin/parities 8$^-$, 10$^-$, 11$^-$, 13$^-$, 15$^-$ in [230], but the E-band requires two. Of more concern is that two 12$^-$ states, and three 14$^-$ and 16$^-$ states are expected, but none of these have been observed or inferred.

The plot of energy $E$ against spin $J$ is approximately linear for the E-band, as it is in Oxygen-16. It is nothing like a linear relation between $E$ and $J(J+1)$. Presumably the E-phonon softens the nucleus, and centrifugal stretching produces an effective moment of inertia that increases with $J$. This approximate linearity with $J$ allows an extrapolation of the E-band down in energy towards spin 0. In this way, the 2$^\pm$ and 4$^\pm$ states in the E-band can be identified, and the E-phonon frequency can be read off as approximately 2.5 MeV.

The negative and positive parity states are fairly well interleaved in the E-band, implying that tunnelling from the tetrahedral Skyrmion to its dual is suppressed. An outlier, below the general trend, is the 5$^-$ yrast state at 4.491 MeV. This is interpreted as in the E-band but with lowered energy because of its partial 1p1h character. The 4$^-$ and 2$^-$ states in the E-band are also yrast states, and are interpreted as having a partial 1p1h character, but a less pronounced lowering of energy. The final 1p1h excitation, with spin/parity 3$^-$ cannot be in the E-band at all. It probably contributes to the 3$^-$ yrast state at 3.737 MeV, assigned to the ground state band.

Unifying several of the rotational bands from refs.[132] and [230] into one tetrahedral, rovibrational E-band is theoretically attractive. It explains not only some of the observed jumps of multiplicity as the spin increases, but also appears to be supported by the many cross-band gamma ray transitions that are observed. For example, transitions of roughly equal strength connect the multiple states with spin above 10 (see Fig. 1 of ref.[132]).

Only a few experimental states can be identified as being in the ground state rotational band. The vibrationally-unexcited Skyrmion appears to be rather rigid, so the rotational energy $E$ rises rapidly, scaling more closely with $J(J+1)$. The $0^+$ ground state and $3^-$ state at 3.737 MeV are clearly in this band, but it appears that above the $3^-$ state the ground state band crosses the E-band, so that its $6^\pm$, $7^-$ and $9^-$ states are well above those in the E-band. (The assignment of the $6^-$ state at 9.860 MeV to the ground state band is tentative, and the expected $8^+$ state is still missing.) This band crossing is similar to what occurs in Oxygen-16, where the ground state band crosses the E-band between spin 4 and spin 6 [117, 206].

The $4^+$ state at 5.279 MeV is a mystery. It occurs at the crossover of the ground state band and the E-band. Uniquely, this state is tentatively assigned to both these bands. Really, two independent states are needed. Possibly the observed state at 6.160 MeV, which has an uncertain $3^-$ spin/parity assignment in [77], and which was the lowest-energy state dropped earlier, is the missing $4^+$ state. In any case, between 5 and 7 MeV there are apparently just three $4^+$ states observed, but the band structure needs four.

The E-band does not accommodate all the known positive parity states with high spin. In particular, there are higher-energy observed states with spin/parities $8^+, 10^+, 12^+, 14^+, 16^+$ in a separate band, shown as band 1 in [132]. They are assigned here to the $E^2$-band. As noted previously, the $E^2$-band splits into an $A_1$ part and an E part. In modelling Oxygen-16 we discovered that the $A_1$ part has clearly lower energy, and we assume the same holds for Calcium-40. So the high-spin states listed above are in the $A_1$ part of the $E^2$-band, which has the same spin/parities as the ground state band. It is assumed that this is again a rather soft band, with an approximately linear relationship between $E$ and $J$. The energies are therefore below those of the ground state band for spins decreasing from 16 to 6, but then the bands cross. By extrapolating down to spin 0, we identify the second-excited $0^+$ state at 5.212 MeV as the $E^2$-bandhead (the lowest state in the $E^2$-band). Its energy matches the estimate for the energy of two E-phonons rather well.

The high-spin states in the E part of the $E^2$-band have not been identified – they would include, for example, further $14^+$ states. But a few low-spin states in this part of the band can be identified, in particular the states with spin/parity $2^\pm$.

The first-excited $0^+$ state at energy 3.353 MeV is interpreted as the bandhead of the A-band, a state with one A-phonon. The A-phonon energy is therefore approximately 3.4 MeV, somewhat higher than the E-phonon energy, but less than the energy of two E-phonons. The A-band is assumed to be less soft than the bands with one or two E-phonons, because the nuclear shape stays closer to spherical. Like the $A_1$ part of the $E^2$-band, the A-band has the spin/parities of the ground state band, and its $3^-$ and $4^+$ states, and possibly its $6^+$ state, can all be identified with observed states having energies about 3 MeV higher than those in the ground state band. However, compared with the model predictions for all the bands combined, there is a shortage of observed $6^+$ states in the required energy range, and a worse shortage of $6^-$ states.

A further band that can be recognised is the F-band, with spin/parities as in the list (12.3). This includes the $1^-$ and $3^+$ yrast states at 5.903 and 6.030 MeV, which cannot occur in other bands. Using the energies of these two states, the F-phonon energy is estimated to be about 5 MeV, higher than the E-phonon and A-phonon energies. States in the F-band up to spin 5 are seen, but with considerable uncertainty.

Above the F-band there is evidence for an $A \times E$ band, an $A \times F$ band, and an $E \times F$ band. The energies of their states are estimated to be the sum of the energies of the two contributing phonons and the rotational energy. Suitable states to place in these bands can only be tentatively assigned, and are shown in Fig. 12.3. A few further states with low spins are tentatively assigned to other multiphonon bands. The band with three E-phonons is the lowest that can accommodate a $0^-$ state. One candidate $0^-$ state is observed at 8.359 MeV, but its spin is not certain.

## 12.4 Summary

More than 100 observed isospin 0 states of Calcium-40 can be assigned to rovibrational bands of an intrinsically tetrahedral nuclear structure. This makes Calcium-40 similar to the smaller doubly-magic nucleus Oxygen-16. A tetrahedrally-symmetric $B = 40$ Skyrmion has been found that makes such a structure plausible. Vibrational A-, F- and E-phonons can be excited, and have approximate energies 3.4, 5 and 2.5 MeV, respectively. States with up to three vibrational phonons can be identified, and their rotational excitations lie in several tetrahedral rovibrational bands. In particular, 31 states are assigned to the band with one E-phonon. Relative to the F-phonon and E-phonon energies, the A-phonon energy appears to be considerably lower than it is in Oxygen-16. Possibly this is because in Oxygen-16 the A-mode is a true breather mode of the tetrahedron of four $\alpha$-particles, requiring an energetically-expensive compression and expansion of the $B = 16$ Skyrmion, whereas in Calcium-40 and its corresponding Skyrmion, it is more likely to be a tetrahedrally-symmetric, volume-preserving vibration, where six $\alpha$-particles move outward and four inward, and vice versa.

The band scheme accommodates almost all observed states with excitation energy below 8 MeV, as well as all the known states of higher energy with spins from 8 up to 16. It explains why there are few observed $0^-$, $1^+$ and $3^+$ states – these can only occur in multiphonon bands (with the exception of one $3^+$ state in the F-band). It also explains why $2^+$ states are more numerous than $2^-$ states – the F-band and $A \times F$ bands exclude the latter. Similarly, $4^+$ states are more numerous than $4^-$ states because the ground state band and A-band exclude $4^-$ states.

One more $4^+$ state between 5 and 7 MeV is predicted, and a few extra states of spin 5 and spin 6 between 8 and 11 MeV. Also, a number of extra states with spin 8 or more, particularly of negative parity, are expected from 11 MeV upwards. One of these is the $8^+$ state in the ground state band, with a predicted energy between 12 and 13 MeV.

The analysis could be made more robust if the spins and parities of 20 observed states between about 7 and 10 MeV were better known. At present, it is necessary to make choices where there is experimental uncertainty, in order to fill some of the higher bands, but these spin/parity choices cannot be regarded as firm predictions, as there are too many of them. On the theoretical side, further information about the $B = 40$ Skyrmion and its low-lying vibrational modes is needed, and it would be useful to investigate more generally the vibrational dynamics of tetrahedral clusters of ten $\alpha$-particles.

## Chapter 13

# Electromagnetic Transition Strengths

Radiative electromagnetic transitions between nuclear states depend on the underlying intrinsic nuclear shapes, and can be used to test models of the structure [45,46,142]. In a nucleus with an even number of both protons and neutrons – an even-even nucleus – the reduced electric quadrupole transition strength $B(E2 : 0^+ \mapsto 2^+)$ (also denoted $B(E2)\uparrow$) from the $0^+$ ground state to the yrast (lowest) $2^+$ state is particularly important. It is used to determine the magnitude of the intrinsic electric quadrupole moment, and hence the deformation of the nucleus away from spherical [199, 202]. Large quadrupole moments and transition strengths indicate collective effects in which many nucleons participate.

Quadrupole moments and $B(E2)\uparrow$ transition strengths for nuclei with baryon numbers $B = 8, 12, 16, 20, 24$ and $32$ have been calculated using Skyrmions [111], and in this chapter we compare the results with the nuclear transition strengths inferred from experiment. Excited $0^+$ states usually have different quadrupole moments, because they involve Skyrmions with distinct shapes. For example, for the substantially prolate $0^+$ Hoyle state of Carbon-12, modelled by the $B = 12$ chain Skyrmion, a large quadrupole moment is predicted; and for Oxygen-16, one can relate the quadrupole moment of the first-excited $0^+$ state to the predicted deformation of the tetrahedral $B = 16$ Skyrmion into a flat or bent square.

Isovector magnetic dipole $M1$ transitions between quantized Skyrmion states have also been considered, but only for baryon numbers 1 and 2. These change the spin by one unit, and preserve parity. The $M1$ transition from a delta resonance to a nucleon has been calculated [7], and also the transition from a deuteron to the $^1S_0$ isovector two-nucleon resonance [52].

The Skyrmions we consider here have pion mass parameter $m = 1$, and energies and symmetries as in Table 8.1. Table 13.1 lists the diagonal elements of the inertia tensors $U_{ij}$, $V_{ij}$ and $W_{ij}$ for these Skyrmions[1], calculated using the formulae (7.4). The Skyrmions are oriented such that all off-diagonal elements vanish. The elements $V_{11}$, $V_{22}$ and $V_{33}$ indicate the prolateness or oblateness of the Skyrmion, and are qualitatively correlated with the quadrupole moments.

The root-mean-square matter radius of a Skyrmion is

$$\langle r^2 \rangle^{\frac{1}{2}}_{\text{matter}} = \left( \frac{\int r^2 \mathcal{E}(\mathbf{x}) \, d^3x}{\int \mathcal{E}(\mathbf{x}) \, d^3x} \right)^{\frac{1}{2}}, \quad (13.1)$$

where $\mathcal{E}(\mathbf{x})$ is the static energy density. Energies and matter radii are listed in Table 13.2. In this chapter, the energy and length scale factors are fixed to match the mass 11178 MeV and matter radius 2.43 fm of Carbon-12 in its ground state [162], and take the values

$$\frac{F_\pi}{4e} = 6.154 \, \text{MeV}, \quad \frac{2}{eF_\pi} = 1.061 \, \text{fm}, \quad (13.2)$$

giving parameter values and a physical pion mass

$$F_\pi = 95.6 \, \text{MeV}, \quad e = 3.89, \quad \hbar = 30.2 \quad \text{and} \quad m_\pi = 186 \, \text{MeV}, \quad (13.3)$$

where $\hbar = 2e^2$. We have seen earlier that calibrating Skyrme theory using the Carbon-12 nucleus proves successful for describing the spectrum of rotational excitations of Carbon-12, including those in the Hoyle band. The predicted nuclear masses and matter radii of selected other nuclei with isospin 0 are also in reasonable agreement with experimental data, as can be seen in Table 13.2.

In the following, we mainly focus on the $B(E2)\uparrow$ quadrupole transition strength between the $0^+$ ground state and the lowest $2^+$ state in isospin 0 nuclei. The transition strength in the relatively short-lived Beryllium-12 nucleus will be calculated later as an example with non-zero isospin.

## 13.1  $B(E2)$ Transition Strengths in Skyrme Theory

In Skyrme theory, the electric charge density $\rho(\mathbf{x})$ is [251]

$$\rho(\mathbf{x}) = \frac{1}{2}\mathcal{B}(\mathbf{x}) + \mathcal{I}_3(\mathbf{x}), \quad (13.4)$$

where $\mathcal{B}(\mathbf{x})$ is the baryon density, the integrand of (4.26), and $\mathcal{I}_3(\mathbf{x})$ is the third component of the $su(2)$ isospin charge density $J^0_{\text{isospin}}$. For quantum

---
[1]These include more recently calculated values than those in Tables 9.6 and 9.10, and there are small numerical differences.

Table 13.1: Skyrmions with baryon numbers $B = 8, 12, 16, 20, 24$ and $32$ for $m = 1$, and the isospin 0 nuclei they model. We list the symmetry group $K$, scaled energy $E$, and the diagonal elements of the inertia tensors $U_{ij}$, $V_{ij}$, $W_{ij}$ (in Skyrme units).

| $B$ | $K$ | Nucleus | $E/12\pi^2 B$ | $U_{11}$ | $U_{22}$ | $U_{33}$ | $V_{11}$ | $V_{22}$ | $V_{33}$ | $W_{11}$ | $W_{22}$ | $W_{33}$ |
|---|---|---|---|---|---|---|---|---|---|---|---|---|
| 8 | $D_{4h}^{\text{twist}}$ | $^8$Be | 1.279 | 298 | 292 | 326 | 4093 | 4094 | 1381 | 0 | 0 | 0 |
|   | $D_{4h}^{\text{no twist}}$ |  | 1.283 | 287 | 291 | 350 | 4615 | 4615 | 1296 | 0 | 0 | 0 |
| 12 | $D_{4h}$ | Hoyle | 1.274 | 440 | 449 | 456 | 12137 | 12137 | 2139 | 0 | 0 | 0 |
|   | $D_{3h}$ | $^{12}$C | 1.278 | 442 | 442 | 497 | 5009 | 5006 | 7627 | -41 | -41 | -38 |
| 16 | $D_{2d}^{\text{bent}}$ |  | 1.271 | 572 | 571 | 674 | 9123 | 9119 | 14602 | 0 | 0 | 0 |
|   | $D_{4h}^{\text{flat}}$ |  | 1.272 | 563 | 567 | 689 | 9143 | 9174 | 15682 | 0 | 0 | 0 |
|   | $T_d$ | $^{16}$O | 1.276 | 586 | 586 | 674 | 9100 | 9101 | 9128 | 0 | 0 | 0 |
| 20 | $T_d$ |  | 1.273 | 757 | 757 | 819 | 12820 | 12820 | 12821 | 0 | 0 | 0 |
|   | $D_{3h}$ | $^{20}$Ne | 1.276 | 857 | 735 | 735 | 18542 | 18591 | 9762 | 15 | -15 | -11 |
| 24 | $D_{3d}$ |  | 1.269 | 879 | 890 | 959 | 19600 | 19600 | 29863 | 0 | 0 | 0 |
|   | $D_{3h}$ | $^{24}$Mg | 1.273 | 869 | 869 | 1006 | 20554 | 20454 | 16226 | -99 | 99 | 99 |
| 32 | $O_h$ | $^{32}$S | 1.264 | 1115 | 1116 | 1367 | 31625 | 31628 | 31704 | 0 | 0 | 0 |

Table 13.2: Skyrmion masses $E$ converted to MeV, and measured nuclear masses $E_{\text{Exp}}$; Skyrmion root-mean-square matter radii $\langle r^2 \rangle^{\frac{1}{2}}$ in Skyrme units, converted to physical units (fm), and the experimental values. The conversion uses the calibration (13.2). The experimental matter radii are from ref.[13]. Due to its instability, the matter radius of Beryllium-8 is unknown, so we give the charge radius for its isobar Lithium-8. For the Hoyle state's experimental matter radius, see ref.[68].

| $B$ | $K$ | Nucleus | $E$ [MeV] | $E_{\text{Exp}}$ [MeV] | $\langle r^2 \rangle^{\frac{1}{2}}$ | $\langle r^2 \rangle^{\frac{1}{2}}$ [fm] | $\langle r^2 \rangle^{\frac{1}{2}}_{\text{Exp}}$ [fm] |
|---|---|---|---|---|---|---|---|
| 8 | $D_{4h}^{\text{twist}}$ | $^8$Be | 7457 | 7452 | 2.05 | 2.18 | 2.34 |
|   | $D_{4h}^{\text{no twist}}$ |  | 7481 |  | 2.15 | 2.28 |  |
| 12 | $D_{4h}$ | Hoyle | 11143 |  | 2.76 | 2.93 | 2.89 |
|  | $D_{3h}$ | $^{12}$C | 11178 | 11178 | 2.29 | 2.43 | 2.43 |
| 16 | $D_{2d}^{\text{bent}}$ |  | 14822 |  | 2.66 | 2.82 |  |
|  | $D_{4h}^{\text{flat}}$ |  | 14833 |  | 2.70 | 2.87 |  |
|  | $T_d$ | $^{16}$O | 14903 | 14904 | 2.35 | 2.50 | 2.70 |
| 20 | $T_d$ |  | 18556 |  | 2.60 | 2.76 |  |
|  | $D_{3h}$ | $^{20}$Ne | 18600 | 18630 | 2.86 | 3.03 | 3.01 |
| 24 | $D_{3d}$ |  | 22198 |  | 3.13 | 3.33 |  |
|  | $D_{3h}$ | $^{24}$Mg | 22212 | 22356 | 2.87 | 3.04 | 3.06 |
| 32 | $O_h$ | $^{32}$S | 29480 | 29808 | 3.17 | 3.36 | 3.26 |

states with isospin 0 the charge density is simply half the baryon density, and the total electric charge is $\frac{1}{2}B$ (in units of the proton charge).

The intrinsic electric quadrupole tensor of a Skyrmion is

$$Q_{ij} = \int \left(3x_i x_j - |\mathbf{x}|^2 \delta_{ij}\right) \rho(\mathbf{x}) \, d^3x \,. \tag{13.5}$$

Usually, when the Skyrmion is in its standard orientation, the quadrupole tensor is diagonal and $Q_{11}, Q_{22}$ and $Q_{33}$ are the quadrupole moments. A quadrupole tensor is traceless so $Q_{11} + Q_{22} + Q_{33} = 0$. Almost all the Skyrmions considered here can be oriented to have a cyclic symmetry greater than $C_2$ along the $x^3$-axis. Then $Q_{11} = Q_{22}$, so $Q_{33}$ has the largest magnitude and determines the other moments.

Table 13.3 lists the quadrupole moments of the Skyrmions with $B = 8, 12, 16, 20, 24$ and 32. These are in Skyrme units and are converted to physical units by multiplying by the square of the length scale factor. The intrinsic electric quadrupole moment of the corresponding isospin 0 nucleus

Table 13.3: Intrinsic quadrupole moments for selected Skyrmions in Skyrme units, and the quadrupole moment $Q_0 = 4Q_{33}/e^2 F_\pi^2$ in physical units. Quadrupole moments that are zero because of the Skyrmion's symmetry are denoted by dashes. The experimental results $Q_0^{\text{Exp}}$ are from ref.[202] and have been derived from experimental $B(E2)\uparrow$ transition strengths using eq.(13.9). The result for Beryllium-8 is not measured, but obtained theoretically by a Green's function Monte Carlo method [69]. The experimental value for Oxygen-16 is bracketed since it has been derived from the transition strength from the $0^+$ ground state to the first-excited $2^+$ state.

| B | K | Nucleus | $Q_{11}$ | $Q_{22}$ | $Q_{33}$ | $Q_0$ [eb] | $Q_0^{\text{Exp}}$ [eb] |
|---|---|---|---|---|---|---|---|
| 8 | $D_{4h}^{\text{twist}}$ | $^8$Be | -8.54 | -8.55 | 17.10 | +0.192 | +0.32 |
|   | $D_{4h}^{\text{no twist}}$ |  | -10.5 | -10.5 | 21.1 | +0.238 |  |
| 12 | $D_{4h}$ | Hoyle | -32.1 | -32.1 | 64.3 | +0.724 |  |
|   | $D_{3h}$ | $^{12}$C | 8.99 | 9.10 | -18.0 | -0.203 | -0.200 |
| 16 | $D_{2d}^{\text{bent}}$ |  | 18.2 | 18.4 | -36.6 | -0.412 | (0.202) |
|   | $D_{4h}^{\text{flat}}$ |  | 21.2 | 21.3 | -42.5 | -0.478 |  |
|   | $T_d$ | $^{16}$O | — | — | — | — |  |
| 20 | $T_d$ |  | — | — | — | — |  |
|   | $D_{3h}$ | $^{20}$Ne | -18.8 | -18.6 | 37.4 | +0.421 | +0.584 |
| 24 | $D_{3d}$ |  | 34.4 | 34.4 | -68.9 | -0.776 |  |
|   | $D_{3h}$ | $^{24}$Mg | -13.6 | -12.5 | 26.1 | +0.294 | +0.659 |
| 32 | $O_h$ | $^{32}$S | — | — | — | — | +0.549 |

is then predicted to be

$$Q_0 = Q_{33} \times \left(\frac{2}{eF_\pi}\right)^2. \tag{13.6}$$

With the calibration (13.2), one obtains $Q_0$ (in units of electron barn, eb) as shown in the penultimate column of Table 13.3. For comparison, the experimental data are also listed, where available. Recall that a factor of $\frac{1}{100}$ is required to convert (fm)$^2$ to barn.

For a nucleus, the reduced electric quadrupole transition strength from an initial state $|J_i, L_3\rangle$ to a final state $|J_f, L_3\rangle$ is obtained from the intrinsic quadrupole moment $Q_0$ using the formula [45, 46]

$$B(E2: J_i, L_3 \mapsto J_f, L_3) = \frac{5}{16\pi} Q_0^2 \langle J_i L_3; 2\,0 | J_f L_3 \rangle^2, \tag{13.7}$$

where $\langle J_i L_3; 2\,0 | J_f L_3\rangle$ is the Clebsch–Gordan coefficient governing the coupling of the angular momenta.

For transitions between states $J_i = J$ and $J_f = J+2$, with $L_3 = 0$, the squared Clebsch–Gordan coefficient simplifies to

$$\langle J\,0;\, 2\,0 | (J+2)\,0 \rangle^2 = \frac{3(J+1)(J+2)}{2(2J+1)(2J+3)}, \qquad (13.8)$$

and this is 1 for $J = 0$. The transition strength from a $0^+$ to a $2^+$ state is therefore

$$B(E2)\!\uparrow = \frac{5}{16\pi} Q_0^2. \qquad (13.9)$$

Note that an $E2$ transition and its reverse are related by [45, 46]

$$B(E2 : J_f \mapsto J_i) = \frac{2J_i+1}{2J_f+1} B(E2 : J_i \mapsto J_f). \qquad (13.10)$$

By substituting the intrinsic quadrupole moments $Q_0$ listed in Table 13.3 in eq.(13.9), we obtain the Skyrme theory predictions for the $B(E2)\!\uparrow$ strengths (in units of $e^2 b^2$). They are shown in the fifth column of Table 13.4, with experimental data in the sixth column. The fourth column gives the $B(E2)\!\uparrow$ strengths in Skyrme units. They are obtained by substituting $Q_{33}^2$ from Table 13.3 into eq.(13.9), and are converted to physical units using the factor $(2/eF_\pi)^4$. Included in Table 13.4 are the calculated $B(E2)\!\uparrow$ transition strengths for the Hoyle state of Carbon-12, and for Beryllium-12, to be discussed below.

Comparison with experimental data is further simplified by comparing the transition strength with Weisskopf's single-particle (s-p) estimate [221]

$$B(E2)\!\uparrow_{\text{s-p}} = 2.97 \times 10^{-5} B^{\frac{4}{3}}\, e^2 b^2, \qquad (13.11)$$

where $B$ is the baryon number. The strength in Weisskopf units,

$$\frac{B(E2)\!\uparrow}{B(E2)\!\uparrow_{\text{s-p}}}\, \text{W}, \qquad (13.12)$$

is a dimensionless measure of the collectivity of the quadrupole effects in a nucleus. A value higher than 5 W indicates substantial collectivity.

The approach used here is limited to Skyrmions with axially symmetric inertia tensors, and having quadrupole moments satisfying $Q_{11} = Q_{22}$. The calculation of $B(E2)$ strengths for transitions between rotational levels in triaxial nuclei [10, 157] would require a different approach.

We now discuss selected nuclei in detail.

Table 13.4: $B(E2)\uparrow$ for selected nuclei. The calibration (13.2) is used to convert Skyrme theory values (4th column) to physical values (5th column). Experimental $B(E2)\uparrow$ values (6th column) are from ref.[202]. Estimated transition strengths for Beryllium-8 are obtained by ($^\mathrm{I}$) a Hartree-Fock+BCS calculation with the Skyrme SIII force [202], ($^\mathrm{II}$) variational Monte Carlo calculation [249], and ($^\mathrm{III}$) Green's function Monte Carlo method [69]. For Beryllium-12, the estimated experimental $B(E2)\uparrow$ value is obtained by multiplying by 5 the measured $B(E2)\downarrow$ value [133].

| B | K | Nucleus | $B(E2)\uparrow$ | $B(E2)\uparrow$ [$e^2b^2$] | $B(E2)\uparrow^\mathrm{Exp}$ [$e^2b^2$] |
|---|---|---|---|---|---|
| 8 | $D_{4h}^\mathrm{twist}$ | $^8$Be | 29.1 | 0.00366 (7.7 W) | 0.003$^\mathrm{I}$ |
|   |   |   |   |   | 0.0074$^\mathrm{II}$ |
|   |   |   |   |   | 0.010$^\mathrm{III}$ |
|   | $D_{4h}^\mathrm{no\,twist}$ |   | 44.6 | 0.00563 (11.8 W) |   |
| 12 | $D_{4h}$ | Hoyle | 411 | 0.0521 (63.9 W) |   |
|   | $D_{3h}$ | $^{12}$C | 32.5 | 0.00409 (5.0 W) | 0.00397 (4.9 W) |
|   | $D_{3h}$ | $^{12}$Be | 14.2 | 0.00181 (2.2 W) | 0.0040 (4.9 W) |
| 16 | $D_{2d}^\mathrm{bent}$ | $^{16}$O | 133 | 0.0168 (14.1 W) |   |
|   | $D_{4h}^\mathrm{flat}$ |   | 179 | 0.0227 (18.9 W) |   |
| 20 | $D_{3h}$ | $^{20}$Ne | 139 | 0.0176 (10.9 W) | 0.0340 (21 W) |
| 24 | $D_{3h}$ | $^{24}$Mg | 68.1 | 0.00864 (4.2 W) | 0.0432 (21 W) |
| 32 | $O_h$ | $^{32}$S | — | — | 0.0300 (9.8 W) |

### 13.1.1 Beryllium-8

For Beryllium-8, there are the twisted and untwisted two-cube Skyrmions, both with $D_{4h}$ symmetry. The $B(E2)\uparrow$ transition strength for the twisted Skyrmion is $0.00366\,e^2b^2$ and for the untwisted Skyrmion $0.00563\,e^2b^2$. Due to the instability of Beryllium-8 to alpha decay, it is not possible to compare these results with actual experimental data. Instead, Table 13.4 includes $B(E2)\uparrow$ values based on Hartree–Fock calculations and on Monte Carlo methods. Note that the theoretical values vary significantly depending on which model is used. This makes it impossible to test the accuracy of the Skyrme theory predictions, but the positive quadrupole moments are consistent with the expected prolate shape.

### 13.1.2 Carbon-12 and the Hoyle state

Rotational excitations of the $D_{3h}$-symmetric, triangular $B = 12$ Skyrmion generate the Carbon-12 ground state band, and excitations of the

$D_{4h}$-symmetric chain solution generate the Hoyle band. The triangular Skyrmion has an oblate shape, and its calculated intrinsic quadrupole moment $Q_0 = -0.203$ eb agrees well with the experimental value $Q_0 = -0.200$ eb [202] extracted from the measured strength of the transition from the ground state to the yrast $2^+$ state, $B(E2 : 0^+ \mapsto 2^+) = 0.00397$.

Measuring the transition strength between the $0^+$ and $2^+$ states in the Hoyle band is experimentally challenging and would require a highly efficient particle-gamma experimental setup [96, 142]. The prediction for the up-transition strength is $B(E2 : 0^+ \mapsto 2^+) = 0.0521 \, e^2 b^2$. This is 63.9 W, a large value arising from the strongly prolate form of the chain Skyrmion, with its intrinsic quadrupole moment $Q_0 = 0.724$ eb.

### 13.1.3 Oxygen-16

Oxygen-16 in its ground state has an intrinsic tetrahedral shape, and therefore zero quadrupole moment. The tetrahedral, ground state rotational band has no $2^+$ state, so there is no $E2$ transition from the ground state within this band. The first-excited $0^+$ state at 6.05 MeV is interpreted as an excited E-manifold state with two E-phonons. Its intrinsic shape is not precisely fixed, because the E-manifold wavefunction has some spread, but the dominant probability is for an oblate flat- or bent-square configuration, which has a negative quadrupole moment.

The $E2$ transition strength from the 6.05 MeV, $0^+$ state to a $2^+$ state with energy near 10 MeV has been calculated using the bent-square Skyrmion. The intrinsic quadrupole moment is $Q_0 = -0.412$ eb, giving $B(E2 : 0^+ \mapsto 2^+) = 0.0168 \, e^2 b^2$. For this transition, there appears to be no experimental $B(E2)$ value available. For completeness, Tables 13.3 and 13.4 include the quadrupole moment and $B(E2)\uparrow$ values found by modelling the $0^+$ state and its $2^+$ excitation by the flat-square Skyrmion. The $2^+$ endpoint of the transition could be the state at 9.66 MeV assigned to the F-band, but is more likely to be the state at 11.26 MeV identified as being in the $E^2$-band.

### 13.1.4 Neon-20, Magnesium-24 and Sulphur-32

In the $\alpha$-particle model, Neon-20 is modelled by five $\alpha$-particles arranged in a triangular bipyramid. The spectrum of Neon-20 can be partly understood in terms of this prolate bipyramidal structure and its vibrational excitations [41, 50]. The analogous bipyramidal cluster arrangement of five cubic $B = 4$ Skyrmions is not the stable $B = 20$ Skyrmion, but it is a nearby saddle point. This bipyramidal Skyrmion would model a prolate Neon-20 ground state. The associated intrinsic quadrupole moment $Q_0 = 0.421$ eb is a little less than the value $Q_0 = 0.584$ eb [202] deduced from the experimental $B(E2)\uparrow$ value. Stable $B = 20$ Skyrmions are shown in Fig. 8.14. The quantized states of the $T_d$-symmetric $B = 20$ Skyrmion have not yet been identified with any observed states of Neon-20, and the $B = 20$ solution formed from two $B = 10$ Skyrmion clusters with $D_{2h}$ symmetry remains to be analysed. The bipyramidal Skyrmion remains the most promising for modelling Neon-20.

For modelling Magnesium-24, two low-energy $B = 24$ Skyrmions are known: the $D_{3d}$-symmetric non-planar ring formed of six $B = 4$ cubes with neighbouring pairs twisted by 90° around the line joining them, shown in Fig. 8.7 (right), and the more compact $D_{3h}$-symmetric solution formed of two parallel triangular $B = 12$ Skyrmions, shown in Fig. 8.8. The ring is calculated to be the Skyrmion of minimal energy, but Magnesium-24 is probably best described by the slightly prolate $D_{3h}$-symmetric Skyrmion, whose quadrupole moment is found to be $Q_0 = 0.294$ eb, still significantly less than the experimental value $Q_0 = 0.659$ eb. The Skyrmion ring solution is oblate, so its quadrupole moment has the wrong sign. A further $B = 24$ solution has been found [111], constructed from six $B = 4$ cubes in a rectangular arrangement, with partial twisting. This has less symmetry, and looks more prolate, but its quantum states and $E2$ transitions need further analysis.

Calculating nuclear properties of Sulphur-32 has proved to be difficult [34], and Hartree–Fock calculations of the rotational spectra in Sulphur-32 have yielded contradictory, model-dependent results. The experimental excitation energies of the $0^+$, $2^+$ and $4^+$ states of Sulphur-32 agree well with expectations for the vibrational excitations of a spherical nucleus [147]. However, Sulphur-32 also has a relatively large positive quadrupole moment [202, 219], requiring a significant prolate nuclear deformation. This can be understood within a nuclear coexistence model [64], in which spherical and prolate rotational bands coexist.

In Skyrme theory, Sulphur-32 is modelled by the cubically-symmetric $B = 32$ Skyrmion shown in Fig. 8.6. This Skyrmion is a candidate for modelling the vibrational excitations of Sulphur-32 [84], but its intrinsic quadrupole moment vanishes. When the Skyrmion spins it deforms, and a non-zero quadrupole moment could be induced. However, a calculation of $E2$ transitions for non-rigidly spinning Skyrmions requires techniques beyond those considered here. Another possibility is that the static Skyrmion may stretch along the cube diagonal, generating a quadrupole moment. Quite a small stretch leads to a configuration of eight $B = 4$ cubes similar to the arrangement of $\alpha$-particles sketched by Wefelmeier and Wheeler in their discussions of $\alpha$-particle molecules. It would be worthwhile to see if this stretch is energetically favoured. The vibrational modes of a slightly stretched, cubic Skyrmion would be little different from those of the unstretched Skyrmion.

## 13.2 Beryllium-12

So far, the discussion of $E2$ transitions has been for isospin 0 nuclei. In this section, we show how the transition strength can be estimated for a nucleus with non-zero isospin.

Beryllium-12 is an $I_3 = -2$ nucleus in an $I = 2$ isoquintet, having four protons and eight neutrons. It exhibits a fairly large quadrupole strength in the transition between the $0^+$ ground state and the $2^+$ state at 2.1 MeV; a down-transition strength $B(E2 : 2^+ \mapsto 0^+) = 0.0008\,e^2b^2$ has been determined through the lifetime measurement of the $2^+$ state [133]. Using eq.(13.10) this results in $B(E2)\uparrow = 0.0040\,e^2b^2$.

Nuclei of baryon number 12, with isospin 1 and 2, are well described within Skyrme theory as quantum states of the $D_{3h}$-symmetric $B = 12$ Skyrmion, shown in Fig. 8.3. This includes the low-lying states of Beryllium-12. The quantum states $\Psi_{J^P,I,|L_3|,|K_3|}$ allowed by the FR constraints are listed in Table 9.11. For both spin/parity $0^+$ and $2^+$ there is a unique $I = 2$ state with $L_3 = K_3 = 0$. Here, we consider $E2$ transitions between them. The relevant isospin state for Beryllium-12 is $|I, K_3, I_3\rangle = |2, 0, -2\rangle$. The new aspect is to include the contribution of the isospin to the electric charge density, and hence to the quadrupole moments. It would be best to do a proper quantum calculation of the expectation value of the quadrupole moments, but this has not yet been possible.

Instead, the isospin state can be interpreted classically, as an analogue of the classically spinning nucleons described in Section 5.5. To achieve the space-fixed isospin projection $I_3 = -2$, the $B = 12$ Skyrmion's red/green/blue colours need to spin in isospace while the black/white colours are fixed. (The spin is about the 3-axis in isospace.) The complete isospin state determines the body-fixed colouring of the Skyrmion having highest probability. If the body-fixed isospin projection in this state had been $K_3 = \pm 2$, then the standard orientation and colouring of the Skyrmion in Fig. 8.3 (as defined by the rational maps (8.5) and (8.6)) would be correct, but as the projection is $K_3 = 0$ it is necessary to reorient the colours first.

The reorientation is by the angles that maximise the magnitude of the isospin state $|2, 0, -2\rangle$. This state's Wigner D-function is $D^2_{0,-2}(\alpha, \beta, \gamma) = e^{-2i\gamma} \sin^2 \beta$, where $\alpha, \beta, \gamma$ are the isorotational Euler angles, whose maximal magnitude is at $\beta = \frac{1}{2}\pi$, Hence, the standard Skyrmion needs to be isorotated by $\beta = \frac{1}{2}\pi$ and (for convenience) $\gamma = 0$, moving the black/white points of the Skyrmion to the faces or edges of the three constituent $B = 4$ cubes, and away from the vertices.

Under this isorotation (with $\alpha$ undetermined) the pion field components $(\pi_1, \pi_2, \pi_3)$ transform to

$$\pi'_1 = -\sin\alpha\, \pi_2 + \cos\alpha\, \pi_3 ,$$
$$\pi'_2 = \cos\alpha\, \pi_2 + \sin\alpha\, \pi_3 ,$$
$$\pi'_3 = -\pi_1 . \tag{13.13}$$

The colours now spin with $I_3 = -2$ about the transformed 3-axis in isospace, which we see is equivalent to spinning about the initial 1-axis. The moment of inertia that determines the angular velocity is therefore $U'_{33} = U_{11}$. For the $D_{3h}$-symmetric Skyrmion, $U_{11} = 442$ (see Table 13.1).

The Beryllium-12 nucleus has physical isospin $-2\hbar$, so in the classical picture its angular velocity $w$ is given by

$$U'_{33}w = U_{11}w = -2\hbar , \tag{13.14}$$

leading to the numerical value

$$w = -0.14 \tag{13.15}$$

as $\hbar = 30.2$ and $U_{11} = 442$. The classical isospin density is then

$$\mathcal{I}_3(\mathbf{x}) = \frac{1}{\hbar}\mathcal{U}_{11}(\mathbf{x})w , \tag{13.16}$$

where $\mathcal{U}_{ij}(\mathbf{x})$ denotes the isorotational moment of inertia density, the integrand in the first of eqs.(7.4). This contributes to $\rho$, the electric charge density (13.4), and hence to the quadrupole tensor (13.5). Note that eq.(13.16) is correctly normalised for Beryllium-12 as its integral is $I_3 = -2$, and this decreases the total electric charge of the $B = 12$ Skyrmion from 6 to 4.

The numerically computed electric quadrupole moments are $Q_{11} = 5.65$, $Q_{22} = 6.32$ and $Q_{33} = -11.9$ for the reoriented $B = 12$ Skyrmion, including the isospin contribution. Thus, in physical units using the calibration (13.2), the intrinsic quadrupole is $Q_0 = -0.135\,\text{eb}$. The corresponding $B(E2)\uparrow$ value for Beryllium-12 is $0.00181\,e^2b^2$, which is approximately half the experimental value.

## 13.3 Further Transitions

We have focussed on $E2$ transitions, mainly to and from the $0^+$ ground state of several nuclei. Further transitions are of interest including, for example, the $E3$ transition between the 6.13 MeV, $3^-$ state of Oxygen-16 and the ground state. Halcrow and Rawlinson have investigated a number of electromagnetic transitions between states in Carbon-12 and in Oxygen-16 using insight into the structure of the states obtained from Skyrme theory [119]. For Carbon-12 the states are defined on Rawlinson's graph of isosceles, triangular configurations of three $B = 4$ Skyrmions. For Oxygen-16 the states are defined on the E-manifold of $D_2$-symmetric configurations of four $B = 4$ Skyrmions. The wavefunctions of these states were discussed and illustrated in Sections 10.4 and 11.3.

The calculations earlier in this chapter used rigid-body rotational states of Skyrmions. In that approach, there can be transitions within a rotational band, but transitions between bands based on different Skyrmions (for example, the ground state band and Hoyle band of Carbon-12) are automatically forbidden. Using wavefunctions that depend non-trivially on the Skyrmion shape gives a more realistic model for electromagnetic transitions, and leads to non-zero transition strengths between states in distinct bands, although these may be small. Halcrow and Rawlinson have calculated some examples involving Skyrmions made from $B = 4$ constituents, and we outline here their method.

The Skyrmion is assumed to be parametrised by shape coordinates **s** and Euler angles $\boldsymbol{\theta}$ specifying the orientation of the body-fixed axes. The wavefunction of the Skyrmion in an isospin 0 state is then of the form

$$|\Psi\rangle = \sum_{L_3} \chi_{L_3}(\mathbf{s})|J, L_3, J_3\rangle, \qquad (13.17)$$

where $J$ and $J_3$ are taken to be fixed, and the state $|J, L_3, J_3\rangle$ depends on the Euler angles $\boldsymbol{\theta}$.

As before, for a quantum state with isospin 0, the charge density of a Skyrmion is half the baryon density. The charge density is simplified to be two units of charge (as for an $\alpha$-particle) concentrated at the centre of each constituent $B = 4$ Skyrmion. In both the graph model for Carbon-12 and the E-manifold model for Oxygen-16, there is a well defined set of locations for these constituents as a function of the shape parameters, as explained earlier. The intrinsic moments of the charge density are most conveniently expressed using the spherical tensor components

$$Q_{lm}(\mathbf{s}) = \int \rho(\mathbf{s}, \mathbf{x}) r^l Y_{lm}^*(\Omega) \, d^3x \,, \tag{13.18}$$

where $r$ and $\Omega$ are radial and angular coordinates in space, and the orientational Euler angles are set to zero. Here, $\rho$ is simply a sum of delta functions.

To calculate an $El$ transition strength, one requires the moments $Q_{lm}(\mathbf{s})$ and then needs to integrate over all possible orientations of the Skyrmion, using the rotated version of the charge density. The moments $Q_{lm}$ transform rather simply under rotations, so this integral gives simply a Clebsch–Gordan coefficient. The rate then depends on the overlap between the initial and final state weighted by the moments, requiring a further integral over the shape parameters. The result is

$$B(El, i \mapsto f) = \frac{2J_f + 1}{2J_i + 1} \tag{13.19}$$

$$\times \left| \int \sum \chi_{L_{3,f}}^*(\mathbf{s}) \chi_{L_{3,i}}(\mathbf{s}) Q_{lm}(\mathbf{s}) \langle J_f L_{3,f}; lm | J_i L_{3,i} \rangle \, d\mathbf{s} \right|^2 ,$$

where the sum is over $L_{3,f}$, $L_{3,i}$ and $m$. This integral will tend to be large if the initial and final shape wavefunctions are similar, i.e. if these states are in a single rotational band of an approximately rigid structure. The integral will be small, but still non-zero, if the shape wavefunctions have a small overlap, provided the moment $Q_{lm}$ varies with shape.

The transition strength satisfies the general reciprocity relation

$$B(El, f \mapsto i) = \frac{2J_i + 1}{2J_f + 1} B(El, i \mapsto f) \,. \tag{13.20}$$

Also, it simplifies if $J_f = 0$. Then the sum over $L_{3,f}$ reduces to the single term with $L_{3,f} = 0$, and more importantly, the strength vanishes unless $l = J_i$. There is a similar simplification if $J_i = 0$.

Using the wavefunctions on the graph of triangles, Halcrow and Rawlinson calculated that $B(E2)\uparrow = 0.00585 \, \text{e}^2\text{b}^2$ in the Carbon-12 ground state band, somewhat higher than the strength found using rigid-body quantization. They also calculated the down-transition strength from the yrast $3^-$ state to the ground state, $B(E3)\downarrow = 0.000062 \, \text{e}^2\text{b}^3$, about 40% below the experimental value. Transition strengths within the Hoyle band were also found, but here there is no experimental data. The difference between $B(E2)$ strengths in the ground state band and Hoyle band is less pronounced than in rigid-body quantization, because the wavefunctions are more spread out, softening the distinction between triangular and chain shapes. Finally, a small, non-zero $B(E2)$ transition strength between the $2^+$ state in the ground state band and the $0^+$ Hoyle state was calculated, a strength that vanishes in the rigid-body approach. The result is $B(E2 : 2^+ \mapsto 0^+_{\text{Hoyle}}) = 0.00011\,\text{e}^2\text{b}^2$, compared to the experimental value $0.00027\,\text{e}^2\text{b}^2$.

In Oxygen-16, the ground state band has $3^-$ and $4^+$ states above the $0^+$ ground state. As noted before, there is no $E2$ transition in this band. The length scale and electric octupole moments in the E-manifold model can be calibrated to fit the observed transition strength $B(E3 : 3^- \mapsto 0^+) = 0.000205\,\text{e}^2\text{b}^3$. The $B(E4)$ transition strength from the $4^+$ state to the ground state is then found to be compatible with the experimental value (which has a substantial uncertainty). The symmetries of the states in the model exclude any $E1$ transition between the $4^+$ and $3^-$ states, as observed experimentally. This result is important evidence for the tetrahedral character of the Oxygen-16 nucleus. Halcrow and Rawlinson also discuss $E2$ transitions between the yrast $2^+$ state at 6.91 MeV and the first-excited $0^+$ state at 6.05 MeV. Both these states have their greatest probability for square shapes of the four $B = 4$ constituents. The transition is experimentally measured to be much stronger than between the yrast $2^+$ state and the $0^+$ ground state, by a factor of about 9, consistent with the similarity in shapes. Using the E-manifold wavefunctions, the ordering of the strengths is the same, but the difference is not so much. This transition is affected by the two states being vibrationally different, since the $2^+$ state has a single E-phonon and the $0^+$ state has two E-phonons.

Epelbaum et al. [80] have also considered the transition between this $2^+$ state and the first-excited $0^+$ state, interpreting both states as rotational excitations of a square configuration of $\alpha$-particles. They calculated the $B(E2)\downarrow$ transition strength using a nuclear lattice effective field theory simulation. The inferred up-transition strength is $B(E2)\uparrow = 0.0110\,\text{e}^2\text{b}^2$, somewhat less than the experimental value $B(E2)\uparrow = 0.0325\,\text{e}^2\text{b}^2$.

## 13.4 Summary

We have reviewed the calculations of electromagnetic transition strengths between the $0^+$ ground state and the nearest accessible $2^+$ state for a range of light nuclei: Beryllium-8, Carbon-12 and Beryllium-12, Neon-20 and Magnesium-24. The calculated $B(E2)\uparrow$ transition strengths have the correct order of magnitude and the intrinsic quadrupole moments have the correct sign. For the Hoyle state of Carbon-12 a large $B(E2)\uparrow$ strength is also predicted, but measurements of the electromagnetic transitions between states of the Hoyle band are technically difficult and have yet to be performed.

For the intrinsically tetrahedral Oxygen-16, the ground state rotational band has no $2^+$ state, so there is no $E2$ transition to consider. However, the $E3$ and $E4$ transitions within the ground state band have been recently calculated. The first-excited $0^+$ state has a $2^+$ rotational excitation. The transition strength has been calculated using bent- and flat-square $B = 16$ Skyrmions in the rigid-body approach, and also using the wavefunctions on the E-manifold.

There remain some challenges. For $B = 20$, the triangular bipyramidal arrangement of five $B = 4$ cubes, used to describe $E2$ transitions in Neon-20, is not a minimal-energy Skyrmion but a saddle point. For $B = 32$, the minimal-energy Skyrmion is cubically symmetric and hence cannot explain the large prolate quadrupole moment of Sulphur-32. However, the Skyrmion may deform under a rotation or be stretched diagonally, and a non-zero quadrupole moment may be induced.

# Chapter 14

# Variants of Skyrme Theory

The Skyrme theory discussed so far has been the original theory proposed by Skyrme, with only the pion mass term treated as a variation. We have seen, however, that this theory does not fit all experimental data. The problem was quantified early on by Adkins, Nappi and Witten [7] and by Adkins and Nappi [5] in their analysis of nucleon properties. This motivates searching for a variant of Skyrme theory, to improve on its undoubted successes.

There is more than one reason for exploring variants. First, it is desirable to extend the theory to include some physics that is omitted from the original version, for example, electromagnetic and weak interactions. We have seen how to calculate electromagnetic transition strengths of some nuclei, but have not investigated the effect of electrostatic Coulomb energy on nuclear structure and stability. Callan and Witten have shown how to couple the Skyrme field to the electromagnetic $U(1)$ gauge field [60], and D'Hoker and Farhi constructed the coupling to the $SU(2)_L$ gauge field of the weak interactions [72], which is responsible for nuclear beta decay among other things. However, the details are quite complicated, and the rates of beta decay of neutrons and larger nuclei have not yet been calculated using Skyrmions, so we will not consider gauged Skyrmions further.

A further limitation of the original Skyrme theory is that it is restricted to the lightest mesons and baryons constructed from u and d quarks – pions and nucleons – but there are well-known mesons and baryons containing the heavier s (strange) quark, and the even heavier c and b quarks, and one would like to consider these. The simplest extension that can accommodate the s quark replaces the $SU(2)$ Skyrme field by an $SU(3)$ Skyrme field [73, 106, 251], and we will discuss this later in the chapter. For the inclusion of heavier quarks and the mesons that contain them, see [212].

However, our main motivation for considering variants is to improve the matching of the theory to nuclear structure and binding energies. A useful idea, very much in the spirit of Skyrme's philosophy, is to add one or more terms to the Skyrme Lagrangian without changing the field content. The simplest addition is a term, sextic (sixth-order) in the derivatives of the field $U$, i.e. sextic in the current $R_\mu = \partial_\mu U U^{-1}$. There is a unique term that is still quadratic in time derivatives, the key property of the quartic Skyrme term. Also adjustable is the form of the pion mass term. Its behaviour close to the vacuum is determined by the pion mass itself, but a more general function than a multiple of $\mathrm{Tr}(1 - U)$ is permitted. The general theory accommodating these variants has a Lagrangian density of the form

$$\mathcal{L} = \mathcal{L}_2 + \mathcal{L}_4 + \mathcal{L}_6 + \mathcal{L}_0, \qquad (14.1)$$

where $\mathcal{L}_2$ and $\mathcal{L}_4$ are the usual quadratic and quartic terms of Skyrme theory. $\mathcal{L}_6$ is the additional sextic term and $\mathcal{L}_0$ a generalised pion mass term. To preserve isospin symmetry, $\mathcal{L}_0$ needs to be a function of $\mathrm{Tr}(U)$.

In this class of theories, particular attention has been given to what is known as the BPS model [2, 3]. This is where $\mathcal{L}$ is dominated by the terms $\mathcal{L}_6 + \mathcal{L}_0$, and minimal-energy Skyrme fields obey a first-order equation that has Skyrmion solutions for all baryon numbers. The energy of a BPS Skyrmion exactly saturates a Bogomolny bound proportional to the baryon number, so there is no classical binding energy. This is seen as helpful phenomenologically, because the observed binding energy of nuclei is approximately 8 MeV per nucleon (for all but the smallest nuclei), less than 1% of the total nuclear mass. If the Coulomb energy is separated off, the strong interaction binding energy is somewhat larger, but still less than 2% of the nuclear mass. Such a small binding energy is not realised in standard Skyrme theory, but it can be in variants close to the BPS model. However, we will see that the BPS model has some conceptual difficulties, because the BPS Skyrmions completely lose their rigidity, and there are too many solutions.

Discussion of binding energies of nucleons into larger nuclei in any version of Skyrme theory is rather complicated, because of the role of the spin energy. A nucleon has spin $\frac{1}{2}$, and in Skyrme theory there is a clear splitting of the nucleon mass into the classical energy of the static $B = 1$ hedgehog Skyrmion and the spin energy, which depends on the moment of inertia. When the Skyrme theory is varied, the classical energy and the moment of inertia both change. Nuclear binding energies are unlikely to be small while the spin energy of a nucleon is substantial. To understand the binding energy in different variants, it is therefore best to focus on the energy of Skyrmions whose baryon number is $B = 4N$, and compare with the energy of $N$ well-separated $B = 4$ Skyrmions. These Skyrmions can all be quantized with zero spin and isospin. Ideally, the energy difference should match the very small experimental binding energy of $N$ $\alpha$-particles into a nucleus with spin and isospin zero (an even, self-conjugate nucleus in its ground state), although here too, the Coulomb energy needs to be considered. For example, the binding energy of three $\alpha$-particles into Carbon-12 is about 7.3 MeV, less than 0.1% of the Carbon-12 mass.

Another approach to improving Skyrme theory phenomenologically is to directly include other mesons in addition to the pions, in particular the heavier $\rho$ and $\omega$ vector mesons [4, 6, 188]. This is somewhat controversial because these mesons, as free particles, are highly unstable and decay into pions; their effect is also partly accounted for by the terms $\mathcal{L}_4$ and $\mathcal{L}_6$ in the standard, pionic Lagrangian density (14.1), as we shall explain. However, using the known physics of the vector mesons, one can build a physically reasonable theory that couples vector meson fields to the Skyrme current $R_\mu$. Such a theory captures the chirally-invariant interactions of the heavier mesons with pions in the massless pion limit, and also their interactions with baryons treated as quantized Skyrmions. A disadvantage is that the theory can have many more parameters than the original Skyrme theory, and it is harder to solve the coupled field equations. However, there has been substantial recent progress in constraining the parameters, and finding Skyrmion solutions that are similar to the familiar solutions discussed earlier, up to baryon number 12 [110, 192]. The binding energies are substantially less than those of the standard theory, which is encouraging.

Perhaps the deepest reason for seeking a variant of standard Skyrme theory is that one would like an effective field theory systematically derived from QCD that can accurately predict any property of mesons, baryons and nuclei, as well as being more tractable than alternative realisations of QCD, like lattice QCD. Of course, one would wish the baryon to remain a soliton. Unfortunately, there is no directly derived theory of this type yet available. However, by an indirect approach involving string theory, a candidate has emerged, and it captures many of the features of the simpler variants of Skyrme theory that we have mentioned. This is *holographic Skyrme theory*, which was pioneered by Sakai and Sugimoto in 2005 [209], developing more general ideas of Witten and 't Hooft for dealing with QCD in the limit of a large number $N_c$ of quark colours. The theory has by now been investigated and developed by many others. It is fascinating from a mathematical perspective, and has considerable phenomenological success. We will sketch this theory in the next chapter.

## 14.1 Adding a Sextic Term

One of the simplest modifications of the original Skyrme theory is to add a term that is sextic in derivatives. This was first investigated by Jackson et al. [138]. There is a unique sextic term that satisfies the requirements of chiral invariance, Lorentz invariance, and being at most quadratic in time derivatives. It is constructed from the baryon current density

$$\mathcal{B}^\mu = -\frac{1}{24\pi^2}\epsilon^{\mu\nu\kappa\lambda}\text{Tr}\left(R_\nu R_\kappa R_\lambda\right), \tag{14.2}$$

a Lorentz vector generalising the baryon density appearing in (4.26). $\epsilon^{\mu\nu\kappa\lambda}$ is the 4-dimensional antisymmetric tensor symbol with $\epsilon^{0123} = 1$. The complete Lagrangian density including this term, for massless pions, is

$$\mathcal{L} = -a_2\text{Tr}(R_\mu R^\mu) + a_4\text{Tr}([R_\mu, R_\nu][R^\mu, R^\nu]) - a_6\mathcal{B}_\mu\mathcal{B}^\mu. \tag{14.3}$$

The coefficients here have different values than previously, and are interpreted somewhat differently, which is why the notation is changed. An improvement would be the addition of the standard pion mass term, or a more general pion mass term $\mathcal{L}_0$.

For static fields, the sextic term's contribution to the energy density is proportional to the square of the baryon density $\mathcal{B} = \mathcal{B}^0$. In the elastic strain formalism, with strain eigenvalues $\lambda_1$, $\lambda_2$ and $\lambda_3$, eq.(4.43) implies that this energy density is a multiple of $\lambda_1^2\lambda_2^2\lambda_3^2$. Like the quartic Skyrme term, it penalises a Skyrmion becoming too small. Floratos and Piette found Skyrmion solutions up to $B = 5$ in this theory [88]. They are rather similar to those of standard Skyrme theory, and have the same symmetries. The rational map approximation is again useful.

The kinetic energy of a rotating $B = 1$ Skyrmion acquires a new contribution from the sextic term, modifying the moment of inertia. Quantization of the rotational motion still leads to spin $\frac{1}{2}$ nucleon and spin $\frac{3}{2}$ delta resonance states that can be used to calibrate the model, and as there is one extra parameter, the model can be slightly closer than the standard theory to the observed properties of baryons and the forces between them. In fact, as noted in [138], it is now acceptable for the quartic term to have a small negative coefficient $a_4$, provided the coefficient $a_6$ is positive. This introduces an attractive component to the force between Skyrmions, although there is still a short-range repulsion.

Jackson *et al.* did not consider further variants of this model, but their idea of adding higher-order terms in the derivatives of the Skyrme field was developed by Marleau and collaborators [36, 48]. In particular, Marleau considered adding an infinite series of higher-order terms, with coefficients depending on just one free parameter [186, 187].

### 14.1.1 The lightly-bound Skyrme model

Among models with a Lagrangian density of the form (14.1), a particular study has been made of a model including a term $\mathcal{L}_0$ that is quartic in $\text{Tr}(1 - U)$ [102]. This disfavours large regions where $U \simeq -1$, even more than the standard pion mass term does. Skyrmion solutions with larger baryon numbers therefore do not have the hollow cores characteristic of Skyrmions with massless pions. The small regions where $U \simeq -1$ do not coalesce, but remain isolated. Consequently, the Skyrmions are collections of $B = 1$ hedgehogs lightly bound together, and the binding energies are also quite small. In particular, the $B = 4$ Skyrmion has a tetrahedral shape.

This lightly-bound model can be further simplified by assuming that the $B = 1$ constituent Skyrmions are undistorted point-particles, with their relative orientations arranged so that they maximally attract. These arrangements are in fact very close to subclusters of the FCC Skyrmion crystal, before it changes phase into the half-Skyrmion crystal. Gillard *et al.* have determined the minimal-energy solutions in the lightly-bound model numerically, up to $B = 8$, and have confirmed the accuracy of the point-particle simplification [101]. Halcrow, Rawlinson and the present author have investigated more broadly the possible subclusters of the FCC lattice with high symmetry – either tetrahedral or octahedral symmetry. Many such clusters, up to $B = 85$, can be identified with weight diagrams (ignoring the weight multiplicities) of the Lie algebra of $SU(4)$, whose weight lattice happens to be the 3-dimensional FCC lattice [118]. The binding energy of a cluster is then estimated to be proportional to the total number of nearest neighbour bonds the cluster has. Using these approximate lightly-bound Skyrmions as a starting point, it has been possible to construct many Skyrmions of standard Skyrme theory that are hard to find in other ways. Notable among these is the tetrahedrally-symmetric Skyrmion with baryon number 40, shown in Fig. 12.1.

For nuclear phenomenology, the lightly-bound model has a challenging problem. Because of the small binding energies, it is important to consider not just the clusters having minimal energy, but different arrangements of the constituents having slightly larger energies. In other words, the quantized model should allow fairly free particle motion, like a traditional model of interacting protons and neutrons. In particular, for the less symmetric clusters, the $B = 1$ constituents in the surface layer should be allowed to hop between sites of the background FCC lattice. The rigid-body approach to quantization, even including some vibrational modes, is not really appropriate.

## 14.2 The BPS Skyrme Model

The BPS Skyrme model is an idealisation where the classical Skyrmion binding energies are exactly zero. It is the limit as $\epsilon_2, \epsilon_4 \to 0$ of a model with Lagrangian density $\mathcal{L} = \epsilon_2\mathcal{L}_2 + \epsilon_4\mathcal{L}_4 + \mathcal{L}_6 + \mathcal{L}_0$ [3]. It is essential to have the sextic term $\mathcal{L}_6$ and some non-zero term $\mathcal{L}_0$. Curiously, in the limit, the terms of the original Skyrme theory both disappear. The B in BPS stands for Bogomolny, who clarified that various field theories with topological solitons have limiting forms where the energy of solitons at rest is just a constant multiple of their topological charge – i.e. the energy is independent of where the solitons are, so there are no static forces between them, and no binding energy [44]. PS stands for Prasad and Sommerfield, who found an exact analytical solution for a non-abelian, unit charge monopole in such a limiting theory [198]. Others, including Prasad, developed this discovery into powerful machinery for constructing and understanding exact monopole solutions for all charges. Skyrmions in the BPS model have energies proportional to their baryon number $B$, so there is no binding energy, and a large (in fact, infinite-dimensional) class of static solutions can be constructed explicitly.

It is assumed that $-\mathcal{L}_0$ (the potential energy density) is a non-negative function of $\mathrm{Tr}(U)$ that vanishes when $U = 1$, and it is convenient to express this as the square of another non-negative function, $\mathcal{V}(\mathrm{Tr}(U))$. The BPS model therefore has an energy for static fields $U(\mathbf{x})$ that reduces, after rescaling, to

$$E = \int_{\mathbb{R}^3} \left(\lambda_1^2\lambda_2^2\lambda_3^2 + \mathcal{V}^2(\mathrm{Tr}(U))\right) d^3x, \tag{14.4}$$

where the sextic term is expressed in terms of the elastic strain eigenvalues. As always in field theories of Bogomolny type, one now completes the square, and rewrites the energy as

$$E = \int_{\mathbb{R}^3} \left(\lambda_1\lambda_2\lambda_3 - \mathcal{V}(\mathrm{Tr}(U))\right)^2 d^3x + 2\int_{\mathbb{R}^3} \lambda_1\lambda_2\lambda_3 \mathcal{V}(\mathrm{Tr}(U)) d^3x. \tag{14.5}$$

The second integral is essentially topological, because $\lambda_1\lambda_2\lambda_3$ is the Jacobian of the map $U$ from the domain $\mathbb{R}^3$ to the target $S^3$. More precisely, its integrand is the pull-back by $U$ of the standard volume form on $S^3$ weighted by the function $\mathcal{V}$, which can be interpreted as a locally-deformed volume form on the target. As $\mathcal{V}$ depends only on $\mathrm{Tr}(U)$, the target effectively becomes a 3-sphere stretched or squashed in one direction (like a 2-sphere being stretched into an egg-shape), preserving $SO(3)$ symmetry. The second integral is therefore the topological degree of the map – the baryon number $B$ – times the total deformed volume of the target. Then, because the first integrand is a complete square, and therefore non-negative, the total energy $E$ is bounded below by $B$ times the deformed 3-sphere volume, multiplied by the explicit factor of 2.

The energy $E$ equals its lower bound, proportional to $B$, provided the first integral vanishes. This imposes the first-order Bogomolny equation on the Skyrme field, $\lambda_1\lambda_2\lambda_3 = \mathcal{V}(\mathrm{Tr}(U))$, which is more transparently written as

$$2\pi^2 \mathcal{B}(\mathbf{x}) = \mathcal{V}(\mathrm{Tr}(U(\mathbf{x}))). \qquad (14.6)$$

The baryon density $\mathcal{B}$ is therefore everywhere locked to the value of $\mathrm{Tr}(U)$. It is easy to satisfy this equation. For $B = 1$ one can assume a field configuration of the usual hedgehog form, depending on a profile function $f(r)$ satisfying the usual boundary conditions, $f(0) = \pi$ and $f(\infty) = 0$. The equation relates the radial derivative of $f$ to $f$ itself, and can be integrated to find the profile. Typically (and this depends on the exact form of $\mathcal{V}$), $f$ reaches 0 at some finite radius. That is because the total spatial volume of the Skyrmion must equal the integral over the target 3-sphere of the standard volume form weighted by $1/\mathcal{V}$. The field outside this radius is simply the vacuum $U = 1$, which contributes neither to the energy nor to the topological charge. Such a $B = 1$ Skyrmion is called a *compacton*, a non-trivial field configuration occupying a finite volume, with vacuum outside. Generally, a solution of (14.6) with baryon number $B$ is a compacton with $B$ times this fundamental volume.

A simple class of examples assumes that $\mathcal{V} = \mu(\mathrm{Tr}(1-U))^n$. The choice $n = \frac{1}{2}$ gives the standard pion mass term, but since there is no standard kinetic term for the pion field in the BPS model, the usual consequences of having a positive pion mass do not hold. For example, the hedgehog Skyrmion does not have an exponentially decaying pion tail. Instead, a better choice is $n = 1$. A standard pion mass term can be added later, accompanying the $\mathcal{L}_2$ term.

For $n = 1$ the Bogomolny equation (14.6) reduces to

$$-\frac{\sin^2 f}{r^2}\frac{df}{dr} = 2\mu(1 - \cos f), \qquad (14.7)$$

where we have used the strain eigenvalues of a hedgehog (4.48). The solution is a compacton of radius $R$, with total volume

$$\frac{4}{3}\pi R^3 = 4\pi \int_0^R r^2 dr = \frac{2\pi}{\mu} \int_0^\pi \frac{\sin^2 f}{1 - \cos f} df = \frac{2\pi^2}{\mu}. \qquad (14.8)$$

Its energy, the second integral in (14.5), is

$$E = -16\pi\mu \int_0^R \frac{\sin^2 f}{r^2}\frac{df}{dr}(1 - \cos f)r^2 \, dr$$

$$= 16\pi\mu \int_0^\pi \sin^2 f(1 - \cos f) \, df$$

$$= 8\pi^2 \mu. \qquad (14.9)$$

The profile $f(r)$ approaches zero linearly near the compacton boundary at radius $R$.

It is characteristic of the BPS model that solutions are far from unique. Because the Bogomolny equation only involves the baryon density, one can use any volume-preserving deformation of space to create a new solution from the hedgehog solution. This can stretch and squeeze the solution in infinitely many ways, in the same way that an incompressible fluid can be deformed by its flow. The fluid here is somewhat unusual in that its density, the baryon density, is not uniform, but depends on the value of $\mathcal{V}(\text{Tr}(U))$.

It is equally straightforward to construct solutions with higher baryon numbers, $B$. A simple and popular method involves replacing $\varphi$ by $B\varphi$ in the hedgehog ansatz [3], where $\varphi$ is the azimuthal polar angle, and adjusting the radial profile, although for large $B$ this orange-segment solution looks rather unnatural. Again, this solution can be deformed in infinitely many ways.

One may identify the locations of each unit of baryon number with the points where $U = -1$. For large $B$, these points behave like a gas of particles in the BPS model. Generically, the points are separated, but for the orange-segment solution they all coalesce.

As we have seen, the BPS model has the attractive feature that both the energy of a Skyrmion and its spatial volume are proportional to the baryon number, and solutions of the Bogomolny equation behave as configurations of an incompressible fluid. But moving away from such solutions, the fluid can be compressed at a finite energy cost. This is similar to what is observed experimentally for larger nuclei. The BPS model is in fact reminiscent of the liquid drop model of larger nuclei, but it seems to be too extreme in this regard. It completely loses the rigid, symmetric shapes of smaller Skyrmions that were established in standard Skyrme theory, and the associated spectra of rigid rotations and vibrations that were key to understanding quantum states. Also, the BPS model is very far from having chiral symmetry, because $\mathcal{L}_0$ plays an essential role. Another rather subtle problem of the BPS model, which depends on the choice of potential $\mathcal{V}$, is that although the solutions as a whole cannot be spatially compressed without a substantial increase of energy, the surface layer can sometimes be compressed rather too easily [2].

Improvement to the BPS model is possible by retaining the terms $\mathcal{L}_2$ and $\mathcal{L}_4$ in the Lagrangian, with their small coefficients, and adding a standard pion mass term with a small coefficient. $\mathcal{L}_2$ is needed to have a conventional kinetic energy for the linearised pion field. The effect of these terms on static solutions is probably largest where $U$ is close to $-1$, the locations of largest energy and baryon densities. For a fixed value of $\lambda_1\lambda_2\lambda_3$, it now becomes energetically favoured for the three strain eigenvalues to be close to equal. This introduces some rigidity to the solutions. Possibly, the extra terms also produce some repulsion between the gas particles mentioned earlier, creating a gas where these particles have an approximately uniform density in space, and do not coalesce to a single point.

Speight has considered treating the $\epsilon_2\mathcal{L}_2$ term as a perturbation [218], and minimising its contribution to the energy among the class of compacton field configurations that satisfy the Bogomolny equation (14.6). This leads to a variant of the harmonic map problem which is not easily solved, even numerically. The true Skyrmion solutions cease to be compactons when $\epsilon_2$ is positive, so the field perturbation is not simple near the boundary, and extends to spatial infinity.

It would interesting if one could derive a positive surface energy density for a compacton from these perturbations. This would suppress the possibility of a compacton of baryon number $B$ simply breaking up into a collection of smaller, spatially-separated compactons, with the same total baryon number. Breakup in the pure BPS model can freely occur, and would be inevitable due to electrostatic repulsion if the Coulomb energy were included. However, spontaneous breakup of nuclei with baryon numbers $B \lesssim 210$ does not occur physically, although beyond this, surface area effects cannot overcome the electrostatic repulsion, and $\alpha$-particle emission or nuclear fission is the result, usually requiring quantum mechanical tunnelling through the Coulomb barrier.

The application of the BPS model that looks most promising is to large nuclei, including neutron stars. The $\mathcal{L}_6$ term is the most important for controlling the energy of compressed nuclear matter, so the BPS model has a characteristic equation of state for nuclear matter at high density and low temperature [2]. When the model is coupled to gravity, the gravitational compression of nuclear matter is balanced by the pressure produced by the $\mathcal{L}_6$ term. The isospin inertia should also be considered, because for a neutron star with no electric charge, the isospin $I$ has to be half the baryon number, and the energetic cost of this needs to be less than the combined Coulomb and isospin energies arising for a different value of $I$.

## 14.3 Including Heavier Mesons

Higher than the pions in mass are the spin 1, vector $\rho$ and $\omega$ mesons, still composed of u and d quark-antiquark pairs. Like the pions, the $\rho$ mesons form an isospin triplet with electric charges $+1, 0, -1$, and the $\omega$ is an isospin singlet with charge 0. The $\rho$ decays primarily to two pions, and is seen as a resonance in processes that produce a pion pair. The $\omega$ decays primarily to three pions – two-pion decay is suppressed by a G-parity constraint[1]. The $\rho$ and $\omega$ have similar masses, 775 and 783 MeV, but their widths are considerably different, being 149 and 8 MeV respectively [256]. The $\rho$ decays rapidly, like the delta baryon resonance, but the three-pion decay of the $\omega$ proceeds relatively slowly.

Both mesons can be modelled by fundamental vector fields coupled to the Skyrme field $U$, with the couplings determined by the experimental data on their decays to pions. In the minimal extensions of Skyrme theory, these vector meson fields each have a single coupling to the Skyrme current $R_\mu = \partial_\mu U U^{-1}$. There are relatively simple Skyrme models including these vector mesons separately, which we review here.

### 14.3.1 *Skyrme model with $\rho$ mesons*

An extension of Skyrme theory including the $\rho$ meson was proposed by Adkins [4]. The $\rho$-meson field is an isospin triplet of real Lorentz vectors, $\rho_a^\mu(x), a = 1, 2, 3$. For realising chiral symmetry, it is convenient to supplement this triplet by a fourth real Lorentz vector $\rho_0^\mu(x)$ and convert to a

---

[1] G-parity is an approximate discrete symmetry of the strong interactions, broken by electromagnetic effects.

$2 \times 2$ matrix field

$$\rho^\mu = \rho_0^\mu + i\rho_a^\mu \tau_a, \tag{14.10}$$

where $\tau_a$ are the Pauli matrices. It is then necessary to impose the constraint

$$\text{Tr}(\rho^\mu U^{-1}) = 0, \tag{14.11}$$

so that in an ambient 4-dimensional target space, with the Skyrme field $U$ restricted to the unit 3-sphere, the field quartet $\rho^\mu$ (for each $\mu$) is tangent to the 3-sphere at $U$.

The kinetic and gradient terms of the Lagrangian density for $\rho^\mu$ are constructed using the antisymmetrised derivative (again a $2 \times 2$ matrix)

$$\rho_{\mu\nu} = \partial_\mu \rho_\nu - \partial_\nu \rho_\mu, \tag{14.12}$$

reminiscent of the field tensor in an abelian gauge theory. (Lorentz indices are lowered and raised using the Minkowski metric.) There is also a mass term quadratic in $\rho^\mu$. Finally, there is a coupling of $\rho_{\mu\nu}$ to the commutator of Skyrme currents, compatible with Lorentz and chiral symmetry, and conservation of parity. The complete Lagrangian density in the case of massless pions is

$$\begin{aligned}\mathcal{L}_{U,\rho} = &-\frac{1}{2}\text{Tr}(R_\mu R^\mu) + \frac{1}{16}\text{Tr}([R_\mu, R_\nu][R^\mu, R^\nu]) + \frac{1}{8}\text{Tr}(\rho^\dagger_{\mu\nu}\rho^{\mu\nu}) \\ &-\frac{1}{4}m_\rho^2 \text{Tr}(\rho^\dagger_\mu \rho^\mu) + g_{\rho\pi\pi}\text{Tr}(\rho_{\mu\nu}U^{-1}[R^\mu, R^\nu]),\end{aligned} \tag{14.13}$$

where $m_\rho$ and $g_{\rho\pi\pi}$ are the $\rho$-meson mass and $\rho\pi\pi$ coupling in Skyrme units. These parameters are calibrated to the physical $\rho$-meson mass and width. The Lagrangian density $\mathcal{L}_{U,\rho}$, together with the constraint (14.11), is invariant under global chiral transformations $(U, \rho^\mu) \mapsto \mathcal{O}_1(U, \rho^\mu)\mathcal{O}_2$, where $\mathcal{O}_1$ and $\mathcal{O}_2$ are arbitrary, constant $SU(2)$ matrices.

The fully coupled field equations are quite complicated, but there is still a static $B = 1$ hedgehog Skyrmion. The Skyrme field has its usual form

$$U(\mathbf{x}) = \cos f(r) + i \sin f(r)\,\hat{\mathbf{x}} \cdot \boldsymbol{\tau}, \tag{14.14}$$

and the spatial part of the $\rho$-meson field is

$$\boldsymbol{\rho}(\mathbf{x}) = i\xi(r)\hat{\mathbf{x}} \times \boldsymbol{\tau}, \tag{14.15}$$

with $\xi(r)$ vanishing at $r = 0$ and as $r \to \infty$, and the time-component $\rho^0$ vanishing everywhere. These fields satisfy the constraint (14.11). Minimising the energy numerically, one finds the profile functions $f(r)$ and $\xi(r)$. Because the mass parameter $m_\rho$ is quite large, $\xi$ is small and $f$ differs little from its usual shape. $\xi(r)$ has its largest amplitude in the core region of the hedgehog Skyrmion, but away from $r = 0$.

Adkins considered the rigid-body quantization of this Skyrmion, allowing for a spinning Skyrmion to have a non-zero $\rho^0$. Some nucleon and delta resonance properties are closer to the data than in standard Skyrme theory, particularly magnetic moments, but the overall improvement is not great.

Naya and Sutcliffe have recently considered a more sophisticated model coupling the Skyrme and $\rho$-meson fields, both without and with a pion mass term [192]. They have added several further interaction terms, including terms cubic and quartic in $\rho^\mu$. Such a model potentially has several new parameters, but Naya and Sutcliffe have fixed all of these by using the version of the couplings that emerges from holographic Skyrme theory. Static Skyrmion solutions have been found for all baryon numbers up to $B = 12$. Qualitatively, in the massless pion case, the Skyrmion shapes are the same as those of standard Skyrme theory, but the classical binding energies are significantly reduced, as expected in a model derived from the holographic perspective. The pion mass term, as usual, disfavours Skyrmions with larger baryon numbers having hollow cores, and leads to Skyrmion solutions with different shapes, having a more pronounced subclustering than occurs in standard Skyrme theory. This starts at $B = 5$, where the solution has clear $B = 4$ and $B = 1$ subclusters. The study of quantum effects for a range of baryon numbers in this variant of Skyrme theory would be interesting.

In Adkins' Lagrangian density, the $\rho$-meson field occurs quadratically, and linearly coupled to the pion fields. It is therefore possible formally to eliminate the $\rho$ field in favour of a series of terms involving the pion field and its derivatives. Since the $\rho$-meson mass is relatively large, much larger than the pion mass and comparable with the nucleon mass, it is quite a good approximation to retain just the leading terms in this series. If the field $\rho^\mu$ coupled directly to the commutator of Skyrme currents, then this procedure could generate the quartic Skyrme term of standard Skyrme theory, but since it is $\rho_{\mu\nu}$ that couples to the commutator, a term with higher derivatives of $U$ is generated.

However, other variants of Skyrme theory involving a $\rho$-meson field have been investigated, and some of these do generate a Skyrme term when this field is eliminated [8]. Typically, the alternative quartic term in the Skyrme current $R_\mu$ occurs too – the term $(\mathrm{Tr}(R_\mu R^\mu))^2$ that Skyrme rejected because part of it is quartic in the time derivative of $U$. One class of models replaces the antisymmetrised derivative $\rho_{\mu\nu}$ by the non-abelian field tensor $\widetilde{\rho}_{\mu\nu} = \partial_\mu \rho_\nu - \partial_\nu \rho_\mu + [\rho_\mu, \rho_\nu]$, although the implied gauge invariance (known as *hidden gauge invariance*) is broken by the $\rho$-meson mass term. The effect of the $\rho$-meson resonance on pion interactions can also be treated phenomenologically, using the method of dispersion relations. In this approach, effective terms in a pion Lagrangian including a Skyrme term are again generated [196].

So, there is some evidence that one could entirely dispense with the Skyrme term, and generate analogous effects from the coupling of the $\rho$ meson to pions via the Skyrme current. This was one reason the Skyrme model with $\rho$-meson fields was proposed. In practice it is better to retain the Skyrme term, possibly with a reduced coefficient (which can be absorbed into the definition of Skyrme units).

In summary, the inclusion of a $\rho$-meson field in Skyrme theory is well motivated because it captures important aspects of meson physics, slightly improves the modelling of baryons, and provides some explanation for the appearance of the Skyrme term in standard Skyrme theory. But further mesons need to be considered too.

### 14.3.2  Skyrme model with $\omega$ mesons

The $\omega$ meson is an isospin singlet Lorentz vector, modelled by a field $\omega^\mu$, and its inclusion in Skyrme theory is slightly easier than for the $\rho$ meson. Again, the kinetic and gradient terms involve the antisymmetrised derivative of the field

$$\omega_{\mu\nu} = \partial_\mu \omega_\nu - \partial_\nu \omega_\mu \,, \tag{14.16}$$

and there is a mass term quadratic in $\omega^\mu$. The coupling to the Skyrme current $R_\mu$ is via the baryon current density $\mathcal{B}^\mu$, defined by eq.(14.2). This coupling is cubic in the derivatives of the pion field, as required to model $\omega$-meson decay. Such a coupling potentially violates parity conservation, because the pions are negative parity scalars, but the 4-index tensor $\epsilon^{\mu\nu\kappa\lambda}$ produces a true vector, so there is overall parity conservation.

The baryon current automatically obeys the conservation law $\partial_\mu \mathcal{B}^\mu = 0$. To see this, note that as $\partial_\mu U^{-1} = -U^{-1}\partial_\mu U U^{-1}$,

$$\begin{aligned}\partial_\mu \mathcal{B}^\mu &\propto \epsilon^{\mu\nu\kappa\lambda} \mathrm{Tr}(\partial_\mu U U^{-1} \partial_\nu U U^{-1} \partial_\kappa U U^{-1} \partial_\lambda U U^{-1})\,, \\ &= \epsilon^{\mu\nu\kappa\lambda} \mathrm{Tr}(R_\mu R_\nu R_\kappa R_\lambda)\,.\end{aligned} \tag{14.17}$$

Under a cyclic permutation of the currents the right hand side is unchanged, because of the trace; at the same time it changes sign because of the $\epsilon$-tensor. So the right hand side must vanish. However, the baryon current is not a Noether current, relying on a field equation for its conservation, and there is no associated symmetry.

As a consequence of baryon current conservation, the corresponding charge, the spatial integral of $\mathcal{B} = \mathcal{B}^0$, is conserved. This is of course the baryon number, although to fix the prefactor in the current, and show that baryon number is an integer, the topological arguments of Chapter 4 are needed.

The simplest Skyrme model that includes both the $\omega$-meson field and massive pions was constructed by Adkins and Nappi [6], and has the Lagrangian density (in convenient units where the $\omega$-meson mass is 1)

$$\mathcal{L}_{U,\omega} = -\frac{1}{2}\text{Tr}(R_\mu R^\mu) + m^2 \text{Tr}(U-1) - \frac{1}{4}\omega_{\mu\nu}\omega^{\mu\nu} + \frac{1}{2}\omega_\mu\omega^\mu - \beta\omega_\mu\mathcal{B}^\mu. \quad (14.18)$$

This has dimensionless parameters $m$, the pion mass relative to the $\omega$ mass, and $\beta$, a dimensionless version of the coupling constant $g_{\omega\pi\pi}$. The quartic Skyrme term is omitted. There is no algebraic constraint as for the $\rho$ field, but something similar emerges from the field equations.

The equation satisfied by the field $\omega^\mu$ is

$$\partial_\nu \omega^{\nu\mu} + \omega^\mu = \beta\mathcal{B}^\mu. \quad (14.19)$$

Conservation of the baryon current and the antisymmetry of $\omega^{\nu\mu}$ imply that $\partial_\mu\omega^\mu = 0$. So the spatial integral of $\omega^0$ is conserved; this is simply $\beta$ times the baryon number. For a static field, $\omega^0$ obeys the Helmholtz equation

$$(-\nabla^2 + 1)\omega^0 = \beta\mathcal{B}, \quad (14.20)$$

which can be solved using the exponentially decaying Green's function for the Helmholtz operator. It is a good approximation to identify the densities and set $\omega^0(\mathbf{x}) = \beta\mathcal{B}(\mathbf{x})$, but more precisely, the field configuration of $\omega^0$ is a smoothed-out version of $\beta\mathcal{B}$.

Static Skyrmion solutions can be found in this model. Adkins and Nappi found the $B=1$ hedgehog solution, quantized it, and determined some of the properties of the nucleons and delta resonances. More recently, the model has been investigated by Sutcliffe [222], who found approximate solutions with baryon numbers up to $B=4$ with the help of rational maps, and by Gudnason and Speight [110], who found Skyrmion solutions numerically for baryon numbers up to $B=8$. The solutions are mostly similar to those of the standard Skyrme theory with massive pions, except in the case of $B=8$, where the expected twisted 2-cube solution is slightly unstable and the lowest-energy solution has $D_{6d}$ symmetry, the symmetry of the standard $B=8$ Skyrmion with massless pions. In all cases, the spatial components of the baryon current and $\omega$ field vanish for a static Skyrmion, though not when the Skyrmion is spinning.

Choosing a calibration of the coupling $\beta$ is tricky, and no choice gives good results in both the meson and baryon sectors. Gudnason and Speight calibrated the model by matching to the masses of the nucleons and Helium-4 nucleus. Because the nucleons have spin $\frac{1}{2}$, there is a quantum contribution to their mass, additional to the $B = 1$ Skyrmion mass, but Helium-4 has spin and isospin 0, so the Skyrmion mass can be used directly. An alternative calibration uses the nucleon radius and Helium-4 mass. Using the nucleon and delta resonance masses is regarded as less reliable. To some extent the overall energy and length scale, determining the physical $\omega$-meson mass, are adjustable too. For the optimal parameters, the binding energy of the Skyrmions is less than in standard Skyrme theory. In fact, for certain ranges of the parameters, the $B = 2$ and $B = 3$ Skyrmions partially unbind into separated $B = 1$ Skyrmions as in the lightly-bound model, and the $B = 4$ Skyrmion ceases to have octahedral symmetry but distorts into a tetrahedral shape. However, parameter values that avoid this distortion are preferable, because the phenomenology of larger Skyrmions and nuclei seems to prefer cubic $B = 4$ subunits.

When the derivatives of the field $\omega^\mu$ can be treated as small, $\omega^\mu$ is just a multiple of the baryon current density $\mathcal{B}^\mu$ and can be eliminated. This generates a term quadratic in $\mathcal{B}^\mu$ in the Lagrangian density, having the form of the sextic term $\mathcal{L}_6$. Eliminating a field in this way relies on the idea that heavier mesons like $\omega$ and $\rho$ do not exist as independent, freely propagating particles in the context of nuclei, but are only present due to Skyrme field interactions in the nuclear cores, affecting the core structure and the short-distance forces between Skyrmions.

In conclusion, a Skyrme model with an $\omega$-meson field and no Skyrme term works quite well, and provides some explanation for the appearance of a sextic term in a purely pionic theory, but better models are likely to require a more systematic treatment of pions and heavier-meson fields. This is what holographic Skyrme theory offers, to be discussed in the next chapter.

## 14.4 The $SU(3)$ Extension of Skyrme Theory

An interesting extension of Skyrme theory, considered since the 1980s, replaces the $SU(2)$-valued Skyrme field by a (flavour) $SU(3)$-valued field, still denoted by $U$ [106, 251]. The purpose is to incorporate the strange quark s, and to model the mesons and baryons with constituent u, d and s quarks and antiquarks. In a theory with exact $SU(3)$ symmetry, where these three quark types have equal masses, mesons and baryons occur in $SU(3)$ multiplets of equal mass. In practice, complete particle multiplets with the expected quark content and electric charges are observed, but the s quark has a mass near 100 MeV, significantly larger than the u and d quarks, so there is breaking of $SU(3)$ symmetry, leading to mass splittings within each meson and baryon multiplet. In comparison, the $SU(2)$ isospin subgroup of $SU(3)$ remains an almost unbroken symmetry.

The u, d and s quarks form a fundamental triplet of $SU(3)$, but quarks are not directly observed. The particles that are observed are mostly quark-antiquark pairs (mesons), three-quark states (baryons) and three-antiquark states (antibaryons). The lowest-energy, spin 0 meson states lie in an $SU(3)$ octet that includes the three pions, four kaons (each containing one s quark or $\bar{s}$ antiquark), and the $\eta$ meson which has significant $s\bar{s}$ content. There is a similar octet of spin 1 mesons, including the $\rho$ and $\omega$ mesons. The lowest-mass baryons also lie in an $SU(3)$ octet, which includes the two nucleons (p and n), and six hyperons ($\Sigma$, $\Lambda$ and $\Xi$) that contain one or two s quarks. These all have spin $\frac{1}{2}$. Next is the baryon decuplet, which combines the four delta resonances, five excited $\Sigma$ and $\Xi$ resonances, and finally the $\Omega^-$ baryon which is made from three s quarks and is much longer-lived than the other particles in this multiplet. These all have spin $\frac{3}{2}$.

The $SU(3)$ Skyrme field $U$ is constructed from the spin 0 octet meson fields in the same way that the $SU(2)$ Skyrme field is constructed from the pion fields. Let us denote the eight meson fields by $\kappa^a(x), 1 \leq a \leq 8$. The eight basis matrices of the $SU(3)$ Lie algebra are the $3 \times 3$ Gell-Mann matrices $\Lambda_a$, generalising the Pauli matrices. An $SU(3)$ Skyrme field has the form

$$U(x) = \exp(i\kappa^a(x)\Lambda_a), \tag{14.21}$$

and the vacuum is $U = 1_3$, the $3 \times 3$ unit matrix. The $SU(3)$ matrix field $U(x)$ is regarded as the fundamental object in Skyrme theory, so the fields $\kappa^a(x)$ need to have a restricted range, otherwise they would be multi-valued.

An $SU(3)$ Skyrme field configuration $U(\mathbf{x})$ at a given time, with $U \to 1_3$ at spatial infinity, is a mapping from $\mathbb{R}^3$ to $SU(3)$ and is classified topologically by an integer baryon number $B$, as in the $SU(2)$ theory. This is because the 8-dimensional manifold of $SU(3)$ is an $S^3$ fibre bundle over $S^5$, locally the product of a 5-sphere and a 3-sphere, and the baryon number is the topological degree of $U(\mathbf{x})$ projected to the $S^3$ factor. (More precisely, the manifold $SU(3)$ has a normalised closed 3-form that is not exact – a cohomology 3-form. Pulling-back this 3-form to spatial $\mathbb{R}^3$ using the Skyrme field $U$, and integrating, gives the baryon number.)

If $SU(3)$ symmetry breaking is ignored at first, then the Lagrangian can be taken to be the standard Skyrme Lagrangian $L$ as in eq.(4.14), supplemented by a Wess–Zumino term [251]. The current is $R_\mu = \partial_\mu U U^{-1}$ as before, but is now a linear combination of ($i$ times) the Gell-Mann matrices. The $B = 1$ hedgehog Skyrmion solution is unchanged. It can be embedded in $SU(3)$ via the $SU(2)$ isospin subgroup, generated by the matrices $\Lambda_1, \Lambda_2$ and $\Lambda_3$. These are essentially the Pauli matrices, so the pion fields are as in the $SU(2)$ Skyrmion, and the remaining five field components are zero. The $SU(2)$ subgroup is one copy of the $S^3$ fibre mentioned above, so the hedgehog Skyrmion still has baryon number $B = 1$.

$SU(3)$ has a second interesting 3-dimensional subgroup. This is the $SO(3)$ subgroup of real, orthogonal $3 \times 3$ matrices with determinant 1, generated by the Gell-Mann matrices $\Lambda_2, \Lambda_5$ and $\Lambda_7$. There is a spherically-symmetric Skyrmion related to this subgroup [19]. Its field configuration $U(\mathbf{x})$ is invariant under combinations of rotations and conjugations by elements in the subgroup, and has a generalised hedgehog form depending on two radial functions. This novel Skyrmion has baryon number $B = 2$, so it is called a *dibaryon*, and its mass is a little less than twice the mass of the $B = 1$ hedgehog.

$SU(3)$ symmetry breaking has a large effect on the Skyrmion solutions. The analogue of the $SU(2)$ pion mass term needs to give a substantial mass to the kaon and $\eta$-meson fields. Such mass terms, preserving isospin symmetry, can be constructed by using $\text{Tr}(U)$ and $\text{Tr}(U\Lambda_8)$. In the presence of these mass terms, the $SU(2)$ hedgehog Skyrmion takes its values in the isospin subgroup and acquires dynamical spin and isospin as usual, but conjugating this solution by a general element of $SU(3)$ now substantially raises its energy. The other type of Skyrmion, the dibaryon, becomes deformed as a result of $SU(3)$ symmetry breaking, losing its spherical symmetry. Possibly it deforms into the familiar toroidal $B = 2$ Skyrmion lying in the isospin subgroup.

Quantization in the $SU(3)$ theory is more complicated than in the $SU(2)$ theory, but interesting. At the perturbative level, in the $B = 0$ sector, there is an octet of spin 0 mesons, but no spin 1 mesons unless further fields are added. Quantized baryons are obtained by rigid-body quantization of the $B = 1$ hedgehog. In the theory retaining full $SU(3)$ symmetry, the internal motion is generated by conjugation of the hedgehog by a time-dependent

$SU(3)$ matrix $A$,

$$U \mapsto A(t)UA(t)^{-1}. \qquad (14.22)$$

Quantization leads to wavefunctions $\Psi(A)$ on this internal $SU(3)$ lying in $SU(3)$ multiplets which must have *triality zero*, because of the Wess–Zumino term. Triality-zero baryon multiplets are those that can be constructed from quarks and antiquarks with net quark number three. The states of lowest energy are the octet baryons and the decuplet baryons, which are three-quark states, followed by an antidecuplet of baryons that is distinct from the usual antidecuplet of antibaryons. All the states in a single $SU(3)$ multiplet have the same energy and the same spin. The relation between spin and isospin for the states with no s quarks (the states with hypercharge $Y = 1$) is the same as in the $SU(2)$ theory, so the octet baryon states (containing the isospin doublet of nucleons) have spin $\frac{1}{2}$ and the decuplet states (containing the isospin quartet of deltas) have spin $\frac{3}{2}$.

The baryon antidecuplet is interesting, and controversial. This multiplet has spin $\frac{1}{2}$, because it has an isospin doublet of states with $Y = 1$. Its particles do not occur as three-quark states, but can occur as *pentaquarks*, constructed from four quarks and one antiquark. In this sense, they are analogous to the isospin $\frac{5}{2}$ baryons of the $SU(2)$ Skyrme theory. Three states in particular, at the vertices of the antidecuplet weight diagram, have quantum numbers that cannot be realised with three quarks. One of these is known as the $\Theta^+$ pentaquark. It is an isospin singlet and has quark content $uudd\bar{s}$. The 'strangeness' of the $\bar{s}$ cannot be realised in a three-quark state. Diakonov, Petrov and Polyakov calibrated the $SU(3)$ Skyrme model against the octet and decuplet baryons. They then obtained fairly precise predictions for the masses of the antidecuplet baryons, including the mass splitting due to $SU(3)$ symmetry breaking [73, 74]. The $\Theta^+$ mass was calculated to be near 1530 MeV. Its width was estimated to be surprisingly narrow for a state having enough energy to decay strongly.

The $\Theta^+$ pentaquark could occur as a two-particle resonance in $nK^+$ or $pK^0$ scattering, as these kaons contain an $\bar{s}$ antiquark. Such a resonance should be observable, for example, in low-energy $K^+$ scattering off a deuterium target (which contains nearly free neutrons). Rather spectacularly, experimenters claimed in 2003 to have seen this resonance [20, 190], but not directly in kaon-nucleon scattering experiments, because kaon beams of sufficiently low energy were not available. Since it was observed in only one bin of a histogram plotting number of observed events against energy, the resonance appeared to be narrow. Unfortunately, further experiments and reanalysis of data from older experiments didn't confirm the initial observations [149]. Not for the first or last time, an interesting 'discovery' in particle physics disappeared. The theorists also showed, after reanalysis, that the $\Theta^+$ is probably considerably broader than initially thought. The $\Theta^+$ pentaquark state may still exist, but being broader, is more difficult to extract from the background, and its existence has not yet been confirmed. This is a pity, because the prediction and apparent observation of the $\Theta^+$ pentaquark generated a great deal of interest in Skyrmions for a few years.

More recently, strong evidence has accumulated for different types of pentaquark baryons, involving heavier quarks than the s quark [149]. Their masses are close to the sum of the masses of a three-quark baryon and a quark-antiquark meson binding together weakly to form a pentaquark.

Many of the particles in the $SU(3)$ baryon octet and decuplet are hyperons, containing at least one strange quark; an example in the octet is the $\Lambda$ baryon with quark content uds. Hyperons are also observed in several exotic nuclei. They are produced when a $K^-$ or $\overline{K}^0$ meson scatters off normal nuclear matter, and turns into a pion. Such exotic nuclei have not been much studied from a Skyrmion perspective. The $SO(3)$ dibaryon, whose quark content is uuddss in its lowest-energy, $SU(3)$-singlet quantum state, is an exception. In more conventional terms, taking into account $SU(3)$ symmetry breaking, this H-dibaryon is doubly exotic, being a bound state or resonance of two $\Lambda$ hyperons. Unfortunately, no such H-dibaryon state has yet been experimentally confirmed.

In summary, the $SU(3)$ Skyrme theory has received considerable attention, and it successfully accounts for mesons and baryons containing s and $\bar{s}$ quarks and antiquarks. Skyrmion quantization can be extended to the additional degrees of freedom, although the effect of $SU(3)$ symmetry breaking is substantial. So far, various exotic baryon states, predicted in this version of Skyrme theory, have not been established experimentally.

## Chapter 15

# The Sakai–Sugimoto Model

The Sakai–Sugimoto model is a holographic variant of Skyrme theory [209], going far beyond the simple variants considered in the last chapter, where we added one or two vector meson fields and a few new terms to the Skyrme Lagrangian. It has its basis in string theory, which is then reduced to a low-energy effective Yang–Mills gauge theory in four space dimensions. This gauge theory is interpreted in ordinary 3-dimensional space as a theory of a Skyrme field coupled to an infinite tower of vector and axial-vector meson fields[1]. Finally, truncation of the infinite tower gives a variant of Skyrme theory similar to one of the variants we considered earlier, with a Skyrme field coupled to a few mesons. Because of the higher-dimensional background geometry, there are powerful constraints on the truncated theory, and the Skyrmions also acquire new collective degrees of freedom.

The Sakai–Sugimoto model cannot be summarised quickly and really deserves more than a chapter. In particular, we cannot properly discuss the string theory underlying the model. String theory is a candidate for a fundamental theory of all of physics, incorporating particles, fields and gravity. But to be fully consistent mathematically, it needs to be formulated in a 10-dimensional spacetime and also be supersymmetric – having a symmetry between bosons and fermions. Low-energy limits of string theory are understood to be well described by quantum field theories coupled to (super)gravity. The field theories that emerge depend on further structures that can be present in the 10-dimensional spacetime – these are 'brane' configurations where strings can end. The brane configurations are not arbitrary, because the string dynamics, combined with the string boundary conditions on the branes, determine equations for the brane locations and

---
[1] For the rest of this chapter, we usually refer to these simply as mesons – noting that pions are included separately in the Skyrme field.

for the generally curved spacetime background in which the branes live.

The branes are hypersurfaces in the 10-dimensional background, behaving like the solitons of string theory. A $p$-brane has $p$-dimensional spatial extent and can evolve in time, so it has a $(p+1)$-dimensional worldsheet dynamics, although in practice branes are usually at rest. If several identical branes are stacked at the same geometrical location, or close together, they can support a non-abelian gauge theory. If the brane separations are zero, then the gauge theory is a massless Yang–Mills theory, whereas non-zero separations can be interpreted as Higgs field expectation values, which produce a Yang–Mills–Higgs theory where some gauge fields are massive.

There have been a number of schemes for constructing brane configurations that support an approximation to QCD, a massless $SU(N_c)$ Yang–Mills gauge theory with $N_c = 3$ colours, coupled to quarks. The most successful so far is that of Sakai and Sugimoto. This has a stack of three 4-branes, which leads to a $U(3)$ gauge theory. There are further 8-branes that intersect the 4-branes, accommodating the $N_f$ flavours of quarks. We shall assume a stack of two 8-branes, to accommodate two light flavours of quarks, u and d. These branes support a $U(N_f)$ gauge theory with $N_f = 2$. This is the most important part of the model for us. The connection between string theory and gauge theory on branes is most convincingly justified in the large-$N_c$ limit of QCD. However, we will suppose, with Sakai and Sugimoto, that it has some validity in the physical case where $N_c = 3$. In principle, corrections of order $1/N_c$ or higher could be calculated, to obtain more accurate results.

The 10-dimensional background spacetime is the product of a 4-sphere and a 6-dimensional spacetime. The 8-branes extend around the 4-sphere whereas the 4-branes do not. However, the 4-sphere is small, so that excitations of the $U(2)$ fields here have high energy and can be neglected (although stringy excitations on this 4-sphere have recently been considered more carefully [126]). The 6-dimensional spacetime has the usual coordinates $x^0, x^1, x^2, x^3$ of Minkowski spacetime and two further spatial coordinates $x^4$ and $x^5$ with more interesting curved geometry. The 4-branes and 8-branes both extend over the Minkowski spacetime; additionally the 4-branes extend in the $x^4$-direction and the 8-branes in the $x^5$-direction. $x^4$ is an angular coordinate on a circle, and 8-branes intersect this at two antipodal points. Actually, one is an 8-brane supporting left-handed quarks and the other an anti-8-brane supporting right-handed quarks. In one of the $x^5$-directions these branes extend to infinity, but in the opposite direction the $x^4$-circle collapses to zero size and the 8-brane and anti-8-brane smoothly join together. The net effect is that there is a single 8-brane parametrised by a single coordinate $-\infty < z < \infty$, with $z = 0$ the location of the join. This curved geometry leads to $z$-dependent warp factors in the Lagrangian for the $U(2)$ fields.

The low-energy dynamics is therefore a $U(2)$ Yang–Mills gauge theory on a $(4+1)$-dimensional warped spacetime. Supersymmetry is no longer present in this gauge theory – it is broken because antiperiodic boundary conditions in the $x^4$-direction need to be imposed on the fermionic, supersymmetric partners of the Yang–Mills field [252]. The inclusion of quark masses is also not easily arranged in this set-up, so the simplest version has massless u and d quarks.

A Yang–Mills gauge theory in $4+1$ dimensions can be interpreted as a field theory in $3+1$ dimensions having an infinite tower of independent fields of increasing mass. This is a modern version of the Kaluza–Klein idea. The tower of fields arises from a suitably chosen expansion in the extra spatial coordinate. For a circular extra dimension, one would expand using a Fourier series, but in the warped Sakai–Sugimoto geometry a different basis of functions of $z$ is required.

One might expect the theory in $3+1$ dimensions to also be a gauge theory, perhaps QCD itself, with additional fields. That is not the case. Instead, the expansion of the $U(2)$ Yang–Mills field is interpreted in terms of gauge-invariant fields. The result is an effective field theory (EFT) in three space dimensions, with the lowest-mass fields being those of the pions, followed by the lightest vector and axial-vector mesons.

In addition to the pion fields and the meson fields, which arise by expanding around the vacuum, the Yang–Mills theory has its own solitons (static solutions in $4+1$ dimensions) [184]. These are *instantons*, which have an integer topological charge $N$ – their second Chern number – also called the instanton number. Instantons are most easily studied in a simplification of the Sakai–Sugimoto model, where the warp factors depending on $z$ are treated approximately, as perturbations of flat space. This simplification is justified, because the Sakai–Sugimoto model has a dimensionless parameter, the 't Hooft coupling constant $\lambda$, whose value can be such that the instanton is smaller than the length scale associated with the warp factors. Instantons in flat 4-dimensional space are relatively simple and rather well known.

For us, the most important thing is that the instantons in four dimensions are closely related to Skyrmions in three dimensions, and the instanton charge $N$ descends to the baryon number $B$ of the Skyrmion. The dynamical theory of the instantons can therefore be interpreted as a variant of Skyrmion dynamics, and it has improved phenomenological features. This interplay of structures in dimensions differing by one is the essence of *holography*. The instanton-Skyrmion connection was actually proposed earlier, by Atiyah and the present author [16], but without reference to string theory, and it was not developed into a quantum theory of mesons and baryons in the way Sakai and Sugimoto have achieved. A particularly interesting aspect of an instanton, in the context of the Sakai–Sugimoto model, is that it has more moduli than a Skyrmion, and quantizing these moduli leads to excited nucleon and delta resonance states that are absent from standard Skyrme theory, but which match observed baryonic states quite well [124, 125].

In flat space, the $N = 1$ instanton has eight moduli in total, compared to six for a $B = 1$ Skyrmion. There are four translational moduli, three orientational moduli (related to global $SU(2)$ gauge transformations), and a scale factor $\ell$. Metrically, the moduli space is flat $\mathbb{R}^8$, slightly modified. The translational moduli produce one $\mathbb{R}^4$ factor. The second $\mathbb{R}^4$ factor has the origin (representing an instanton of zero size) removed. Its radial coordinate is $\ell$, and the orientational moduli are the coordinates on the 3-sphere at each radius. Antipodal points on the 3-sphere are identified, because classically the orientations are parametrised by $SO(3)$ rather than $SU(2)$. The static instanton energy is constant on the moduli space, so the instanton moduli dynamics in flat space is purely kinetic, like the rigid-body motion of a Skyrmion.

However, in the theory where the warp factors are included as a first-order perturbation, the instanton moduli dynamics is more elaborate, and includes a potential energy term. 4-dimensional scale invariance and translational invariance in the $z$-direction are broken. The geometry favours a particular size for the instanton, and for the centre to be at $z = 0$. The 3-dimensional translational and orientational moduli remain those of a Skyrmion, but the two additional moduli, the instanton size and the location of its centre in the $z$-direction, are subject to the potential.

When the 1-instanton moduli are quantized, the wavefunction $\Psi$ is not single-valued on the $SO(3)$ of orientations, for the same reason that the wavefunction of a $B = 1$ Skyrmion changes sign under a $2\pi$ rotation. $\Psi$ is instead well-defined on the covering space, which can be regarded as all of $\mathbb{R}^8$ provided $\Psi$ vanishes for an instanton of zero size. There is, in fact, no interesting quantum mechanics arising purely from the flat $\mathbb{R}^8$ geometry. But with the potential included, the quantum states are similar to those of a $B = 1$ Skyrmion, except that the usual nucleon and delta states are supplemented by states where the two additional moduli are excited.

An alternative treatment of the instanton dynamics and its relation to Skyrme theory has been developed by Sutcliffe [223]. Here one constructs a formal expansion of an $SU(2)$ Yang–Mills gauge field in flat, 4-dimensional space. This involves a Skyrme field, a $\rho$-meson field, and a tower of further meson fields. The fundamental energy function is assumed to be that of pure Yang–Mills theory. By substituting the field expansion into this, one obtains an energy function for the Skyrme field coupled to the meson fields in three space dimensions.

When this expansion is truncated at lowest order, the 3-dimensional theory involves only the Skyrme field $U$ and its spatial current $R_i$, and most remarkably, the energy has precisely the form of standard Skyrme theory, with its quadratic and quartic terms, and no pion mass term. So the relationship between standard Skyrme theory for massless pions and pure Yang–Mills theory in one dimension higher is very direct and simple. In Sutcliffe's scheme, finding an exact Skyrmion solution is reinterpreted as finding the best approximation to an instanton at the lowest order in the expansion. This is complementary to the Atiyah–Manton idea, where the Skyrme energy function was fundamental and exact instantons were used for finding approximate Skyrmions. When the truncation is at next order, the Skyrme field $U$ is coupled to the $\rho$-meson field, and has the energy function mentioned in Chapter 14, with very specific interaction terms [192]. Solving the coupled equations gives a better approximation to an instanton, and a lower energy.

Both the Sakai–Sugimoto approach and Sutcliffe's approach are only approximate. Sakai and Sugimoto's instanton moduli dynamics captures some collective effects of the higher-mass meson fields, but not the full dynamics, and it relies on a flat-space instanton as a starting point. However, it does take into account the warped geometry in $z$. Sutcliffe's scheme potentially captures all the dynamics of the mesons, but uses a flat-space approximation to the Sakai–Sugimoto geometry and needs some truncation. The truncation produces effects analogous to having warp factors, and suppresses mesons more massive than about 1 GeV that are believed to have little effect on nuclei.

Other authors have considered variants of the geometry, without worrying about the string theory justification. For example, Pomarol and Wulzer, in a series of papers reviewed in [197], have studied a background $(4 + 1)$-dimensional anti-de Sitter geometry. There is a similar reduction of the Yang–Mills theory to a theory of coupled mesons and baryons in Minkowski space, with appropriate instantons reinterpreted as Skyrmions. This model has quantized nucleons and deltas whose properties compare favourably with those of Sakai and Sugimoto, and also improve slightly on those obtained by Adkins, Nappi and Witten in the purely pionic theory of Skyrmions.

The most attractive features of the Sakai–Sugimoto model are that it incorporates many mesons, treats the baryons as solitons, and involves very few free parameters. It leads to an EFT whose couplings are almost completely determined. It is probably a mystery to those who are not string theorists why all this should work. It seems to be because string theory provides a mathematical framework for studying gauge theory, and systematically deriving an EFT of mesons and baryons from it, whether or not string theory is correct as a fundamental physical theory. Supersymmetry and the large $N_c$ limit do not seem to be essential, at least qualitatively. The success of this scheme can be assessed by comparing the many predicted meson masses and couplings with those determined phenomenologically from meson and baryon physics. The fit seems rather good. Sakai and Sugimoto have in particular made predictions for the spectrum of baryon resonances – the excited nucleon and delta states – using their model of quantized instanton moduli. This also seems to work well, showing the need for the warp factors, and confirming that the instantons of the model, closely related to Skyrmions, are an essential feature.

Sutcliffe's version of the Sakai–Sugimoto model in 4-dimensional flat space has its own attractive features. In the topological sector with charge $N$, the complete, untruncated Yang–Mills energy is minimised by instantons whose energy is proportional to $N$. Instantons, interpreted as Skyrmions coupled to an infinite tower of meson fields in three dimensions, therefore have zero binding energy, something one would like to achieve exactly or approximately in a Skyrme model. The baryon number of the Skyrmion equals the charge of the instanton. The structure of this model looks better than that of the BPS Skyrme model, because the energy of the truncated model at lowest order is exactly that of standard Skyrme theory, and the next-order terms give a physically reasonable coupling of Skyrmions to $\rho$ mesons.

The sectors with higher instanton numbers are less studied. Many classical instanton solutions in flat space are known, and can be interpreted as Skyrmions with reduced binding energies, but the quantization of the instanton moduli remains to be done. The 16-dimensional geometry of the

2-instanton moduli space is understood quite well [9,17], and a systematic analysis of 2-instanton moduli dynamics has been started by Halcrow and Winyard [120]. The moduli space of charge $N$ instantons is $8N$-dimensional, and beyond $N = 2$ the metrics on the moduli spaces are complicated. Further work is needed to construct quantized models of nuclei with baryon numbers greater than two, using instanton moduli dynamics.

In conclusion, the Sakai–Sugimoto model seems to be the most successful variant of Skyrme theory so far, and can be used to justify several simpler variants including the original Skyrme theory. In the following sections, we will explore some aspects of the Sakai–Sugimoto model in more detail. We start by recalling the Atiyah–Manton construction of Skyrmions from instantons. We then describe Sutcliffe's field expansion for general gauge potentials in flat 4-dimensional space, which clarifies the energetic relation between instantons and Skyrmions, and also leads to the attractive variant of Skyrme theory including $\rho$ mesons. Finally, we discuss quantized 1-instanton moduli dynamics and the resulting baryon states in the Sakai–Sugimoto model, where the geometrical warp factors play an important role.

## 15.1 Skyrmions from Instantons

In 4-dimensional Euclidean space $\mathbb{R}^4$ with coordinates $x^I$, $I = 1, 2, 3, 4$, pure $SU(2)$ Yang–Mills gauge theory has soliton solutions known as instantons, which have an integer topological charge. When the fourth coordinate is thought of as a spatial analogue of time, the instanton is localised in time and space – hence the name.

Let us denote the Yang–Mills gauge potential by $A_I(x)$; it is valued in the Lie algebra $su(2)$ and its field tensor is $F_{IJ} = \partial_I A_J - \partial_J A_I + [A_I, A_J]$. Under a gauge transformation $g(x)$, the gauge potential transforms to $A'_I = gA_I g^{-1} - \partial_I g g^{-1}$ and the field tensor transforms to $F'_{IJ} = gF_{IJ}g^{-1}$. The Yang–Mills energy is the gauge invariant integral

$$E_{\text{YM}} = \int_{\mathbb{R}^4} -\frac{1}{8}\text{Tr}(F_{IJ}F_{IJ})\, d^4x\,, \tag{15.1}$$

and the topological charge is the second Chern number,

$$N = -\frac{1}{32\pi^2} \int_{\mathbb{R}^4} \epsilon_{IJKL}\text{Tr}(F_{IJ}F_{KL})\, d^4x\,, \tag{15.2}$$

where $\epsilon_{IJKL}$ is the antisymmetric 4-index tensor with $\epsilon_{1234} = -1$. For a gauge potential with suitable boundary conditions, $N$ is an integer.

Instantons are gauge potentials satisfying the self-duality condition

$$F_{IJ} = \frac{1}{2}\epsilon_{IJKL}F_{KL}, \tag{15.3}$$

which plays the role of a Bogomolny equation. In the sector with positive charge $N$, the instantons are the minima of the Yang–Mills energy, and from eqs.(15.1) and (15.2) we see that their energy is $E_{\text{YM}} = 2\pi^2 N$. General $N$-instanton solutions can all be obtained using the ADHM construction [15]. This is not completely explicit, but some subfamilies of instantons can be obtained using the simpler 't Hooft ansatz [225] and its generalisation, the JNR ansatz [137]. All 1-instanton and 2-instanton solutions can be obtained in this more explicit way.

The Skyrmions from instantons scheme was proposed by Atiyah and Manton [16] and involves computing the *holonomy* of $SU(2)$ instantons in $\mathbb{R}^4$ along lines parallel to the $x^4$-axis, which are parametrised by points $\mathbf{x}$ in $\mathbb{R}^3$. Interchangably with $x^I$, it is convenient to use the notation $x^i$, $i = 1, 2, 3$ and $z = x^4$ for the coordinates, so the gauge potential has components $A_i$, $i = 1, 2, 3$ and $A_z$, and the self-duality condition (15.3) becomes $F_{ij} = -\epsilon_{ijk}F_{kz}$. The holonomy at $\mathbf{x}$ generated from the instanton, also known as the Wilson line element, is

$$U(\mathbf{x}) = \mathcal{P}\exp\left(\int_{-\infty}^{\infty} A_z(\mathbf{x}, z)\, dz\right), \tag{15.4}$$

where $\mathcal{P}$ denotes path ordering from left to right as $z$ increases. Since $A_z$ takes values in the Lie algebra $su(2)$, $U$ is a map from $\mathbb{R}^3$ to $SU(2)$ that can be interpreted as a Skyrme field configuration.

The holonomy is not in general an elementary integral, but can be found by solving the (matrix) ordinary differential equation

$$\frac{d\tilde{U}}{dz} = \tilde{U}A_z \tag{15.5}$$

along the line in the $z$-direction at $\mathbf{x}$, satisfying the boundary condition $\tilde{U}(\mathbf{x}, z = -\infty) = 1$. The holonomy is then $U(\mathbf{x}) = \tilde{U}(\mathbf{x}, z = \infty)$.

Because of the conformal invariance of 4-dimensional Yang–Mills theory, $\mathbb{R}^4$ can be compactified by adding a point at infinity to obtain a 4-sphere, $S^4$. The instanton extends to a smooth connection (gauge potential) on a non-trivial $SU(2)$ bundle over $S^4$. In (15.4), the endpoints $-\infty$ and $\infty$ now refer to a single point on $S^4$, so the holonomy is along a family of closed loops in $S^4$, and is almost gauge invariant. The only effect of a gauge transformation $g(x)$ is to conjugate $U(\mathbf{x})$ by a fixed element $g(\infty)$. This corresponds to an isospin rotation of the Skyrme field. Also, the boundary condition $U \to 1$ as $|\mathbf{x}| \to \infty$ is satisfied, because the loops on $S^4$ tend to zero size in this limit. In practice, ensuring the holonomy is along a closed loop means that sometimes an additional bundle transition function should be included in the formula (15.4). For an instanton given by the 't Hooft ansatz, the formula (15.4) is complete as it stands, but for the JNR ansatz, an additional factor of $-1$ is required. In the axial gauge, where $A_z = 0$, the holonomy is entirely contained in the transition function, and the simple compactification to $S^4$ is no longer valid. This will be important below.

Starting from any Yang–Mills field with topological charge $N$, the Skyrme field constructed from it has baryon number $B = N$. This can be verified using specific examples, and the general result follows by continuity. By restricting to instantons, one obtains an $(8N-1)$-dimensional family of Skyrme field configurations from the $8N$-dimensional moduli space of charge $N$ instantons; one parameter is lost since a translation of the instanton in the $z$-direction does not change the Skyrme field $U$, due to the integration over $z$. These configurations are never exact solutions of the Skyrme field equation but some can be good approximations to minimal-energy Skyrmions. When we discuss Sutcliffe's approach to Yang–Mills fields in the next section, it will be clearer why one should focus on instantons.

The basic example is the $N = 1$ instanton, discovered by Belavin *et al.* [37]. This is described by the 't Hooft ansatz, which has the general form

$$A_I(x) = \frac{i}{2}\sigma_{IJ}\partial_J \log \chi(x). \tag{15.6}$$

$\chi$ has to satisfy the 4-dimensional Laplace equation $\partial_I \partial_I \chi = 0$, and $\sigma_{IJ}$ is the set of antisymmetric and anti-self-dual Pauli matrices

$$\sigma_{iz} = \tau_i, \quad \sigma_{ij} = \epsilon_{ijk}\tau_k. \tag{15.7}$$

In standard orientation and centred at the origin, the 1-instanton is derived using the solution $\chi = 1 + \frac{\ell^2}{R^2}$ of the Laplace equation, where $R^2 = r^2 + z^2$ is the 4-dimensional squared radial coordinate, with $r = |\mathbf{x}|$ the usual 3-dimensional radial coordinate, and $\ell$ is an arbitrary, positive scale parameter. The gauge potential components are then calculated to be

$$\mathbf{A}(\mathbf{x}, z) = -i(\mathbf{x} \times \boldsymbol{\tau} + z\boldsymbol{\tau})\left(\frac{1}{R^2} - \frac{1}{R^2 + \ell^2}\right),$$
$$A_z(\mathbf{x}, z) = i\mathbf{x} \cdot \boldsymbol{\tau}\left(\frac{1}{R^2} - \frac{1}{R^2 + \ell^2}\right). \tag{15.8}$$

The holonomy (15.4) becomes an elementary integral here, because $A_z$ is proportional to a fixed element $i\mathbf{x}\cdot\boldsymbol{\tau}$ of $su(2)$ for given $\mathbf{x}$, and the integration results in a hedgehog Skyrme field $U(\mathbf{x}) = \cos f(r) + i\sin f(r)\hat{\mathbf{x}}\cdot\boldsymbol{\tau}$ with profile function [16]

$$f(r) = \pi\left(1 - \frac{r}{\sqrt{r^2+\ell^2}}\right). \tag{15.9}$$

$\ell$ can be chosen so as to minimise the standard energy function (4.49) for a hedgehog Skyrmion with massless pions. The optimal value is $\ell^2 = 2.11$, for which the energy is $E = 1.243 \times 12\pi^2$, only 1% above that of the true $B = 1$ Skyrmion solution.

The holonomies of $N = 2$ instantons generate a 15-dimensional family of $B = 2$ Skyrme field configurations, including configurations with two well-separated $B = 1$ Skyrmions having arbitrary positions, orientations and scales. This accounts for fourteen of the instanton moduli and the final one – the $z$-separation of the instanton pair – has little effect. A particularly symmetric $N = 2$ instanton gives rise to a good approximation to the toroidal $B = 2$ Skyrmion. A 10-dimensional submanifold of the instanton-generated Skyrme fields, corresponding to two $B = 1$ Skyrmions of optimal size in the attractive channel, that includes the toroidal Skyrmion, has been used to study the quantization of the $B = 2$ Skyrmion in [166], producing a reasonable model of the deuteron.

There are symmetric instantons that can be used to generate approximations to the tetrahedral $B = 3$ Skyrmion, the cubic $B = 4$ Skyrmion, and the icosahedral $B = 7$ Skyrmion [165, 214]. The first has a description in terms of the JNR ansatz, but the others require the ADHM formalism. By considering infinitesimal variations of the ADHM data away from the tetrahedral $N = 3$ instanton, it is possible to classify many of the vibrational modes of the $B = 3$ Skyrmion (in fact, 23 modes in total, including nine zero modes). This calculation reproduces all 18 vibrational modes up to the breather, and a few higher-frequency modes too [108]. The modes up to the breather are also captured by the more restricted variations within the JNR ansatz [238]. A similar analysis should be possible for the $B = 4$ and $B = 7$ Skyrmions, although here it is essential to use the ADHM data. A large range of further approximate Skyrmions with various baryon numbers have recently been constructed using a novel numerical implementation of both the ADHM formalism and the instanton holonomy calculation [67].

It is important to understand what happens to holonomies if one transforms to the axial gauge, where the $z$-component of the gauge potential vanishes. The Skyrme field shouldn't change. We have assumed so far that the gauge potential approaches zero at $z = \pm\infty$, but that $A_z$ is non-zero. After a 4-dimensional gauge transformation $g(x)$, the transformed $z$-component of the gauge potential is

$$A'_z = gA_z g^{-1} - \partial_z g g^{-1}, \qquad (15.10)$$

so $A'_z = 0$ if $\partial_z g = gA_z$. This last equation is the same as the matrix differential equation (15.5) defining the holonomy, and its solution satisfying the boundary condition $g(\mathbf{x}, z = -\infty) = 1$ is the path-ordered exponential with upper endpoint $z$,

$$g(\mathbf{x}, z) = \mathcal{P} \exp\left( \int_{-\infty}^{z} A_z(\mathbf{x}, z) \, dz \right). \qquad (15.11)$$

This therefore is the gauge transformation that leads to the axial gauge. The bundle transition function from $z = \infty$ to $z = -\infty$ was initially unity, but in the axial gauge it becomes $g(\mathbf{x}, z = \infty)$. Let us denote this again by $U(\mathbf{x})$. The complete holonomy at $\mathbf{x}$ (along the closed loop based at $z = \infty$) is now the transition function $U(\mathbf{x})$ that connects $z = \infty$ to $z = -\infty$, multiplied on the right by the path-ordered exponential of $A'_z$ along the line at $\mathbf{x}$, which is now unity because $A'_z = 0$. So the holonomy – the Skyrme field – is $U(\mathbf{x})$, as it was before.

The spatial part $A_i$ of the gauge potential at $z = \pm\infty$ was initially zero. In the axial gauge, $A'_i$ is still zero at $z = -\infty$, but at $z = \infty$

$$A'_i = -\partial_i U U^{-1}. \qquad (15.12)$$

$A'_i(\mathbf{x})$ at $z = \infty$ is therefore the 3-dimensional gauge potential with zero field strength (a pure gauge) generated by the gauge transformation $U(\mathbf{x})$. Interpreting $U$ as a Skyrme field, we see that $A'_i$ is the negative of the Skyrme current $R_i$.

It is clearer in the axial gauge why the baryon number $B$ of a Skyrme field, obtained as an instanton holonomy, equals the instanton charge $N$. This charge is defined in terms of the topology of the $SU(2)$ bundle on which the instanton is defined, which is more fundamental than the second Chern number, although the two agree. In the axial gauge, the instanton charge is the topological degree of the transition function connecting $z = \infty$ to $z = -\infty$. This is precisely the topological degree of $U$, which by definition is also the baryon number of the Skyrme field $U$.

## 15.2 An Expansion for 4-Dimensional Gauge Potentials

Sutcliffe has introduced an expansion for $su(2)$ gauge potentials in flat 4-dimensional space, in terms of a tower of 3-dimensional meson fields [223]. This is a simplification of the expansion considered by Sakai and Sugimoto. To make the meson fields unambiguous, the axial gauge $A_z = 0$ is chosen (we now drop the primes when working in this gauge). Only the spatial components $A_i(\mathbf{x}, z)$ of the gauge potential remain. Let us first consider a gauge potential in the topologically trivial sector with instanton number zero, satisfying the boundary conditions $A_i = 0$ at $z = \pm\infty$.

The expansion depends on choosing a basis of orthonormal functions of $z$ that decay as $z \to \pm\infty$. Sutcliffe's choice was the orthonormal sequence of Hermite functions (Hermite polynomials multiplied by a Gaussian),

$$Q_n(z) = \frac{(-1)^n}{\sqrt{n!\, 2^n \sqrt{\pi}}} e^{\frac{1}{2}z^2} \frac{d^n}{dz^n} e^{-z^2}, \quad 0 \leq n < \infty. \tag{15.13}$$

These are energy eigenstates of a quantized harmonic oscillator, although that interpretation is not relevant here. The expansion of the gauge potential is

$$A_i(\mathbf{x}, z) = \sum_{n=0}^{\infty} V_i^n(\mathbf{x}) Q_n(z), \tag{15.14}$$

where $V_i^n(\mathbf{x})$ are interpreted as $su(2)$-valued meson fields. Under the remaining $SU(2)$ global gauge transformations, which preserve the gauge choice $A_z = 0$ and the boundary conditions for $A_i$, the mesons transform as $SU(2)$ isospin triplets. In principle, the field $A_i$ and all the fields $V_i^n$ could be dynamical, although here we focus on time-independent fields.

The dynamics of the mesons is invariant under the 4-dimensional inversion that reverses the signs of $\mathbf{x}$ and $z$, and this inversion symmetry is preserved by quantization. The functions $Q_n$ are alternately even and odd in $z$ (with $Q_0$ even). The effect is that the mesons with even $n$ are vector mesons (with odd parity) in 3-dimensional space, and those with odd $n$ are axial-vector mesons (with even parity). Parity is a good quantum number, and constrains the interactions between the mesons.

We now want to extend this scheme to include more general gauge potentials in the axial gauge. As explained above, it can be arranged that $A_i = 0$ at $z = -\infty$ and $A_i = -\partial_i U U^{-1}$ at $z = \infty$, where $U$ is the holonomy of the gauge potential. In the topologically trivial sector, $U(\mathbf{x})$ can be expressed in terms of pion fields with no baryons, but if the gauge potential has instanton number $N$, then $U$ is a Skyrme field with baryon number $B = N$.

Let us introduce a smooth step function $Q_{-1}(z)$ interpolating from 0 at $z = -\infty$ to 1 at $z = \infty$. Then $A_i + (\partial_i U U^{-1}) Q_{-1}$ vanishes at $z = \pm\infty$ so we can make use of the expansion (15.14). $A_i$ therefore has the modified expansion

$$A_i(\mathbf{x}, z) = (-\partial_i U U^{-1})(\mathbf{x}) Q_{-1}(z) + \sum_{n=0}^{\infty} V_i^n(\mathbf{x}) Q_n(z). \tag{15.15}$$

There is some freedom in the choice of $Q_{-1}$ to be used here, but it is convenient to arrange that $\frac{dQ_{-1}}{dz}$ is proportional to $Q_0$, so that a small shift by $Z$ in the $z$-direction, replacing $Q_{-1}(z)$ by $Q_{-1}(z - Z)$, is equivalent to exciting $V_i^0$ by an amount proportional to $\partial_i U U^{-1}$. With this choice, $Q_{-1}(z) = \frac{1}{2}(1 + \mathrm{erf}(z/\sqrt{2}))$ in Sutcliffe's scheme, as the error function $\mathrm{erf}(z/\sqrt{2})$ is the integral of the relevant Gaussian.

Any gauge potential can be expressed exactly using the expansion (15.15), but a good though crude approximation truncates the series at the leading term. The gauge potential then has the factorised form

$$A_i(\mathbf{x}, z) = (-\partial_i U U^{-1})(\mathbf{x}) \, \psi(z), \tag{15.16}$$

which can be regarded as a separation of variables. In order to find the best approximation to an instanton in this form, we allow $U : \mathbb{R}^3 \mapsto SU(2)$ to be any map of degree $B$, and $\psi$ to be any step function interpolating from 0 to 1, not necessarily the function $Q_{-1}$ as above. With $U$ interpreted as a Skyrme field, $R_i = \partial_i U U^{-1}$ is the spatial Skyrme current, and $A_i = -R_i \psi$.

For this factorised gauge potential in the axial gauge, the field tensor components are $F_{iz} = -\partial_z A_i$ and $F_{ij} = \partial_i A_j - \partial_j A_i + [A_i, A_j]$. Differentiating (15.16) and including the commutator term, we find

$$F_{iz} = R_i \frac{d\psi}{dz} \quad \text{and} \quad F_{ij} = -[R_i, R_j]\psi(1 - \psi). \tag{15.17}$$

The instanton number now simplifies to

$$\begin{aligned}
N &= \frac{1}{8\pi^2} \int_{\mathbb{R}^4} \epsilon_{ijk} \mathrm{Tr}(F_{ij} F_{kz}) \, d^3x \, dz, \\
&= -\frac{1}{4\pi^2} \int_{\mathbb{R}^3} \epsilon_{ijk} \mathrm{Tr}(R_i R_j R_k) \, d^3x \times \int_{-\infty}^{\infty} \psi(1-\psi) \frac{d\psi}{dz} \, dz, \\
&= -\frac{1}{4\pi^2} \int_{\mathbb{R}^3} \epsilon_{ijk} \mathrm{Tr}(R_i R_j R_k) \, d^3x \times \left[\frac{1}{2}\psi^2 - \frac{1}{3}\psi^3\right]_0^1, \\
&= -\frac{1}{24\pi^2} \int_{\mathbb{R}^3} \epsilon_{ijk} \mathrm{Tr}(R_i R_j R_k) \, d^3x, \tag{15.18}
\end{aligned}$$

confirming that for this class of gauge potentials, the instanton number again equals the baryon number of the Skyrme field $U(\mathbf{x})$.

Let us optimise the factorised gauge potential by minimising the 4-dimensional Yang–Mills energy (15.1), which we rewrite as

$$E_{\rm YM} = \int_{\mathbb{R}^4} \left\{ -\frac{1}{4}{\rm Tr}(F_{iz}F_{iz}) - \frac{1}{8}{\rm Tr}(F_{ij}F_{ij}) \right\} d^4x. \tag{15.19}$$

Using the field tensor components (15.17) this simplifies to

$$E_{\rm YM} = \int_{\mathbb{R}^4} \left\{ -\frac{1}{4}{\rm Tr}(R_i R_i) \left(\frac{d\psi}{dz}\right)^2 \right.$$
$$\left. -\frac{1}{8}{\rm Tr}([R_i, R_j][R_i, R_j])\psi^2(1-\psi)^2 \right\} d^4x, \tag{15.20}$$

where the integrals over $\mathbb{R}^3$ and over $z$ are independent. Integrating over $z$ we get almost the standard energy for a static Skyrme field $U$ with massless pions, the only difference being that the coefficients of the quadratic and quartic terms, which are both positive, depend on the function $\psi(z)$, which is not yet determined. The minimal-energy configuration for any $B$ is therefore an exact Skyrmion, but it does not have the standard energy or length scale, in general. Integrating instead over $\mathbb{R}^3$, the result is

$$E_{\rm YM} = \int_{-\infty}^{\infty} \left\{ \frac{1}{2}E_2 \left(\frac{d\psi}{dz}\right)^2 + 2E_4 \, \psi^2(1-\psi)^2 \right\} dz. \tag{15.21}$$

Here

$$E_2 = \int_{\mathbb{R}^3} -\frac{1}{2}{\rm Tr}(R_i R_i) \, d^3x \quad \text{and} \quad E_4 = \int_{\mathbb{R}^3} -\frac{1}{16}{\rm Tr}([R_i, R_j][R_i, R_j]) \, d^3x \tag{15.22}$$

are the standard quadratic and quartic terms in the Skyrme energy, and $E_{\rm YM}$, in the form (15.21), is the energy for a 1-dimensional scalar $\psi^4$ theory with a double-well potential [184]. For configurations $\psi(z)$ interpolating between 0 and 1, $E_{\rm YM}$ is minimised by a $\psi^4$ kink. Together, these arguments show that the optimised factorised gauge potential combines a Skyrmion with a $\psi^4$ kink.

The length scales of the Skyrmion and kink are correlated, but they are not fixed, because an instanton has an arbitrary size. By choosing a scale, however, we can always arrange that $E_4 = E_2$, the relation we anticipate for a standard Skyrmion. Then the energy (15.21) has the normalised form

$$E_{\rm YM} = \frac{1}{2}E_2 \int_{-\infty}^{\infty} \left\{ \left(\frac{d\psi}{dz}\right)^2 + 4\psi^2(1-\psi)^2 \right\} dz, \tag{15.23}$$

and its minimum occurs for the kink solution

$$\psi(z) = \frac{1}{2}(1 + \tanh(z)), \tag{15.24}$$

which satisfies the Bogomolny equation $\frac{d\psi}{dz} = 2\psi(1-\psi)$. Both $z$-integrals in (15.23) have value $\frac{1}{3}$, so $E_{\text{YM}} = \frac{1}{3}E_2$.

With $\psi(z)$ determined, and after integrating over $z$, the expression (15.20) for $E_{\text{YM}}$ reduces to one sixth of the standard Skyrme energy function. This is optimised by the standard, minimal-energy Skyrmion with baryon number $B$. Denoting its energy by $E_B$ as usual, we see that the factorised approximate instanton has energy $E_{\text{YM}} = \frac{1}{6}E_B$. The same result is obtained from the kink energy $E_{\text{YM}} = \frac{1}{3}E_2$, as $E_B = E_2 + E_4 = 2E_2$.

If the Skyrmion energy had attained the Faddeev–Bogomolny bound $E_B = 12\pi^2 B$, then the instanton constructed from the Skyrmion and kink would have had energy $2\pi^2 B$, the energy of a true instanton with charge $N = B$. Similarly, it is straightforward to check that if the Skyrmion had satisfied the Faddeev–Bogomolny equation (4.40), then the factorised gauge potential would have satisfied the self-dual Yang–Mills equation (15.3) exactly. We have therefore found a new interpretation of the failure of Skyrmions to satisfy the Faddeev–Bogomolny equation. Previously, the interpretation was that the mapping $U : \mathbb{R}^3 \mapsto S^3$ could not be an isometry. Now we see that it is a failure to be able to factorise an instanton gauge potential into functions depending separately on $\mathbf{x}$ and on $z$. It is nevertheless remarkable that the optimal approximation to an instanton, as a factorised gauge potential in the axial gauge, combines precisely a standard Skyrmion with a $\psi^4$ kink. This is a novel justification for the original version of the Skyrme energy function.

As an aside, recall that the rational map approximation for Skyrmions also exploits a separation of variables, between the radial and angular dependence in three dimensions, so it may make sense to exploit these separations of variables together in the context of approximate instantons.

An improved approximation to an instanton requires further terms in the expansion (15.15). Let us consider the expansion truncated at the second term. More details are given in [192, 223]. The gauge potential now takes the form

$$A_i(\mathbf{x}, z) = (-\partial_i U U^{-1})(\mathbf{x}) Q_{-1}(z) + \rho_i(\mathbf{x}) Q_0(z) \tag{15.25}$$

with $A_z = 0$. $\rho_i$ is new notation for $V_i^0$. $U$ is again interpreted as a Skyrme field, and $\rho_i$ as a $\rho$-meson field. Naya and Sutcliffe investigated two versions of this truncated expansion; in the first, $Q_{-1}$ is the kink of error-function type, and in the second, $Q_{-1}$ is the $\psi^4$ kink. In both, $Q_0 \propto \frac{dQ_{-1}}{dz}$. The results are very similar. In either case, the field tensor components $F_{iz}$ and $F_{ij}$ are needed. Their $z$-dependence involves the functions $Q_{-1}$, $Q_0$ and their derivatives. In the instanton energy $E_{\text{YM}}$, integrals over $z$ of powers and products of these functions occur, which can all be done numerically, and some exactly. When $Q_{-1}$ is the $\psi^4$ kink, the $z$-integrals can all be done exactly.

The resulting 3-dimensional energy function is more complicated than that of the one-term truncation; it generalises the Skyrme energy for $U$ by including the $\rho$-meson field. The coupled system has standard derivative and mass terms for the $\rho$ field, $\rho$ self-couplings up to quartic order, and several terms coupling the Skyrme current to $\rho$ (as mentioned in Chapter 14). The coefficients are phenomenologically reasonable, despite there being essentially no free parameters in this scheme apart from the overall length and energy scales.

Naya and Sutcliffe found the Skyrmion solutions minimising the energy of the coupled system, up to $B = 12$ [192]. They look very similar to standard Skyrmions, but their energies $E_B$ are much closer to the topological lower bound $12\pi^2 B$. For the $B = 1$ Skyrmion, the coupled system's energy exceeds the bound by a factor 1.06, compared with the one-term truncation where the factor is 1.23. This factor decreases as the baryon number increases, reaching 1.02 for $B = 12$.

Overall, this is an appealing version of the coupled Skyrmion/$\rho$-meson system, because the energy has a lower bound $12\pi^2 B$ (six times the instanton energy) that is close to being attained, so Skyrmion binding energies are small. These Skyrmions have not yet been quantized, but because their shapes and symmetries are qualitatively like those of standard Skyrmions, the spectrum of quantum states should be similar too. It would be interesting to investigate this in detail. Also interesting would be to see how the configurations of the $\rho$-meson field bring the 4-dimensional gauge potentials closer to satisfying the self-dual Yang–Mills equation than can be achieved by pure, factorised gauge potentials. The $\rho$ field is likely to be largest where the underlying Skyrmion $U$ most strongly violates the Faddeev–Bogomolny equation, probably where there are holes in the Skyrmion's baryon density.

Sutcliffe has briefly discussed the truncation at the next order [223]. There are now two mesons included, the vector meson $\rho$ and the axial-vector meson known to particle physicists as the $a_1$ (with mass 1230 MeV). Using the properties of the Hermite functions for $n = 0$ and $n = 1$, the ratio of these meson masses is predicted to be

$$\frac{m_{a_1}}{m_\rho} = \sqrt{3} \simeq 1.73, \tag{15.26}$$

which is close to the experimentally determined ratio of 1.59.

Finally, we mention a situation where factorisation of an instanton works exactly. Recall that the Faddeev–Bogomolny energy bound is attained by the $B = 1$ Skyrmion on a unit 3-sphere. This Skyrmion is, rather trivially, the identity map between 3-spheres, suggesting that if we separate variables in $\mathbb{R}^4$, using the 4-dimensional radial coordinate $R$ and angular coordinates on the 3-sphere, then the 1-instanton solution centred at the origin, in radial gauge, should have the factorised form $A_I = (-\partial_I U U^{-1})\psi$, where $U$ depends only on the angles and $\psi$ only on $R$.

This is confirmed as follows. The 1-instanton gauge potential components were given previously in (15.8), and they are in the required radial gauge $A_R = 0$, as $x^I A_I = 0$. Let $U$ be the identity map between 3-spheres (for all $R$),

$$U = \frac{1}{R}(z + i\mathbf{x}\cdot\boldsymbol{\tau}). \tag{15.27}$$

From this, we calculate that $-\partial_I U U^{-1}$ has components

$$-\boldsymbol{\nabla} U U^{-1} = -\frac{i}{R^2}(\mathbf{x}\times\boldsymbol{\tau} + z\boldsymbol{\tau}),$$

$$-\partial_z U U^{-1} = \frac{i}{R^2}\mathbf{x}\cdot\boldsymbol{\tau}, \tag{15.28}$$

and we see that the instanton gauge potential is the product of this angular part (which has no radial component, because $\partial_R U = 0$) and the radial profile

$$\psi(R) = 1 - \frac{R^2}{R^2 + \ell^2}, \tag{15.29}$$

a kink interpolating between $\psi = 0$ at $R = \infty$ and $\psi = 1$ at $R = 0$.

Curiously, if we set $R = \ell e^{-y}$ then this kink, as a function of $y$, has the previous form (15.24). This is a consequence of the conformal equivalence $\mathbb{R}^4 \simeq S^3 \times \mathbb{R}$, and the conformal invariance of Yang–Mills theory and its self-dual equation.

## 15.3 Baryon Resonances in the Sakai–Sugimoto Model

The Sakai–Sugimoto model, reduced to a $(4+1)$-dimensional gauge theory, is more elaborate than what we have been studying so far. Its gauge group is $U(2)$ rather than $SU(2)$, and there are geometrical warp factors. Also, the dynamical Yang–Mills action $S$ combines the usual action of such a theory, quadratic in the field tensor, with a Chern–Simons term. Following Sutcliffe, we use spacetime coordinates $x^\Gamma$ with capital greek indices in $4+1$ dimensions, with $x^0$ the time coordinate and $x^I$ the spatial coordinates. As before, the spatial coordinates split into $x^i$ for ordinary 3-dimensional space, and $z$ for the additional coordinate on which the warp factors depend. The spacetime metric is

$$ds^2 = (1+z^2)^{\frac{2}{3}}(-(dx^0)^2+(dx^1)^2+(dx^2)^2+(dx^3)^2)+(1+z^2)^{-\frac{2}{3}}dz^2. \quad (15.30)$$

The $u(2)$ gauge potential $\tilde{A}_\Gamma$ is a $2\times 2$ matrix that splits into an antihermitian, traceless $su(2)$ part $A_\Gamma$ and a real $u(1)$ part $\hat{A}_\Gamma$ as

$$\tilde{A}_\Gamma = A_\Gamma + \frac{i}{2}\hat{A}_\Gamma. \quad (15.31)$$

The action is

$$S = \lambda \int k(z)\mathrm{Tr}(\tilde{F}_{\Gamma\Delta}\tilde{F}^{\Gamma\Delta})\,d^5x + i\int \Omega_{\mathrm{CS}} \quad (15.32)$$

where, in the first term, $\lambda$ is a rescaled 't Hooft coupling, indices are raised using the inverse of the metric (15.30), and $k(z) = 1+z^2$ is the square root of (minus) the determinant of the metric. In the second term, $\Omega_{\mathrm{CS}}$ is the imaginary Chern–Simons 5-form constructed from the 1-form gauge potential $\tilde{A} = \tilde{A}_\Gamma dx^\Gamma$ and its 2-form field tensor $\tilde{F} = d\tilde{A} + \tilde{A}\wedge\tilde{A}$. Both contributions to the action $S$ are real.

As a differential form, and suppressing the coordinate indices,

$$\Omega_{\mathrm{CS}} = \mathrm{Tr}\left(\tilde{F}\wedge\tilde{F}\wedge\tilde{A} - \frac{1}{2}\tilde{F}\wedge\tilde{A}\wedge\tilde{A}\wedge\tilde{A} + \frac{1}{10}\tilde{A}\wedge\tilde{A}\wedge\tilde{A}\wedge\tilde{A}\wedge\tilde{A}\right), \quad (15.33)$$

which has the property that $d\Omega_{\mathrm{CS}} = \mathrm{Tr}(\tilde{F}\wedge\tilde{F}\wedge\tilde{F})$. In the above expressions, $\tilde{A}$, $d\tilde{A}$ and $\tilde{F}$ are $2\times 2$ matrices whose entries are differential forms. The matrix multiplication and the trace are performed as usual, with the entries multiplied as differential forms using the wedge product.

A $u(2)$ gauge potential has just four independent 1-form components, so $\tilde{A}\wedge\tilde{A}\wedge\tilde{A}\wedge\tilde{A}\wedge\tilde{A}$ vanishes. The Chern–Simons 5-form for a pure $su(2)$ gauge potential also vanishes. A further combination of terms can be expressed as a total derivative and discarded. It follows that the action simplifies to

$$S = \lambda\int k(z)\mathrm{Tr}(\tilde{F}_{\Gamma\Delta}\tilde{F}^{\Gamma\Delta})\,d^5x$$
$$- \int \epsilon_{\Gamma\Delta\Sigma\Xi\Upsilon}\left(\frac{3}{8}\hat{A}_\Gamma\mathrm{Tr}(F_{\Delta\Sigma}F_{\Xi\Upsilon}) + \frac{1}{16}\hat{A}_\Gamma\hat{F}_{\Delta\Sigma}\hat{F}_{\Xi\Upsilon}\right)d^5x. \quad (15.34)$$

We are primarily interested in static fields – the instantons in the warped 4-dimensional space. The spatial part of the $u(1)$ gauge potential has no instanton solution, so this part can be set to zero, as can the time component of the $su(2)$ potential. Remaining are the spatial $su(2)$ potential $A_I$ and the time component of the $u(1)$ potential $\hat{A}_0$. The energy for static fields in the Sakai–Sugimoto model is therefore

$$E_{\text{SS}} = \lambda \int k(z)\left(-\text{Tr}(F_{IJ}F^{IJ}) + \partial_I \hat{A}^0 \partial^I \hat{A}^0\right) d^3x\, dz$$
$$-\frac{3}{8}\int \hat{A}_0\, \epsilon_{IJKL} \text{Tr}(F_{IJ}F_{KL})\, d^3x\, dz. \qquad (15.35)$$

Using the field tensor components $F_{ij}$ and $F_{iz}$, and the explicit metric factors, this reduces to

$$E_{\text{SS}} = -\lambda \int \left(h(z)\text{Tr}(F_{ij}F_{ij}) + 2k(z)\text{Tr}(F_{iz}F_{iz})\right)d^3x\, dz$$
$$+\lambda \int \left(h(z)\partial_i \hat{A}_0 \partial_i \hat{A}_0 + k(z)\partial_z \hat{A}_0 \partial_z \hat{A}_0\right)d^3x\, dz$$
$$-\frac{3}{2}\int \hat{A}_0\, \epsilon_{ijk}\text{Tr}(F_{ij}F_{kz})\, d^3x\, dz, \qquad (15.36)$$

where $h(z) = (1+z^2)^{-\frac{1}{3}}$, and $k(z) = 1 + z^2$ as before. In the final term, $\hat{A}_0$ is coupled to the $su(2)$ field in a similar way as $\omega_0$ was coupled to the baryon density in the theory discussed in Section 14.3.2.

There is an exact 1-instanton solution minimising the energy $E_{\text{SS}}$, having a similar structure to the 1-instanton in flat space [47]. The instanton has $SO(3)$ rotational symmetry but not the larger $SO(4)$ symmetry of the flat-space instanton, because of the warp factors, and the centre is at $z = 0$. The $u(1)$ field is a 4-dimensional analogue of an electromagnetic field, and from the last integral contributing to $E_{\text{SS}}$ the instanton acquires the analogue of an electric charge density proportional to $\epsilon_{ijk}\text{Tr}(F_{ij}F_{kz})$, a multiple of the $su(2)$ topological charge density. The increasing energy in the resulting electric field as the instanton gets smaller is balanced by the increasing energy due to the warp factors as the instanton gets larger. Equilibrium occurs at a finite instanton size depending on $\lambda$.

At large $\lambda$, the instanton is smaller than the length scale associated with the warp factors, and it is a good approximation to treat its field components as those of a flat-space instanton with $SO(4)$ symmetry. The electric field has this symmetry too, and can be calculated by solving a radial equation. Using these flat-space fields, it is now possible to reintroduce an arbitrary scale parameter $\ell$ and centre $Z$ in the $z$-direction, and recalculate the energy $E_{\text{SS}}$ including the warp factors. The result is a potential energy of the form

$$V(\ell, Z) = M_0 + \frac{A}{\ell^2} + B\ell^2 + CZ^2, \qquad (15.37)$$

with some calculable positive constants $M_0, A, B, C$. The minimum of $V$ occurs at $Z = 0$ and some positive $\ell$, as expected.

By using the flat-space instanton in this way, with its eight dynamical moduli, Sakai and Sugimoto have determined the approximate dynamics of a slowly-moving instanton in the warped spacetime. This takes place on flat $\mathbb{R}^8$, the 1-instanton moduli space mentioned earlier, in the presence of the potential $V(\ell, Z)$. $Z$ is a linear coordinate in one $\mathbb{R}^4$ factor of $\mathbb{R}^8$, and $\ell$ is the radial coordinate in the second $\mathbb{R}^4$ factor. Note that the coordinate $Z$ previously dropped out from the Skyrmion obtained as the holonomy of an instanton, but it doesn't drop out here, because a shift of $Z$ affects the meson fields and is therefore interpreted as a collective excitation of these fields in the Skyrmion background.

The 1-instanton dynamics can be quantized, resulting in an extension of the rigid-body quantization of a $B = 1$ Skyrmion with its six moduli, and the states can again be interpreted as those of a baryon. If $\ell$ and $Z$ are fixed at the minimum of the potential $V$, then the instanton and Skyrmion dynamics are equivalent. The quantum states are labelled by translational momentum in $\mathbb{R}^3$ (which we ignore from here on), spin $J$ and isospin $I$. An improvement is to allow for harmonic vibrations of $\ell$ and $Z$ around the minimum of the potential $V$. This results in a sequence of higher-energy baryon excitations with equal energy spacings.

In fact, the energy spectrum can be found more precisely. The quantized $Z$-dynamics is exactly that of a linear harmonic oscillator. If the constant $A$ in (15.37) vanished, the dynamics on the second $\mathbb{R}^4$ factor would be that of a 4-dimensional, spherical harmonic oscillator, whose energy levels and degeneracies are easily found using Cartesian coordinates. The same states can also be obtained by transforming to 4-dimensional spherical polars. In this formulation, part of the Hamiltonian is a centrifugal term proportional to $1/\ell^2$. The term $A/\ell^2$ in the potential is therefore an adjustment to this centrifugal term, and its effect is to alter the dependence of the energy on the spin $J$.

The final expression for the mass (energy) of the baryon, before calibrating the energy and length scale, is then [209]

$$M = M_0 + \sqrt{\frac{(2J+1)^2}{6} + \frac{6}{5} + \frac{2(n_\ell + n_Z) + 2}{\sqrt{6}}}. \quad (15.38)$$

$n_\ell$ and $n_Z$ are the excitation numbers of the radial and linear oscillators, and $M_0$ is the classical energy of the Sakai–Sugimoto instanton. In the same way as for the Skyrmion, the isospin and spin quantum numbers $I$ and $J$ have to be equal, and only half-integer values are allowed. Because the warp factors do not break 4-dimensional inversion symmetry, quantum states that are even/odd under $Z$-reflection, i.e. those with $n_Z$ even/odd, have positive/negative 3-dimensional parity, and parity is conserved in the interactions between baryons and mesons.

The lowest-energy states are those with $n_\ell = n_Z = 0$. They are the $J^P = \frac{1}{2}^+$ nucleon states and the $J^P = \frac{3}{2}^+$ delta resonance states. Higher-energy states arise as $n_\ell$ and $n_Z$ increase, having the same spins but either parity. The predicted spin $\frac{1}{2}$ states with $(n_\ell, n_Z) = (1,0)$, $(0,1)$, $(1,1)$ and $(2,0)$ match observed excited nucleons well. The first is the Roper resonance, a $\frac{1}{2}^+$ radial excitation of a nucleon; the next three are higher-energy $\frac{1}{2}^-$ and $\frac{1}{2}^+$ nucleon excitations. Similarly, the $(n_\ell, n_Z) = (1,0)$, $(0,1)$ and $(2,0)$ excitations of the delta match observed $\frac{3}{2}^+$ and $\frac{3}{2}^-$ states. For yet higher states, the observational evidence is poor.

We have focussed here on the quantum states in the 1-baryon sector, using the 1-instanton collective coordinates, but there is also a complete expansion of the Sakai–Sugimoto $u(2)$ gauge potential in four dimensions in terms of functions of $z$. The $su(2)$ and $u(1)$ parts models isovector and isoscalar meson degrees of freedom, even in the absence of baryons. The natural basis functions of $z$ to use are the normalised eigenfunctions satisfying the Sturm–Liouville equation

$$\frac{d}{dz}\left(k(z)\frac{d\psi_n}{dz}\right) = \lambda_n h(z)\psi_n, \quad (15.39)$$

with $h(z)$ and $k(z)$ the warp factors defined above. The eigenvalues $\lambda_n$ are positive and the eigenfunctions $\psi_n(z)$ are orthonormal in the sense that

$$\int_{-\infty}^{\infty} h(z)\psi_m \psi_n \, dz = \delta_{mn}. \quad (15.40)$$

Multiplying eq.(15.39) by $\psi_m$ and integrating, one finds the further useful relations

$$\int_{-\infty}^{\infty} k(z)\frac{d\psi_m}{dz}\frac{d\psi_n}{dz} \, dz = \lambda_n \delta_{mn}. \quad (15.41)$$

The field expansions of the $su(2)$ and $u(1)$ potentials are similar to those in Sutcliffe's scheme, but the functions $Q_n$ are replaced by $\psi_n$. The lowest-mass, $n = 0$ mesons in the Sakai–Sugimoto model are therefore isotriplet and isosinglet vector mesons – the $\rho$ and $\omega$ – followed at $n = 1$ by isotriplet and isosinglet axial-vector mesons – the $a_1$ and $f_1$ with measured masses 1230 and 1282 MeV.

After carrying out the $z$-integrations, the Sakai–Sugimoto model becomes a theory of interacting mesons in $3+1$ dimensions. The predicted squared meson masses are a multiple of the eigenvalues $\lambda_n$. Using the numerical values of $\lambda_0$ and $\lambda_1$ one finds equal masses for the $\rho$ and $\omega$, equal masses for the $a_1$ and $f_1$, and

$$\frac{m_{a_1}}{m_\rho} = 1.54, \tag{15.42}$$

a ratio very close to what is measured experimentally.

The field expansion can be extended to the baryon sector by introducing the appropriate step function

$$\psi_{-1}(z) \propto \int_{-\infty}^{z} \psi_0(z') \, dz'. \tag{15.43}$$

Truncating the expansion at $n = 0$ then gives a model for a Skyrme field coupled to $\rho$ and $\omega$ vector mesons, and truncating at $n = 1$ gives a more elaborate model including $a_1$ and $f_1$ axial-vector mesons, with a rich meson and baryon phenomenology.

# Bibliography*

[1] C. Adam, M. Haberichter, T. Romanczukiewicz and A. Wereszczynski, Roper resonances and quasi-normal modes of Skyrmions, *JHEP* **03** (2018) 023.

[2] C. Adam, C. Naya, J. Sánchez-Guillén, R. Vazquez and A. Wereszczyński, The Skyrme model in the BPS limit, in *The Multifaceted Skyrmion, 2nd ed.*, eds. M. Rho and I. Zahed, World Scientific: Singapore, Hackensack NJ, 2016.

[3] C. Adam, J. Sánchez-Guillén and A. Wereszczyński, A Skyrme-type proposal for baryonic matter, *Phys. Lett.* **B691**, 105 (2010); A BPS Skyrme model and baryons at large $N_c$, *Phys. Rev.* **D82**, 085015 (2010).

[4] G. S. Adkins, Rho mesons in the Skyrme model, *Phys. Rev.* **D33**, 193 (1986).

[5] G. S. Adkins and C. R. Nappi, The Skyrme model with pion masses, *Nucl. Phys.* **B233**, 109 (1984).

[6] G. S. Adkins and C. R. Nappi, Stabilization of chiral solitons via vector mesons, *Phys. Lett.* **B137**, 251 (1984).

[7] G. S. Adkins, C. R. Nappi and E. Witten, Static properties of nucleons in the Skyrme model, *Nucl. Phys.* **B228**, 552 (1983).

[8] I. J. R. Aitchison, C. M. Fraser and P. J. Miron, Effective Lagrangian for Skyrmion physics, *Phys. Rev.* **D33**, 1994 (1986).

[9] J. P. Allen and D. J. Smith, The low energy dynamics of charge two dyonic instantons, *JHEP* **02** (2013) 113.

[10] J. M. Allmond *et al.*, Triaxial rotor model description of $E2$ properties in $^{186,188,190,192}$Os, *Phys. Rev.* **C78**, 014302 (2008).

[11] R. Álvarez-Rodriguez, E. Garrido, A. S. Jensen, D. V. Fedorov and H. O. U. Fynbo, Structure of low-lying $^{12}$C-resonances, *Eur. Phys. J.* **A31**, 303 (2007).

[12] L. L. Ames, Natural parity levels in $^{16}$O, *Phys. Rev.* **C25**, 729 (1982).

[13] I. Angeli and K. P. Marinova, Table of experimental nuclear ground state

---

*Journal reference style is generally **volume**, page/article (year); *JHEP* style is **issue/month** (year) article.

charge radii: An update, *At. Data Nucl. Data Tables* **99**, 69 (2013).
[14] M. F. Atiyah and N. J. Hitchin, *The Geometry and Dynamics of Magnetic Monopoles*, Princeton University Press: Princeton NJ, 1988.
[15] M. F. Atiyah, N. J. Hitchin, V. G. Drinfeld and Yu. I. Manin, Construction of instantons, *Phys. Lett.* **A65**, 185 (1978).
[16] M. F. Atiyah and N. S. Manton, Skyrmions from instantons, *Phys. Lett.* **B222**, 438 (1989).
[17] M. F. Atiyah and N. S. Manton, Geometry and kinematics of two Skyrmions, *Commun. Math. Phys.* **153**, 391 (1993).
[18] B. Aubert et al., Observation of the bottomonium ground state in the decay $\Upsilon(3S) \to \gamma\eta_b$, *Phys. Rev. Lett.* **101**, 071801 (2008).
[19] A. P. Balachandran, A. Barducci, F. Lizzi, V. G. J. Rodgers and A. Stern, Doubly strange dibaryon in the chiral model, *Phys. Rev. Lett.* **52**, 887 (1984).
[20] V. V. Barmin et al., Observation of a baryon resonance with positive strangeness in K+ collisions with Xe nuclei, *Phys. Atom. Nucl.* **66**, 1715 (2003).
[21] C. Barnes, K. Baskerville and N. Turok, Normal modes of the $B = 4$ Skyrme soliton, *Phys. Rev. Lett.* **79**, 367 (1997); Normal mode spectrum of the deuteron in the Skyrme model, *Phys. Lett.* **B411**, 180 (1997).
[22] W. K. Baskerville, Making nuclei out of the Skyrme crystal, *Nucl. Phys.* **A596**, 611 (1996).
[23] W. K. Baskerville, Vibrational spectrum of the $B = 7$ Skyrme soliton, arXiv:hep-th/9906063 (1999).
[24] W. K. Baskerville, C. Barnes and N. G. Turok, Normal mode spectra of multi-Skyrmions, in *Solitons: Properties, Dynamics, Interactions, Applications, p.11*, eds. R. MacKenzie, M. B. Paranjape and W. J. Zakrzewski, Springer: New York, 2000.
[25] W. H. Bassichis and G. Ripka, A Hartree–Fock calculation of excited states of $O^{16}$, *Phys. Lett.* **15**, 320 (1965).
[26] R. A. Battye, C. J. Houghton and P. M. Sutcliffe, Icosahedral Skyrmions, *J. Math. Phys.* **44**, 3543 (2003).
[27] R. A. Battye, S. Krusch and P. M. Sutcliffe, Spinning Skyrmions and the Skyrme parameters, *Phys. Lett.* **B626**, 120 (2005).
[28] R. A. Battye, N. S. Manton and P. M. Sutcliffe, Skyrmions and the $\alpha$-particle model of nuclei, *Proc. Roy. Soc.* **A463**, 261 (2007).
[29] R. A. Battye, N. S. Manton, P. M. Sutcliffe and S. W. Wood, Light nuclei of even mass number in the Skyrme model, *Phys. Rev.* **C80**, 034323 (2009).
[30] R. A. Battye and P. M. Sutcliffe, Symmetric Skyrmions, *Phys. Rev. Lett.* **79**, 363 (1997); Solitonic fullerene structures in light atomic nuclei, *Phys. Rev. Lett.* **86**, 3989 (2001); Skyrmions, fullerenes and rational maps, *Rev. Math. Phys.* **14**, 29 (2002).
[31] R. A. Battye and P. M. Sutcliffe, A Skyrme lattice with hexagonal symmetry, *Phys. Lett.* **B416**, 385 (1998).
[32] R. A. Battye and P. M. Sutcliffe, Skyrmions and the pion mass, *Nucl. Phys.* **B705**, 384 (2005).

[33] R. A. Battye and P. M. Sutcliffe, Skyrmions with massive pions, *Phys. Rev.* **C73**, 055205 (2006).

[34] W. Bauhoff, H. Schultheis and R. Schultheis, Alpha-cluster structure of $^{32}$S, *Phys. Rev.* **C22**, 861 (1980).

[35] W. Bauhoff, H. Schultheis and R. Schultheis, Alpha cluster model and the spectrum of $^{16}$O, *Phys. Rev.* **C29**, 1046 (1984).

[36] M.-O. Beaudoin and L. Marleau, Near-BPS Skyrmions: Constant baryon density, *Nucl. Phys.* **B883**, 328 (2014).

[37] A. A. Belavin, A. M. Polyakov, A. S. Schwarz and Yu. S. Tyupkin, Pseudoparticle solutions of the Yang–Mills equations, *Phys. Lett.* **B59**, 85 (1975).

[38] M. V. Berry and J. M. Robbins, Indistinguishability for quantum particles: spin, statistics and the geometric phase, *Proc. R. Soc. Lond.* **A453**, 1771 (1997).

[39] H. A. Bethe and R. F. Bacher, Nuclear physics A: Stationary states of nuclei, *Rev. Mod. Phys.* **8**, 82 (1936).

[40] R. Bijker and F. Iachello, Evidence for tetrahedral symmetry in $^{16}$O, *Phys. Rev. Lett.* **112** 152501 (2014).

[41] R. Bijker and F. Iachello, Cluster structure of $^{20}$Ne: Evidence for $D_{3h}$ symmetry, *Nucl. Phys.* **A1006**, 122077 (2021).

[42] J. W. Bittner and R. D. Moffat, Elastic scattering of alpha particles by carbon, *Phys. Rev.* **96**, 374 (1954).

[43] J. M. Blatt and V. F. Weisskopf, *Theoretical Nuclear Physics*, Wiley: New York, 1952.

[44] E. B. Bogomol'nyi, The stability of classical solutions, *Sov. J. Nucl. Phys.* **24**, 449 (1976).

[45] A. Bohr and B. R. Mottelson, *Nuclear Structure, Vol. I: Single Particle Motion*, World Scientific: Singapore, 1998.

[46] A. Bohr and B. R. Mottelson, *Nuclear Structure, Vol. II: Nuclear Deformations*, World Scientific: Singapore, 1998.

[47] S. Bolognesi and P. Sutcliffe, The Sakai–Sugimoto soliton, *JHEP* **01** (2014) 078.

[48] E. Bonenfant and L. Marleau, Nuclei as near BPS Skyrmions, *Phys. Rev.* **D82**, 054023 (2010).

[49] R. Bott and L. W. Tu, *Differential Forms in Algebraic Topology*, Springer: New York, 1982.

[50] M. Bouten, $\alpha$-particle model of $^{20}$Ne and $^{24}$Mg, *Nuovo Cim.* **26**, 63 (1962).

[51] E. Braaten and L. Carson, Deuteron as a toroidal Skyrmion, *Phys. Rev.* **D38**, 3525 (1988).

[52] E. Braaten and L. Carson, Deuteron as a toroidal Skyrmion: Electromagnetic form factors, *Phys. Rev.* **D39**, 838 (1989).

[53] E. Braaten, S. Townsend and L. Carson, Novel structure of static multi-soliton solutions in the Skyrme model, *Phys. Lett.* **B235**, 147 (1990).

[54] E. Braaten, S.-M. Tse and C. Willcox, Electroweak form factors of the Skyrmion, *Phys. Rev.* **D34**, 1482 (1986).

[55] D. M. Brink, H. Friedrich, A. Weiguny and C. W. Wong, Investigation of

the alpha-particle model for light nuclei, *Phys. Lett.* **B33**, 143 (1970).
[56] D. M. Brink and G. F. Nash, Excited states in Oxygen 16, *Nucl. Phys.* **40**, 608 (1963).
[57] G. E. Brown (ed.), *Selected Papers, with Commentary, of Tony Hilton Royle Skyrme*, World Scientific: Singapore, 1994.
[58] G. E. Brown and M. Rho (eds.), *The Multifaceted Skyrmion*, World Scientific: Singapore, 2010.
[59] E. Browne, R. B. Firestone and V. S. Shirley, *Table of Radioactive Isotopes*, Wiley-Interscience: New York, 1986.
[60] C. G. Callan and E. Witten, Monopole catalysis of Skyrmion decay, *Nucl. Phys.* **B239**, 161 (1984).
[61] L. Carson, $B = 3$ nuclei as quantized multi-Skyrmions, *Phys. Rev. Lett.* **66**, 1406 (1991).
[62] L. Castillejo, P. S. J. Jones, A. D. Jackson, J. J. M. Verbaarschot and A. Jackson, Dense Skyrmion systems, *Nucl. Phys.* **A501**, 801 (1989).
[63] L. Castillejo and M. Kugler, The interaction of Skyrmions, Weizmann Institute report WIS-87/66-PH (1987).
[64] Y. T. Cheng, A. Goswami, M. J. Throop and D. K. McDaniels, Status of nuclear coexistence for $^{32}$S, *Phys. Rev.* **C9**, 1192 (1974).
[65] M. Chernykh *et al.*, Structure of the Hoyle state in $^{12}$C, *Phys. Rev. Lett.* **98**, 032501 (2007).
[66] H. Clement, On the history of dibaryons and their final observation, *Prog. Part. Nucl. Phys.* **93**, 195 (2017).
[67] J. Cork and C. Halcrow, ADHM Skyrmions, arXiv:2110.15190 (2021).
[68] A. N. Danilov *et al.*, Determination of nuclear radii for unstable states in $^{12}$C with diffraction inelastic scattering, *Phys. Rev.* **C80**, 054603 (2009).
[69] V. M. Datar *et al.*, Electromagnetic transition from the $4^+$ to $2^+$ resonance in $^8$Be measured via the radiative capture in $^4$He + $^4$He, *Phys. Rev. Lett.* **111**, 062502 (2013).
[70] D. M. Dennison, Excited states of the O$^{16}$ nucleus, *Phys. Rev.* **57**, 454 (1940); Energy levels of the O$^{16}$ nucleus, *Phys. Rev.* **96**, 378 (1954).
[71] G. H. Derrick, Comments on nonlinear wave equations as models for elementary particles, *J. Math. Phys.* **5**, 1252 (1964).
[72] E. D'Hoker and E. Farhi, Skyrmions and/in the weak interactions, *Nucl. Phys.* **B241**, 109 (1984).
[73] D. Diakonov and V. Petrov, Exotic baryon resonances in the Skyrme model, in *The Multifaceted Skyrmion*, eds. G. E. Brown and M. Rho, World Scientific: Singapore, 2010.
[74] D. Diakonov, V. Petrov and M. Polyakov, Exotic anti-decuplet of baryons: Prediction from chiral solitons, *Z. Phys.* **A359**, 305 (1997).
[75] N. Dorey, J. Hughes and M. P. Mattis, Skyrmion quantization and the decay of the $\Delta$, *Phys. Rev.* **D50**, 5816 (1994).
[76] I. Duck and E. C. G. Sudarshan, *Pauli and the Spin-Statistics Theorem*, World Scientific: Singapore, 1997.
[77] ENSDF: Evaluated Nuclear Structure Data File, https://www.nndc.bnl.gov/ensdf/index.jsp .

[78] E. Epelbaum, Few-nucleon forces and systems in chiral effective field theory, *Prog. Part. Nucl. Phys.* **57**, 654 (2006).
[79] E. Epelbaum et al., Structure and rotations of the Hoyle state, *Phys. Rev. Lett.* **109**, 252501 (2012).
[80] E. Epelbaum et al., Ab initio calculation of the spectrum and structure of $^{16}$O, *Phys. Rev. Lett.* **112**, 102501 (2014).
[81] M. J. Esteban, A direct variational approach to Skyrme's model for meson fields, *Commun. Math. Phys.* **105**, 571 (1986); Erratum: Existence of 3D Skyrmions, complete version, *Commun. Math. Phys.* **251**, 209 (2004).
[82] L. D. Faddeev, Some comments on the many dimensional solitons, *Lett. Math. Phys.* **1**, 289 (1976).
[83] D. T. J. Feist, Interactions of $B = 4$ Skyrmions, *JHEP* **02** (2012) 100.
[84] D. T. J. Feist, *Skyrmion Interactions and Vibrations*, Ph.D. thesis, Cambridge University, 2013.
[85] D. T. J. Feist, P. H. C. Lau and N. S. Manton, Skyrmions up to baryon number 108, *Phys. Rev.* **D87**, 085034 (2013).
[86] A. Fert, N. Reyren and V. Cros, Magnetic Skyrmions: Advances in physics and potential applications, *Nat. Rev. Mater.* **2**, 17031 (2017).
[87] D. Finkelstein and J. Rubinstein, Connection between spin, statistics and kinks, *J. Math. Phys.* **9**, 1762 (1968).
[88] I. Floratos and B. Piette, Multi-Skyrmion solutions for the sixth order Skyrme model, *Phys. Rev.* **D64**, 045009 (2001).
[89] D. Foster and S. Krusch, Scattering of Skyrmions, *Nucl. Phys.* **B897**, 697 (2015).
[90] D. Foster and N. S. Manton, Scattering of nucleons in the classical Skyrme model, *Nucl. Phys.* **B899**, 513 (2015).
[91] D. S. Freed, Pions and generalized cohomology, *J. Diff. Geom.* **80**, 45 (2008).
[92] M. Freer et al., $^8$Be and $\alpha$ decay of $^{16}$O, *Phys. Rev.* **C51**, 1682 (1995).
[93] M. Freer et al., $^8$Be+$^8$Be decay of excited states in $^{16}$O, *Phys. Rev.* **C70**, 064311 (2004).
[94] M. Freer et al., Reexamination of the excited states of $^{12}$C, *Phys. Rev.* **C76**, 034320 (2007).
[95] M. Freer, H. Horiuchi, Y. Kanada-En'yo, D. Lee and U-.G. Meißner, Microscopic clustering in light nuclei, *Rev. Mod. Phys.* **90**, 035004 (2018).
[96] M. Freer and H. O. U. Fynbo, The Hoyle state in $^{12}$C, *Prog. Part. Nucl. Phys.* **78**, 1 (2014).
[97] Y. Fujiwara et al., Chapter 2: Comprehensive study of alpha-nuclei, *Prog. Theor. Phys. Suppl.* **68**, 29 (1980).
[98] T. Fukui, Y. Kanada-En'yo, K. Ogata, T. Suhara and Y. Taniguchi, Investigation of spatial manifestation of $\alpha$ clusters in $^{16}$O via $\alpha$-transfer reactions, *Nucl. Phys.* **A983**, 38 (2019).
[99] N. Furutachi, M. Kimura, A. Doté, Y. Kanada-En'yo and S. Oryu, Cluster structures in oxygen isotopes, *Prog. Theor. Phys.* **119**, 403 (2008).
[100] G. W. Gibbons, C. M. Warnick and W. W. Wong, Non-existence of Skyrmion-Skyrmion and Skyrmion-anti-Skyrmion static equilibria, *J.*

Math. Phys. **52**, 012905 (2011).

[101] M. Gillard, D. Harland, E. Kirk, B. Maybee and M. Speight, A point particle model of lightly bound Skyrmions, *Nucl. Phys.* **B917**, 286 (2017).

[102] M. Gillard, D. Harland and M. Speight, Skyrmions with low binding energies, *Nucl. Phys.* **B895**, 272 (2015).

[103] T. Gisiger and M. B. Paranjape, Recent mathematical developments in the Skyrme model, *Phys. Rep.* **306**, 109 (1998).

[104] D. Giulini, On the possibility of spinorial quantization in the Skyrme model, *Mod. Phys. Lett.* **A8**, 1917 (1993).

[105] J. Goldstone, Field theories with "superconductor" solutions, *Nuovo Cim.* **19**, 154 (1961).

[106] E. Guadagnini, Baryons as solitons and mass formulae, *Nucl. Phys.* **B236**, 35 (1984).

[107] S. B. Gudnason and C. Halcrow, $B = 5$ Skyrmion as a two-cluster system, *Phys. Rev.* **D97**, 125004 (2018).

[108] S. B. Gudnason and C. Halcrow, Vibrational modes of Skyrmions, *Phys. Rev.* **D98**, 125010 (2018).

[109] S. B. Gudnason and C. Halcrow, Database of Skyrmion vibrations, www1.maths.leeds.ac.uk/pure/geometry/SkyrmionVibrations/ .

[110] S. B. Gudnason and J. M. Speight, Realistic classical binding energies in the $\omega$-Skyrme model, *JHEP* **07** (2020) 184.

[111] M. Haberichter, P. H. C. Lau and N. S. Manton, Electromagnetic transition strengths for light nuclei in the Skyrme model, *Phys. Rev.* **C93**, 034304 (2016).

[112] L. R. Hafstad and E. Teller, The alpha-particle model of the nucleus, *Phys. Rev.* **54**, 681 (1938).

[113] C. J. Halcrow, Vibrational quantisation of the $B = 7$ Skyrmion, *Nucl. Phys.* **B904**, 106 (2016).

[114] C. J. Halcrow, unpublished.

[115] C. Halcrow and D. Harland, An attractive spin-orbit potential from the Skyrme model, *Phys. Rev. Lett.* **125**, 042501 (2020).

[116] C. J. Halcrow, C. King and N. S. Manton, Dynamical $\alpha$-cluster model of $^{16}$O, *Phys. Rev.* **C95**, 031303(R) (2017).

[117] C. J. Halcrow, C. King and N. S. Manton, Oxygen-16 spectrum from tetrahedral vibrations and their rotational excitations, *Int. J. Mod. Phys.* **E28**, 1950026 (2019).

[118] C. J. Halcrow, N. S. Manton and J. I. Rawlinson, Quantized Skyrmions from $SU(4)$ weight diagrams, *Phys. Rev.* **C97**, 034307 (2018).

[119] C. J. Halcrow and J. I. Rawlinson, Electromagnetic transition rates of $^{12}$C and $^{16}$O in rotational-vibrational models, *Phys. Rev.* **C102**, 014314 (2020).

[120] C. Halcrow and T. Winyard, A consistent two-Skyrmion configuration space from instantons, arXiv:2103.15669 (2021).

[121] D. Harland and C. J. Halcrow, Nucleon-nucleon potential from Skyrmion dipole interactions, *Nucl. Phys.* **B967**, 115430 (2021).

[122] D. Harland and N. S. Manton, Rolling Skyrmions and the nuclear spin-orbit force, *Nucl. Phys.* **B935**, 210 (2018).

[123] D. Harland and R. S. Ward, Chains of Skyrmions, *JHEP* **12** (2008) 093.
[124] K. Hashimoto, T. Sakai and S. Sugimoto, Holographic baryons: Static properties and form factors from gauge/string duality, *Prog. Theor. Phys.* **120**, 1093 (2008).
[125] H. Hata, T. Sakai, S. Sugimoto and S. Yamato, Baryons from instantons in holographic QCD, *Prog. Theor. Phys.* **117**, 1157 (2007).
[126] Y. Hayashi, T. Ogino, T. Sakai and S. Sugimoto, Stringy excited baryons in holographic quantum chromodynamics, *Prog. Theor. Exp. Phys.* **2020**, 053B04 (2020).
[127] G. Herzberg, *Molecular Spectra and Molecular Structure: II. Infrared and Raman Spectra of Polyatomic Molecules*, Van Nostrand: Princeton NJ, 1945.
[128] A. Heusler, High spin states in the heavy nucleus $^{208}$Pb and the coupling of one-particle one-hole states to platonic shapes, *J. Phys.: Conf. Ser.* **1643**, 012137 (2020); Rotational bands with tetrahedral and icosahedral symmetry in $^{208}$Pb, preprint (2021).
[129] C. Houghton and S. Magee, The effect of pion mass on Skyrme configurations, *EPL* **77**, 11001 (2007).
[130] C. J. Houghton, N. S. Manton and P. M. Sutcliffe, Rational maps, monopoles and Skyrmions, *Nucl. Phys.* **B510**, 507 (1998).
[131] F. Hoyle, On nuclear reactions occuring in very hot stars. I. The synthesis of elements from carbon to nickel, *Astrophys. J. Suppl.* **1**, 121 (1954).
[132] E. Ideguchi *et al.*, Superdeformation in the doubly magic nucleus $^{40}_{20}$Ca$_{20}$, *Phys. Rev. Lett.* **87**, 222501 (2001).
[133] N. Imai *et al.*, First lifetime measurement of $2_1^+$ state in $^{12}$Be, *Phys. Lett.* **B673**, 179 (2009).
[134] J. M. Irvine, C. D. Latorre and V. F. E. Pucknell, The structure of $^{16}$O; a review of the theory, *Adv. Phys.* **20:88**, 661 (1971).
[135] P. Irwin, Zero mode quantization of multi-Skyrmions, *Phys. Rev.* **D61**, 114024 (2000).
[136] R. Jackiw and N. S. Manton, Symmetries and conservation laws in gauge theories, *Ann. Phys.* **127**, 257 (1980).
[137] R. Jackiw, C. Nohl and C. Rebbi, Conformal properties of pseudoparticle configurations, *Phys. Rev.* **D15**, 1642 (1977).
[138] A. Jackson, A. D. Jackson, A. S. Goldhaber, G. E. Brown and L. C. Castillejo, A modified Skyrmion, *Phys. Lett.* **B154**, 101 (1985).
[139] A. Jackson, A. D. Jackson and V. Pasquier, The Skyrmion-Skyrmion interaction, *Nucl. Phys.* **A432**, 567 (1985).
[140] A. D. Jackson and M. Rho, Baryons as chiral solitons, *Phys. Rev. Lett.* **51**, 751 (1983).
[141] S. Jarvis, A rational map for Euclidean monopoles via radial scattering, *J. reine angew. Math.* **524**, 17 (2000).
[142] D. G. Jenkins, Electromagnetic transitions as a probe of nuclear clustering, in *Clusters in Nuclei, Vol. 3, Lecture Notes in Physics 875, p.25*, ed. C. Beck, Springer International: Cham, 2014.
[143] M. Johnston and D. M. Dennison, The interaction between vibration and

rotation for symmetrical molecules, *Phys. Rev.* **48**, 868 (1935).
[144] S. L. Kameny, α-particle model of O$^{16}$, *Phys. Rev.* **103**, 358 (1956).
[145] Y. Kanada-En'yo, Tetrahedral 4α and $^{12}$C + α cluster structures in $^{16}$O, *Phys. Rev.* **C96**, 034306 (2017).
[146] Y. Kanada-En'yo and Y. Hidaka, Tetrahedral shape and surface density wave of $^{16}$O caused by α-cluster correlations, arXiv:1608.03642 (2016).
[147] A. Kangasmäki *et al.*, Lifetime of $^{32}$S levels, *Phys. Rev.* **C58**, 699 (1998).
[148] L. B. Kapitanski and O. A. Ladyzenskaia, On the Coleman's principle concerning the stationary points of invariant functionals, *Zap. Nauchn. Semin., LOMI* **127**, 84 (1983).
[149] M. Karliner and T. Skwarnicki, Pentaquarks, in P. A. Zyla *et al.* (Particle Data Group), Review of Particle Physics, *Prog. Theor. Exp. Phys.* **2020**, 083C01 (2020).
[150] D. G. Kendall, Shape manifolds, procrustean metrics, and complex projective spaces, *Bull. Lond. Math. Soc.* **16**, 81 (1984).
[151] O. S. Kirsebom *et al.*, Breakup of $^{12}$C resonances into three α particles, *Phys. Rev.* **C81**, 064313 (2010).
[152] I. Klebanov, Nuclear matter in the Skyrme model, *Nucl. Phys.* **B262**, 133 (1985).
[153] F. Klein, *Lectures on the Icosahedron*, Dover Publications: Mineola NY, 2003.
[154] V. B. Kopeliovich, Quantization of the axially-symmetric systems' rotations in the Skyrme model (in Russian), *Yad. Fiz.* **47**, 1495 (1988).
[155] V. B. Kopeliovich and B. E. Stern, Exotic Skyrmions, *JETP Lett.* **45**, 203 (1987).
[156] S. Krusch, Homotopy of rational maps and the quantization of Skyrmions, *Ann. Phys.* **304**, 103 (2003); Finkelstein–Rubinstein constraints for the Skyrme model with pion masses, *Proc. R. Soc.* **A462**, 2001 (2006).
[157] V. A. Krutov and N. V. Zackrevsky, On the theory of rotation of non-axial nuclei, *J. Phys.* **A2**, 448 (1969).
[158] M. Kugler and S. Shtrikman, A new Skyrmion crystal, *Phys. Lett.* **B208**, 491 (1988); Skyrmion crystals and their symmetries, *Phys. Rev.* **D40**, 3421 (1989).
[159] M. Lacombe *et al.*, Parametrization of the Paris $N$-$N$ potential, *Phys. Rev.* **C21**, 861 (1980).
[160] L. D. Landau and E. M. Lifschitz, *Quantum Mechanics - Course of Theoretical Physics Vol. 3, 3rd ed.*, Pergamon: Oxford, 1977.
[161] P. H. C. Lau, unpublished.
[162] P. H. C. Lau and N. S. Manton, States of Carbon-12 in the Skyrme model, *Phys. Rev. Lett.* **113**, 232503 (2014).
[163] P. H. C. Lau and N. S. Manton, Quantization of $T_d$- and $O_h$-symmetric Skyrmions, *Phys. Rev.* **D89**, 125012 (2014).
[164] R. Lazauskas and J. Carbonell, Three-neutron resonance trajectories for realistic interaction models, *Phys. Rev.* **C71**, 044004 (2005).
[165] R. A. Leese and N. S. Manton, Stable instanton-generated Skyrme fields with baryon numbers three and four, *Nucl. Phys.* **A572**, 575 (1994).

[166] R. A. Leese, N. S. Manton and B. J. Schroers, Attractive channel Skyrmions and the deuteron, *Nucl. Phys.* **B442**, 228 (1995).

[167] H. Leutwyler, On the foundations of chiral perturbation theory, *Ann. Phys.* **235**, 165 (1994).

[168] K. Lezuo, Ground state rotational bands in $^{16}$O, $^{40}$Ca and $^{208}$Pb?, *Z. Naturforsch.* **30a**, 158 (1975).

[169] W. T. Lin and B. Piette, Skyrmion vibration modes within the rational map ansatz, *Phys. Rev.* **D77**, 125028 (2008).

[170] M. MacCormick and G. Audi, Evaluated experimental isobaric analogue states from $T = 1/2$ to $T = 3$ and associated IMME coefficients, *Nucl. Phys.* **A925**, 61 (2014).

[171] V. G. Makhankov, Y. P. Rybakov and V. I. Sanyuk, *The Skyrme Model: Fundamentals, Methods, Applications*, Springer: Berlin Heidelberg, 1993.

[172] O. V. Manko and N. S. Manton, Angularly localized Skyrmions, *J. Phys.* **A39**, 1507 (2006).

[173] O. V. Manko and N. S. Manton, On the Spin of the $B = 7$ Skyrmion, *J. Phys.* **A40**, 3683 (2007).

[174] O. V. Manko, N. S. Manton and S. W. Wood, Light nuclei as quantized Skyrmions, *Phys. Rev.* **C76**, 055203 (2007).

[175] N. S. Manton, Is the $B = 2$ Skyrmion axially symmetric?, *Phys. Lett.* **B192**, 177 (1987).

[176] N. S. Manton, Geometry of Skyrmions, *Commun. Math. Phys.* **111**, 469 (1987).

[177] N. S. Manton, Skyrmions and their pion multipole moments, *Acta Phys. Pol.* **B25**, 1757 (1994).

[178] N. S. Manton, Classical Skyrmions - static solutions and dynamics, *Math. Meth. Appl. Sci.* **35**, 1188 (2012).

[179] N. S. Manton, Evidence for tetrahedral structure of Calcium-40, *Int. J. Mod. Phys.* **E29**, 2050018 (2020).

[180] N. S. Manton, Skyrmions, tetrahedra and magic numbers, *Quart. J. Math.* **72**, 735 (2021).

[181] N. S. Manton and B. M. A. G. Piette, Understanding Skyrmions using rational maps, *Prog. Math.* **201**, 469 (2001).

[182] N. S. Manton and P. J. Ruback, Skyrmions in flat space and curved space, *Phys. Lett.* **B181**, 137 (1986).

[183] N. S. Manton, B. J. Schroers and M. A. Singer, The interaction energy of well-separated Skyrme solitons, *Commun. Math. Phys.* **245**, 123 (2004).

[184] N. Manton and P. Sutcliffe, *Topological Solitons*, Cambridge University Press: Cambridge, 2004.

[185] D. J. Marín-Lámbarri et al., Evidence for triangular $D_{3h}$ symmetry in $^{12}$C, *Phys. Rev. Lett.* **113**, 012502 (2014).

[186] L. Marleau, All-orders Skyrmions, *Phys. Rev.* **D45**, 1776 (1992).

[187] L. Marleau and J.-F. Rivard, Generating function for all-orders Skyrmions, *Phys. Rev.* **D63**, 036007 (2001).

[188] U.-G. Meissner and I. Zahed, Skyrmions in the presence of vector mesons, *Phys. Rev. Lett.* **56**, 1035 (1986).

[189] H. Morinaga, Interpretation of some of the excited states of 4n self-conjugate nuclei, *Phys. Rev.* **101**, 254 (1956).
[190] T. Nakano et al., Evidence for a narrow $S = +1$ baryon resonance in photoproduction from the neutron, *Phys. Rev. Lett.* **91**, 012002 (2003).
[191] P. Navrátil and E. Caurier, Nuclear structure with accurate chiral perturbation theory nucleon-nucleon potential: Application to $^6$Li and $^{10}$B, *Phys. Rev.* **C69**, 014311 (2004).
[192] C. Naya and P. Sutcliffe, Skyrmions in models with pions and rho mesons, *JHEP* **05** (2018) 174; Skyrmions and clustering in light nuclei, *Phys. Rev. Lett.* **121**, 232002 (2018).
[193] E. M. Nyman and D. O. Riska, Low-energy properties of baryons in the Skyrme model, *Rep. Prog. Phys.* **53**, 1137 (1990).
[194] R. Partridge et al., Observation of an $\eta_c$ candidate state with mass $2978 \pm 9$ MeV, *Phys. Rev. Lett.* **45**, 1150 (1980).
[195] J. K. Perring and T. H. R. Skyrme, The alpha-particle and shell models of the nucleus, *Proc. Phys. Soc.* **A69**, 600 (1956).
[196] T. N. Pham and T. N. Truong, Evaluation of the derivative quartic terms of the meson chiral Lagrangian from forward dispersion relations, *Phys. Rev.* **D31**, 3027(R) (1985).
[197] A. Pomarol and A. Wulzer, Baryon physics in a five-dimensional model of hadrons, in *The Multifaceted Skyrmion*, eds. G. E. Brown and M. Rho, World Scientific: Singapore, 2010.
[198] M. K. Prasad and C. M. Sommerfield, Exact classical solution for the 't Hooft monopole and the Julia–Zee dyon, *Phys. Rev. Lett.* **35**, 760 (1975).
[199] B. Pritychenko, M. Birch, M. Horoi and B. Singh, B(E2) evaluation for $0_1^+ \to 2_1^+$ transitions in even-even nuclei, *Nucl. Data Sheets* **120**, 112 (2014).
[200] W. D. M. Rae, S. C. Allcock and J. Zhang, Spin assignments for states in $^{16}$O using the $^{12}$C($^{12}$C, $^{16}$O*)$^8$Be transfer reaction, *Nucl. Phys.* **A568**, 287 (1994).
[201] R. Rajaraman, *Solitons and Instantons*, Elsevier: Amsterdam, 1982.
[202] S. Raman, C. W. Nestor Jr. and P. Tikkanen, Transition probability from the ground to the first-excited $2^+$ state of even-even nuclides, *At. Data Nucl. Data Tables* **78**, 1 (2001).
[203] J. I. Rawlinson, An alpha particle model for Carbon-12, *Nucl. Phys.* **A975**, 122 (2018).
[204] J. I. Rawlinson, Coriolis terms in Skyrmion quantization, *Nucl. Phys.* **B949**, 114800 (2019).
[205] M. Rho and I. Zahed (eds.), *The Multifaceted Skyrmion, 2nd ed.*, World Scientific: Singapore, Hackensack NJ, 2016.
[206] D. Robson, Evidence for the tetrahedral nature of $^{16}$O, *Phys. Rev. Lett.* **42**, 876 (1979); Test of tetrahedral symmetry in the $^{16}$O nucleus, *Phys. Rev.* **C25**, 1108 (1982).
[207] D. Robson, Many-body interactions from quark exchanges and the tetrahedral crystal structure of nuclei, *Nucl. Phys.* **A308**, 381 (1978).
[208] Yu. P. Rybakov and V. I. Sanyuk, Methods for studying $3 + 1$ localized structures: The Skyrmion as the absolute minimizer of energy, *Int. J. Mod.*

Phys. **A7**, 3235 (1992).
[209] T. Sakai and S. Sugimoto, Low energy hadron physics in holographic QCD, *Prog. Theor. Phys.* **113**, 843 (2005); More on a holographic dual of QCD, *Prog. Theor. Phys.* **114**, 1083 (2005).
[210] B. J. Schroers, Dynamics of moving and spinning Skyrmions, *Z. Phys.* **C61**, 479 (1994).
[211] B. J. Schroers, On the existence of minima in the Skyrme model, *JHEP* proceedings PRHEP-unesp2002/034 (2002).
[212] N. Scoccola, Heavy-quark Skyrmions, in *The Multifaceted Skyrmion*, eds. G. E. Brown and M. Rho, World Scientific: Singapore, 2010.
[213] G. Segal, The topology of spaces of rational maps, *Acta Math.* **143**, 39 (1979).
[214] M. A. Singer and P. M. Sutcliffe, Symmetric instantons and Skyrme fields, *Nonlinearity* **12**, 987 (1999).
[215] T. H. R. Skyrme, A nonlinear field theory, *Proc. R. Soc. Lond.* **A260**, 127 (1961).
[216] T. H. R. Skyrme, A unified field theory of mesons and baryons, *Nucl. Phys.* **31**, 556 (1962).
[217] P. C. Sood, Centrifugal stretching of a classical rotator and collective motions in nuclei, *Can. J. Phys.* **46**, 1419 (1968).
[218] J. M. Speight, Near BPS Skyrmions and restricted harmonic maps, *J. Geom. Phys.* **92**, 30 (2015).
[219] N. J. Stone, Table of nuclear magnetic dipole and electric quadrupole moments, *At. Data Nucl. Data Tables* **90**, 75 (2005).
[220] R. F. Streater and A. S. Wightman, *PCT, Spin and Statistics, and all that*, Benjamin: New York, 1964.
[221] J. Suhonen, *From Nucleons to Nucleus*, Springer: Berlin Heidelberg, 2007.
[222] P. Sutcliffe, Multi-Skyrmions with vector mesons, *Phys. Rev.* **D79**, 085014 (2009).
[223] P. Sutcliffe, Skyrmions, instantons and holography, *JHEP* **08** (2010) 19; Skyrmions in a truncated BPS theory, *JHEP* **04** (2011) 45.
[224] I. Tanihata *et al.*, Measurements of interaction cross sections and nuclear radii in the light $p$-shell region, *Phys. Rev. Lett.* **55**, 2676 (1985).
[225] G. 't Hooft, unpublished.
[226] D. R. Tilley *et al.*, Energy levels of light nuclei $A = 5, 6, 7$, *Nucl. Phys.* **A708**, 3 (2002).
[227] D. R. Tilley *et al.*, Energy levels of light nuclei $A = 8, 9, 10$, *Nucl. Phys.* **A745**, 155 (2004).
[228] D. R. Tilley, H. R. Weller and C. M. Cheves, Energy levels of light nuclei $A = 16 - 17$, *Nucl. Phys.* **A564**, 1 (1993).
[229] D. R. Tilley, H. R. Weller and G. M. Hale, Energy levels of light nuclei $A = 4$, *Nucl. Phys.* **A541**, 1 (1992).
[230] S. Torilov *et al.*, Spectroscopy of $^{40}$Ca and negative-parity bands, *Eur. Phys. J.* **A19**, 307 (2004).
[231] J. J. M. Verbaarschot, Axial symmetry of bound baryon number two solution of the Skyrme model, *Phys. Lett.* **B195**, 235 (1987).

[232] J. J. M. Verbaarschot, T. S. Walhout, J. Wambach and H. W. Wyld, Symmetry and quantization of the two-Skyrmion system: The case of the deuteron, *Nucl. Phys.* **A468**, 520 (1987).

[233] R. Vinh Mau, M. Lacombe, B. Loiseau, W. N. Cottingham and P. Lisboa, The static baryon-baryon potential in the Skyrme model, *Phys. Lett.* **B150**, 259 (1985).

[234] A. Volya and Y. M. Tchuvil'sky, Nuclear clustering using a modern shell-model approach, *Phys. Rev.* **C91**, 044319 (2015).

[235] M. Vorabbi, P. Navrátil, S. Quaglioni and G. Hupin, $^7$Be and $^7$Li nuclei within the no-core shell model with continuum, *Phys. Rev.* **C100**, 024304 (2019).

[236] T. Waindzoch and J. Wambach, Stability of the $B = 2$ hedgehog in the Skyrme model, *Nucl. Phys.* **A602**, 347 (1996).

[237] T. Waindzoch and J. Wambach, Skyrmion dynamics on the unstable manifold and the nucleon-nucleon interaction, arXiv:nucl-th/9705040 (1997).

[238] N. R. Walet, Quantising the $B = 2$ and $B = 3$ Skyrmion systems, *Nucl. Phys.* **A606**, 429 (1996).

[239] T. S. Walhout, MultiSkyrmions as nuclei, *Nucl. Phys.* **A531**, 596 (1991).

[240] T. S. Walhout, Quantizing the four-baryon Skyrmion, *Nucl. Phys.* **A547**, 423 (1992).

[241] W. Wefelmeier, Ein geometrisches Modell des Atomkerns (A geometrical model of the nucleus), *Zeit. f. Phys.* **A107**, 332 (1937).

[242] S. Weinberg, Phenomenological Lagrangians, *Physica* **96A**, 327 (1979).

[243] H. Weigel, *Chiral Soliton Models for Baryons (Lecture Notes in Physics 743)*, Springer: Berlin Heidelberg, 2008.

[244] C. F. von Weizsäcker, Zur Theorie der Kernmassen (On the theory of nuclear masses), *Z. Phys.* **96**, 431 (1935).

[245] J. Wess and B. Zumino, Consequences of anomalous Ward identities, *Phys. Lett.* **B37**, 95 (1971).

[246] J. A. Wheeler, Molecular viewpoints in nuclear structure, *Phys. Rev.* **52**, 1083 (1937).

[247] C. Wheldon *et al.*, States at high excitation in $^{12}$C from the $^{12}$C($^3$He,$^3$He)3$\alpha$ reaction, *Phys. Rev.* **C90**, 014319 (2014).

[248] J. G. Williams, Topological analysis of a nonlinear field theory, *J. Math. Phys.* **11**, 2611 (1970).

[249] R. B. Wiringa, S. C. Pieper, J. Carlson and V. R. Pandharipande, Quantum Monte Carlo calculations of $A = 8$ nuclei, *Phys. Rev.* **C62**, 014001 (2000).

[250] A. Wirzba and H. Bang, The mode spectrum and the stability analysis of Skyrmions on a 3-sphere, *Nucl. Phys.* **A515**, 571 (1990).

[251] E. Witten, Global aspects of current algebra, *Nucl. Phys.* **B223**, 422 (1983); Current algebra, baryons, and quark confinement, *Nucl. Phys.* **B223**, 433 (1983).

[252] E. Witten, Anti-de Sitter space, thermal phase transition, and confinement in gauge theories, *Adv. Theor. Math. Phys.* **2**, 505 (1998).

[253] E. Wold, The centrifugal stretching model and the deviation from the $L(L+1)$ rule for rare-earth nuclei, *Nucl. Phys.* **A130**, 650 (1969).

[254] S. W. Wood, *Skyrmions and Nuclei*, Ph.D. thesis, Cambridge University, 2009.
[255] A. H. Wuosmaa, R. R. Betts, M. Freer and B. R. Fulton, Recent advances in the study of nuclear clusters, *Annu. Rev. Nucl. Part. Sci.* **45**, 89 (1995).
[256] P. A. Zyla *et al.* (Particle Data Group), Review of Particle Physics, *Prog. Theor. Exp. Phys.* **2020**, 083C01 (2020).

# Index

$\alpha - \alpha$ bonds, 136, 143, 199
$\alpha$-particle, 20, 124, 132, 139, 154, 160, 171, 185, 199, 223
$\omega$ meson, 251, 258, 261, 290
$\rho$ meson, 251, 258, 283, 290

ADHM formalism, 276, 278
Adkins–Nappi calibration, 73
Adkins–Nappi–Witten calibration, 72
algebraic degree, 87
angular integral $\mathcal{I}$, 89, 99, 144
antivacuum, 54, 60
asymptotic field, 103
asymptotic interactions, 58
Atiyah, M.F., 5, 272, 276
attractive channel, 59
axial gauge, 279
axial-vector mesons, 280, 284, 290

$B = 2$ Skyrmion, 81, 119, 278
baryon, 18
baryon current, 252, 261
baryon number, 48, 87, 279
baryon number conservation, 19, 48
baryon resonances, 285, 289
bent square, 140, 200
Beryllium-12, 176, 234, 242
Beryllium-7/Lithium-7, 21, 128, 189
Beryllium-8, 136, 159, 164, 219, 239
beta decay, 18
Bethe–Weizsäcker mass formula, 22

Boron-10, 169
BPS model, 250, 254, 274
brane, 269
breather mode, 182, 186, 200, 231

Calcium-40, 221
Carbon-12, 136, 159, 171, 173, 178, 190, 234, 239
Carbon-12 graph model, 191, 244
$C_{60}$ Buckyball, 85, 102
charge density, 234
charge radius, 157
Chern number, 275
Chern–Simons term, 286
chiral symmetry, 46, 60
chiral symmetry restoration, 108
classical field, 12, 33
classically spinning Skyrmion, 74, 242
Clebsch–Gordan coefficient, 237, 245
collective coordinates, 65, 111
compacton, 255
conservation laws, 32
Coriolis coupling, 185, 186
Coriolis parameter, 211, 213
corner cutting, 147, 152
current conservation, 37, 46
cutoff, 63

delta resonance, 18, 70
deuteron, 121
dibaryon, 265, 267

dihedral symmetry, 83
double rational map ansatz, 133, 137, 145, 150, 172
doubly-magic nucleus, 23, 185, 221, 230

E-manifold, 201, 221, 244
E-phonons, 207
E2 transitions, 233
E3 transition, 244, 246
E4 transition, 246
effective field theory (EFT), 6, 24, 274
elastic strain formalism, 52, 88, 252, 254
electromagnetic transitions, 233, 244
electromagnetism, 12, 20
energy-momentum tensor, 39
Esteban inequality, 105
Esteban, M., 56
Euler–Lagrange equation, 30, 34, 45

factorised gauge potential, 281, 285
Faddeev–Bogomolny bound, 52, 55, 108, 283
Faddeev–Bogomolny equation, 52, 283
FCC lattice, 107, 253
fermionic quantization, 66, 77
Feynman diagrams, 15, 26, 63
Finkelstein–Rubinstein constraints, 114, 123, 125, 128, 168, 172
Finkelstein–Rubinstein sign, 116, 117, 120, 123, 125
flat square, 141, 204
fluid, 12, 256
fullerene, 85

gamma rays, 20, 223, 228
Gell-Mann matrices, 264, 265
Goldstone particles, 36
Goldstone's theorem, 36
gravity, 13, 269

half-Skyrmion, 107, 144, 253
half-Skyrmion grid, 145, 152
hedgehog, 54, 61, 69, 83, 91, 265, 278

Helium-3/Hydrogen-3, 123
Helium-6, 128, 163
holography, 251, 269, 272
holonomy, 276, 278
homotopy group, 66
Hoyle state, 138, 171, 176, 192, 233, 239
hypercharge, 266
hyperon, 264, 267

icosahedral symmetry, 84, 95, 102, 126, 134, 153
inertia tensors, 112, 158, 234
instanton, 271, 272, 275
internal symmetry, 35
intrinsic symmetry, 68, 82, 114, 132
inversion, 91, 117
isorotation, 17, 82
isospace, 82
isospin, 17, 21, 39, 47, 68, 113, 123, 125, 126, 129, 234, 242

JNR ansatz, 276

kink, 64, 282, 283, 285
Klein polynomials, 92, 139
Klein–Gordon theory, 38

Lagrangian dynamics, 29
Lagrangian field theory, 33
Lead-208, 153, 221
left current, 46
Lie derivative, 32
lightly-bound model, 222, 253
Lithium-5/Helium-5, 126
Lithium-6, 128, 129, 159, 162
Lithium-8, 166
Lorentz invariance, 34

M1 transitions, 233
Möbius transformation, 90, 91, 115
Magnesium-24, 241
Magnetic Skyrmions, 8
matter radius, 179, 234
moduli space, 56, 275

Neon-20, 240
neutron, 17, 69, 154
Noether charge, 38
Noether's theorem, 32, 37, 46
non-contractible loop, 44, 66, 67
nonlinear sigma model, 35
nucleon, 17, 19, 70
numerical methods, 154

octahedral symmetry, 81, 83, 95, 124, 142, 146, 149, 203
Oxygen-16, 136, 140, 199, 214, 221, 240

parity, 117, 127, 130, 204, 223, 280
parity doubling, 215
Pauli matrices, 42, 87
Pauli principle, 18
pentaquark, 266
phase transition, 107
pion, 17, 39, 63
pion dipoles, 55, 58, 97
pion fields, 44
pion mass term, 60, 131
pion multipole, 104
pion-Skyrmion interaction, 64
Planck's constant, 3, 73, 157, 234
Platonic symmetry, 92
point particle, 11
preimage counting, 50, 83, 87, 134
principle of least action, 30, 34
product ansatz, 59
proton, 17, 69, 154

QCD, 4, 15, 18, 79, 251, 270
quadrupole moment, 233, 236
quadrupole vibrations, 186
quantum field theory, 11, 14
quantum theory, 13
quarks, 4, 15, 27, 264

rational map ansatz, 81, 86, 87, 96, 115, 168
reflection symmetry, 91, 103
rescaling, 51
right current, 46

right-angle scattering, 77
rigid-body Hilbert space, 113
rigid-body quantization, 67, 111, 157, 173, 288
rigorous results, 103
Roper resonance, 183, 289
rovibrational states, 185, 195, 196, 203, 209, 225
Runge colour sphere, 56

s quark, 249, 264
Sakai–Sugimoto model, 251, 269
Schrödinger equation, 25, 64, 203
sextic term, 252
shell model, 23, 140, 200, 224
sigma field, 44
simulated annealing, 99
Skyrme field, 29, 41, 44
Skyrme field energy, 51, 282
Skyrme field equation, 45
Skyrme field quantization, 63
Skyrme field topology, 47, 66
Skyrme Lagrangian, 45
Skyrme term, 52
Skyrme theory, 27, 44
Skyrme, T.H.R., 2
Skyrmion crystal, 82, 106, 142, 144
Skyrmion deformation, 141, 181, 189, 242, 244
Skyrmion quantization, 157
Skyrmion vibrations, 181, 278
Skyrmions from instantons, 275
spin, 17, 68, 113, 123, 125, 126, 129, 189, 223
spin-orbit coupling, 74, 126
spin-statistics relation, 4, 67
spontaneous symmetry breaking, 36
Standard Model, 6, 15
string theory, 269
$SU(2)$, 41, 276
$su(2)$ Lie algebra, 42, 276
Sulphur-32, 241
$SU(2)/SO(3)$ relationship, 43, 111
$SU(3)$, 249, 263
Sutcliffe expansion, 273, 280, 281
symmetry, 32, 54, 90

tensor force, 74
tetrahedral group characters, 207, 210
tetrahedral symmetry, 81, 83, 94, 122, 139, 149, 199, 221
't Hooft, G., 251
't Hooft ansatz, 276, 277
't Hooft coupling, 271, 286
topological degree, 48, 87, 255
topological soliton, 3, 27
topological suspension, 88
triton, 123
two-nucleon resonance, 121, 122

vacuum, 46, 60

warp factors, 271, 272, 286, 287, 289
Weisskopf unit, 238
Wess–Zumino term, 78, 265
Witten, E., 4, 78, 251
Wronskian, 91, 95

Yang–Mills theory, 269, 271, 275
yrast state, 215, 218, 228, 233
Yukawa dipole, 131

zero modes, 182

# About the Author

**Nicholas S Manton** is emeritus Professor of Mathematical Physics at the University of Cambridge, and is a Fellow of St John's College, Cambridge. He was elected a Fellow of the Royal Society in 1996. His research has mainly been on a variety of topological solitons, including magnetic monopoles, vortices and kinks. With Frans Klinkhamer he discovered the unstable sphaleron solution in the electroweak theory of W-, Z- and Higgs bosons, proposed as a mediator for the creation of the matter-antimatter asymmetry of the universe. He has been particularly interested in Skyrmions, the topological solitons featured in this book, because of their physical application to the modelling of protons, neutrons and larger atomic nuclei, and because of their striking mathematical beauty. He has published numerous research papers, including a few in collaboration with the late, distinguished mathematician Sir Michael Atiyah, and has coauthored two earlier books – the monograph *Topological Solitons* in collaboration with Paul Sutcliffe, now Professor at Durham University, and the broad-ranging survey of physics *The Physical World* in collaboration with Nicholas Mee, author of several popular books on physics.